Monographs in Mathematics
Vol. 90

Managing Editors:
H. Amann
Universität Zürich, Switzerland
J.-P. Bourguignon
IHES, Bures-sur-Yvette, France
K. Grove
University of Maryland, College Park, USA
P.-L. Lions
Université de Paris-Dauphine, France

Associate Editors:
H. Araki, Kyoto University
J. Ball, Heriot-Watt University, Edinburgh
F. Brezzi, Università di Pavia
K.C. Chang, Peking University
N. Hitchin, University of Warwick
H. Hofer, Universität Bochum
H. Knörrer, ETH Zürich
K. Masuda, University of Tokyo
D. Zagier, Max-Planck-Institut Bonn

Philippe Tondeur

Geometry of Foliations

Birkhäuser Verlag
Basel · Boston · Berlin

Author:
Department of Mathematics
University of Illinois at Urbana-Champaign
273 Altgeld Hall
1409 West Green Street
Urbana, IL 61801
USA

1991 Mathematics Subject Classification SC 53C12, 57R30, 58A14

A CIP catalogue record for this book is available from the Library of Congress, Washington D.C., USA

Deutsche Bibliothek Cataloging-in-Publication Data
Tondeur Philippe:
Geometry of foliations / Philippe Tondeur. - Basel ; Boston ; Berlin : Birkhäuser, 1997
  (Monographs in mathematics ; Vol. 90)
   ISBN 3-7643-5741-X (Basel ...)
   ISBN 0-8176-5741-X (Boston)

This work is subject to copyright. All rights are reserved, whether the whole or part of the material is concerned, specifically the rights of translation, reprinting, re-use of illustrations, recitation, broadcasting, reproduction on microfilms or in other ways, and storage in data banks. For any kind of use the permission of the copyright owner must be obtained.

© 1997 Birkhäuser Verlag, P.O. Box 133, CH-4010 Basel, Switzerland
Printed on acid-free paper produced from chlorine-free pulp. TCF ∞
Printed in Germany
ISBN 3-7643-5741-X
ISBN 3-8176-5741-X

9 8 7 6 5 4 3 2 1

Contents

Introduction	....	vii
1	Examples and Definition of Foliations	1
2	Foliations of Codimension One	7
3	Holonomy, Second Fundamental Form, Mean Curvature	17
4	Basic Forms, Spectral Sequence, Characteristic Form	33
5	Transversal Riemannian Geometry	43
6	Flows	69
7	Hodge Theory for the Transversal Laplacian	81
8	Cohomology Vanishing and Tautness	99
9	Lie Foliations	107
10	Structure of Riemannian Foliations	113
11	Spectral Geometry of Riemannian Foliations	117
12	Foliations as Noncommutative Spaces	129
13	Infinite-Dimensional Riemannian Foliations	133
References on Riemannian Foliations	....	139
Appendix A	Books and Surveys on Particular Aspects of Foliations	167
Appendix B	Proceedings of Conferences and Symposia devoted to Foliations	171
Appendix C	Bibliography on Foliations. This should be a reasonably complete list of all papers on the subject of foliations up to 1995	175
Appendix D	Numbers of papers on foliations published during consecutive five year periods up to 1995	303
Index of Subjects	....	304
Index of Notations	....	305

Introduction

The topics in this survey volume concern research done on the differential geometry of foliations over the last few years. After a discussion of the basic concepts in the theory of foliations in the first four chapters, the subject is narrowed down to Riemannian foliations on closed manifolds beginning with Chapter 5. Following the discussion of the special case of flows in Chapter 6, Chapters 7 and 8 are devoted to Hodge theory for the transversal Laplacian and applications of the heat equation method to Riemannian foliations. Chapter 9 on Lie foliations is a preparation for the statement of Molino's Structure Theorem for Riemannian foliations in Chapter 10. Some aspects of the spectral theory for Riemannian foliations are discussed in Chapter 11. Connes' point of view of foliations as examples of noncommutative spaces is briefly described in Chapter 12. Chapter 13 applies ideas of Riemannian foliation theory to an infinite-dimensional context.

Aside from the list of references on Riemannian foliations (items on this list are referred to in the text by [ ]), we have included several appendices as follows.

Appendix A is a list of books and surveys on particular aspects of foliations. Appendix B is a list of proceedings of conferences and symposia devoted partially or entirely to foliations. Appendix C is a bibliography on foliations, which attempts to be a reasonably complete list of papers and preprints on the subject of foliations up to 1995, and contains approximately 2500 titles.

Piqued by curiosity, we have counted the number of published papers on this subject during consecutive 5-year periods, and came up with the figures in Appendix D. This indicates that since the early 1970's, there is a steady flow of about one hundred papers per year. This volume is a record of activities centered in the 1980's. The last chapters of this book are a bit more prospective in character, and might stimulate research in somewhat different directions. If this occurs, the author would be delighted, and the work on this volume amply justified.

It remains to thank the persons who have contributed to this volume. First of all my thanks go to Franz Kamber, for an intense and fruitful collaboration over a twenty year period. Then to the many other collaborators over the years, Jesus Alvarez López, Mike Hvidsten, Hobum Kim, Yoshiaki Maeda, Maung Min-Oo, Seiki Nishikawa, Mohan Ramachandran, Steve Rosenberg, Ernst Ruh, Gabor Toth, and Lieven Vanhecke. Careful readers who have spotted inaccuracies, and who have helped to eliminate them, include Eric Boeckx and Louis Kerofsky. Hilda Britt has

TEXed so many versions of this manuscript with grace and a smile, that my life has been the better for it.

Finally I thank the University of Illinois and the National Science Foundation for the financial support which made this intellectual journey possible.

Champaign-Urbana, May 1996
Philippe Tondeur

Chapter 1
Examples and Definition of Foliations

Simple examples of foliations are given by submersions. A (smooth) submersion $f : M \to B$ is a map of manifolds with a surjective derivative map at every point of M. The inverse images of points in the target space form a family of closed submanifolds of M, the leaves of the foliation $\mathcal{F}$ on M defined by f. All these submanifolds have the same dimension p. If n denotes the dimension of M, and q the dimension of B, then $n = p + q$. A particularly simple situation is the case of a foliation of M by the level hypersurfaces of a smooth function $f : M \to \mathbb{R}$. The submersion condition is the requirement of the absence of critical points for f (this is of course only possible if M is not compact).

What is special about these submersion examples is that all leaves are closed, and all leaves are of the same topological (diffeomorphism) type, at least if M is closed. The latter fact can be seen as follows. Let g be a Riemannian metric on M, and $H \to M$ the bundle of vectors orthogonal to the fibers of f. A (smooth) path $\gamma : I \to B$ can be lifted to a horizontal path $\tilde{\gamma} : I \to M$ covering γ, i.e. such that $f \circ \tilde{\gamma} = \gamma$. This defines a parallel transport of the fibers of f along γ, giving rise to diffeomorphisms of $f^{-1}(\gamma(0))$ to $f^{-1}(\gamma(t))$ for all $t \in I$.

An approximate definition of a foliation $\mathcal{F}$ on M is that it is given by a partition of M into p-dimensional submanifolds of M, called leaves, such that the quotient map of M onto the space B of leaves has locally the properties of a submersion. A precise definition follows later in this chapter.

Another interesting class of examples is given by certain group actions on smooth manifolds. Let G be a Lie group, M a manifold, and $G \times M \to M$ a (smooth) G-action, i.e. a smooth map $(g, x) \mapsto gx$ such that

$$ex = x, \qquad (g_1 g_2)x = g_1(g_2 x)$$

for all $x \in M$ and $g_1, g_2 \in G, e \in G$ the identity element. This defines a group homomorphism $G \to \mathrm{Diff}(M)$ by $g \to (x \mapsto gx)$. For a point $x \in M$, the orbit $Gx = \{gx \in M | g \in G\}$ appears as the image of the orbit map $G \to M$ given by $g \to gx$. This map is constant on the isotropy group $G_x = \{g \in G | gx = x\}$, and hence defines an identification $G/G_x \cong Gx$. Carrying over the smooth homogeneous space structure, each orbit appears as a smoothly immersed submanifold of M. That the manifold structure on Gx need not coincide with the topology induced from M is already clear from the linear action $\mathbb{R} \times T^2 \to T^2$ with straight flow lines of irrational slope.

The isotropy groups for different points of the same orbit are conjugate, precisely $G_{gx} = gG_x g^{-1}$. Thus an orbit is described by the conjugacy class (H) of the isotropy group H of any of its points, its G-isotropy type. Two orbits are of the same type, if they have the same isotropy types. This is the case if and only if the two orbits F_1, F_2 are equivariantly diffeomorphic, i.e. there exists a diffeomorphism $\varphi : F_1 \to F_2$ such that $\varphi(gx) = g\varphi(x)$ for all $g \in G, x \in F_1$. An orbit Gx is of principal type if there is a neighborhood U of x such that for all points of U the orbits are of the same type.

Let $B = M/G$ be the set of all orbits, and $f : M \to B$ the orbit map $x \to Gx$. If all orbits are of the same principal type, the situation looks like in the initial submersion situation. In general, the orbits not of principal type will play the role of singular fibers or leaves.

A case of considerable interest is the situation when all isotropy groups are discrete. In this case all orbits have the same dimension (the dimension of G) and are of principal type.

Next we consider the model foliation of dimension p on $\mathbb{R}^n$ given by the family of affine subspaces parallel to $\mathbb{R}^p$. The codimension is $q = n - p$. The leaves of this foliation are given by the fibers of the projection $\mathbb{R}^n = \mathbb{R}^p \times \mathbb{R}^q \to \mathbb{R}^q$.

Let $(x, y) = (x_1, \ldots, x_p; y_1, \ldots, y_q)$ denote the coordinates in $\mathbb{R}^p \times \mathbb{R}^q$. A local automorphism of this model foliation is a diffeomorphism $\Phi : U \to U'$ of open subsets of $\mathbb{R}^n$ of the form

$$(x', y') = (\Phi_1(x, y), \Phi_2(y)).$$

This condition can be equivalently expressed by

$$\frac{\partial y'_\alpha}{\partial x_i} = 0$$

for $\alpha = 1, \ldots, q$ and $i = 1, \ldots, p$. The map $y' = \Phi_2(y)$ induces a local diffeomorphism of $\mathbb{R}^q$.

The family of local automorphisms of this model foliation is a pseudogroup of transformations $\Gamma = \Gamma_{n,q}$. This means the following:

(i) if $\Phi \in \Gamma$, then $\Phi^{-1} \in \Gamma$;
(ii) if $\Phi : U \to U'$ and $\Phi' : U' \to U''$ belong to Γ, then $\Phi' \circ \Phi \in \Gamma$;
(iii) if $\Phi \in \Gamma$, then its restriction to any open subset belongs to Γ;
(iv) if $\Phi : U \to U'$ is a local diffeomorphism which coincides on a neighborhood of each point of U with an element of Γ, then $\Phi \in \Gamma$.

A foliated or distinguished chart on M^n is a diffeomorphism of an open subset of M^n with an open subset of a local model on $\mathbb{R}^n$. A foliated atlas of codimension q

on M^n is an open covering of M by the domains of foliated charts, with coordinate transitions belonging to the pseudogroup $\Gamma_{n,q}$. A foliation $\mathcal{F}$ of codimension q on M is a maximal foliated atlas of codimension q on M (Reeb, 1947).

The connected components of the sets $y = $ constant in a distinguished chart are called plaques (plates) of $\mathcal{F}$. Fixing y, the map $x \to (x,y)$ is a smooth embedding. Thus the plaques are connected p-dimensional submanifolds of M. Given that the coordinate transitions respect the sets $y = $ constant, the plaques have a well-defined meaning. The maximal foliated atlas represents all possible microscopic views of the foliation $\mathcal{F}$.

To describe a macroscopic view, consider the subbundle $L \subset TM$ given by vectors tangent to plaques. In a distinguished chart U, the bundle $L \mid U$ is framed by the vector fields $\frac{\partial}{\partial x_1}, \ldots, \frac{\partial}{\partial x_p}$. Since these have pairwise vanishing brackets, the p-dimensional bundle $L \to M$ is integrable, i.e. for sections $V, W \in \Gamma(U, L)$ we have $[V, W] \in \Gamma(U, L)$.

Frobenius' Theorem establishes that given such an integrable subbundle $L \subset TM$, there exists a foliated atlas of codimension q on M. The leaves of $\mathcal{F}$ are defined as the maximal integral submanifolds of L. The leaf through x is constructed as the sets of points of M accessible from x by piecewise smooth paths for which the tangent vectors belong to L. In a distinguished chart U, a leaf $\mathcal{L}$ intersects with U in a union of plaques. This turns $\mathcal{L}$ into an immersed p-dimensional submanifold of M. The topology on $\mathcal{L}$ is defined by the plaques in $\mathcal{L}$ as an open basis.

The global view of a foliation $\mathcal{F}$ of codimension q on M^n is thus a partition $\{\mathcal{L}_\alpha\}_{\alpha \in A}$ of M into connected subsets with the following properties. There is a foliated atlas of codimension q on M, such that for each leaf $\mathcal{L}_\alpha$ the connected components of $U \cap \mathcal{L}_\alpha$ in a foliated chart are defined by the equation $y = $ constant. Next we need to describe the coordinate transitions. Let $\varphi_\alpha = (x_\alpha, y_\alpha) : U_\alpha \to \mathbb{R}^n$ be a distinguished chart, and $f_\alpha : U_\alpha \to \mathbb{R}^n \to \mathbb{R}^q$ the natural composition. It is a submersion defining the plaques of $\mathcal{F}$ on U_α. For two such overlapping distinguished charts, the corresponding submersions f_α, f_β are related by

$$f_\beta = \tau_{\beta\alpha} f_\alpha \quad \text{on} \quad U_\alpha \cap U_\beta,$$

where $\tau_{\beta\alpha} : f_\alpha(U_\alpha \cap U_\beta) \to f_\beta(U_\beta \cap U_\alpha)$. This leads to the equivalent definition of a foliation by a family of submersions $f_\alpha : U_\alpha \to \mathbb{R}^q$ for an open covering $\mathcal{U} = \{U_\alpha, U_\beta, \ldots\}$, together with a family of local diffeomorphisms (transition functions) $\tau_{\beta\alpha}$ of $\mathbb{R}^q$, for each (α, β) with $U_\alpha \cap U_\beta \neq \emptyset$, satisfying the relations above. For these relations to be compatible on triple intersections, one needs the cocycle conditions

$$\tau_{\gamma\alpha} = \tau_{\gamma\beta} \circ \tau_{\beta\alpha} \quad \text{on} \quad U_\alpha \cap U_\beta \cap U_\gamma \neq \emptyset.$$

This definition of foliations by a cocycle was the point of view adopted by Haefliger (1958) in [Ha 1] and throughout his later work on foliations.

Note that the transition functions relate open sets of q-dimensional spaces, in contrast to the coordinate transitions in $\Gamma_{n,q}$ discussed before. The relation is that a coordinate transition in $\Gamma_{n,q}$ projects to a transition function as just described.

One way to rephrase the situation is to consider the disjoint union $T = \bigcup_\alpha f_\alpha(U_\alpha)$ of all the targets. It is a q-dimensional manifold with a pseudogroup Γ generated by the transition functions $\tau_{\alpha\beta}$. This is a particularly useful point of view if one wants to consider more specific transversal geometries for a foliation $\mathcal{F}$. For the example of greatest interest in these notes, the manifold T is equipped with a Riemannian metric, and Γ is the pseudogroup of all local isometries of this metric. A foliation $\mathcal{F}$ is Riemannian, if the local transition functions belong to this pseudogroup. Similarly, the manifold T can be equipped with a symplectic structure and Γ is then the pseudogroup of all local diffeomorphisms preserving this symplectic structure. A foliation is symplectic, if the local transition functions belong to this pseudogroup. A foliation is Kählerian, if the local transition functions belong to the pseudogroup of local diffeomorphisms preserving a given Kähler structure on T. If T is viewed as a measure space with respect to a measure μ, then we can consider the pseudogroup of local measure preserving transformations. A foliation is (transversely) measurable, if the local transition functions are measure preserving.

There is a wide spectrum of interest in particular classes of foliations. The class of measurable foliations ties in with questions of ergodic theory, while the flavour of Riemannian and Kähler foliations is strongly differential geometric. There is a hierarchy of increasingly complex transversal structures. The transversal topology ties in with connectivity and compactness questions, the transversal smooth structure provides transversal infinitesimal linearity, while a transversal Riemannian structure provides transversal curvature data. This book is mostly in the latter spirit.

The normal bundle Q of a codimension q foliation $\mathcal{F}$ on M is the quotient bundle $Q = TM/L$. Equivalently, this is expressed by the exact sequence of vector bundles

$$0 \to L \to TM \xrightarrow{\pi} Q \to 0.$$

On the domain U_α of a distinguished chart $f_\alpha : U_\alpha \to f_\alpha(U_\alpha)$, the bundle $Q \mid U_\alpha$ is represented by (i.e. is isomorphic to) the pullback $f_\alpha^* Tf_\alpha(U_\alpha)$.

If $(x_1, \ldots, x_p; y_1, \ldots, y_q)$ are local coordinates in a distinguished chart U, the bundle $Q \mid U$ is framed by the vector fields $\pi \frac{\partial}{\partial y_1}, \ldots, \pi \frac{\partial}{\partial y_q}$. For a vector field $Y \in \Gamma TM$, we denote also $\overline{Y} = \pi Y \in \Gamma Q$. A vector field Y on U is projectable, if $Y = \sum_i a_i \frac{\partial}{\partial x_i} + \sum_\alpha b_\alpha \frac{\partial}{\partial y_\alpha}$ with $\frac{\partial b_\alpha}{\partial x_i} = 0$ for all $\alpha = 1 \ldots, q$ and $i = 1, \ldots, p$. This means that the functions $b_\alpha = b_\alpha(y)$ are independent of x. Then $\overline{Y} = \sum_\alpha b_\alpha \frac{\overline{\partial}}{\partial y_\alpha}$ with b_α independent of x. This property is preserved under change of distinguished charts, hence makes intrinsic sense. An equivalent property to the projectability of Y is that for all vector fields $V \in \Gamma L$ the bracket $[V, Y] \in \Gamma L$. The set of projectable vector fields $V(\mathcal{F})$ is closed under Lie brackets, and $\Gamma L \subset V(\mathcal{F})$ is an ideal. This yields an exact sequence of Lie algebras

$$0 \longrightarrow \Gamma L \longrightarrow V(\mathcal{F}) \longrightarrow \Gamma Q^L \longrightarrow 0.$$
$$Y \longmapsto \overline{Y}$$

Here we have denoted by $\Gamma Q^L \subset \Gamma Q$ the sections invariant under the ΓL-action

$$\theta(V)s = \overline{[V, Y_s]} \quad \text{for} \quad V \in \Gamma L, s \in \Gamma Q,$$

where $Y_s \in \Gamma TM$ is any choice with $\overline{Y}_s = s$.

Let $G = GL(q)$. The bundle P of linear frames of Q is a principal G-bundle. For a distinguished chart, the bundle $P \mid U_\alpha$ is isomorphic to the pullback of the linear frame bundle of $Tf_\alpha(U_\alpha)$.

The bundle P over the foliated manifold $(M, \mathcal{F})$ carries a natural foliation $\tilde{\mathcal{F}}$. We describe it by defining its tangent bundle $\tilde{L} \subset TP$. Let $V \in \Gamma(U, L)$ with local flow φ_t. Then the derivative $\varphi_{t*} : TM \to TM$ carries L into itself, hence induces an endomorphism $Q \to Q$. The corresponding flow of bundle isomorphisms of Q induces a flow $\tilde{\varphi}_{t*}$ of isomorphisms of the normal bundle P. The vector field $\tilde{V} \in \Gamma(U, P)$ generating $\tilde{\varphi}_{t*}$ is a lift of V with respect to the bundle projection $P \to M$. The vector fields $\tilde{V}$ generate in this fashion a p-dimensional subbundle $\tilde{L} \subset TP$. It is integrable, and defines the foliation $\tilde{\mathcal{F}}$ on P. Moreover the subbundle $\tilde{L} \subset TP$ is G-invariant. This is what is meant by a foliated bundle structure on a G-bundle $P \to M$ over a foliated manifold $(M, \mathcal{F})$, namely a G-invariant p-dimensional foliation on P projecting to the given p-dimensional foliation on M.

In the next chapter we discuss examples and properties of foliations of codimension one.

Chapter 2
Foliations of Codimension One

The simplest examples of foliations of codimension one are the level surfaces of a function $f : M^{n+1} \to \mathbb{R}$ with no critical points. This is of course only possible for a noncompact manifold M. The one-form $\omega = df$ is thus assumed nonsingular, i.e. $\omega_x \neq 0$ for all $x \in M$.

A necessary condition for a one-form ω to occur in the form $\omega = df$ is the condition $d\omega = 0$. Now if the De Rham cohomology class $[\omega] \in H^1_{DR}(M)$ is nontrivial, there is no global solution of $\omega = df$. What is the situation if M is simply connected? Then $H^1_{DR}(M) = 0$, the equation $\omega = df$ has a global solution $f : M \to \mathbb{R}$, and the field of hyperplanes defined by $\omega = 0$ is tangent to the foliation by the level surfaces of the function f.

When does a nonsingular one-form ω define a foliation? The question is the integrability of the subbundle $L \subset TM$ defined by $L_x = \ker \omega_x$ for all $x \in M$.

The multiplication of the one-form ω by a nonzero function does not modify the subbundle $L \subset TM$. A natural variation of the integration problem is to ask for the local solvability of the equation $\omega = g\, df$ with $f, g : U \to \mathbb{R}$, g nonzero, on an open subset U of M. Exterior differentiation yields the necessary integrability condition

$$\omega \wedge d\omega = (g\, df) \wedge d(g\, df) = g\, df \wedge dg \wedge df = 0.$$

This condition is simultaneously satisfied for ω and $h\omega$, since

$$h\omega \wedge d(h\omega) = h\omega \wedge (dh \wedge \omega + h\, d\omega) = h^2 \omega \wedge d\omega.$$

The theorem of Frobenius states conversely, that the integrability condition $\omega \wedge d\omega = 0$ for a nonsingular one-form ω guarantees the local solvability of the equation $\omega = g\, df$ with $f, g : U \to \mathbb{R}$ and g nonzero (see Proposition 2.1 below).

An example of a nonintegrable one-form on $\mathbb{R}^3 - \{0\}$ is given by $\omega = x\, dy + y\, dz + z\, dx$. Then $\omega \wedge d\omega = (x + y + z)\, dx \wedge dy \wedge dz$.

For $n = 2$ the 3-form $\omega \wedge d\omega$ is necessarily 0. A nonsingular one-form defines by $\omega = 0$ a line field. If M^2 is compact, it follows that the Euler characteristic $\chi(M^2) = 0$. Thus M^2 is the torus T^2 in the orientable case, the Klein bottle K^2 in the nonorientable case.

In cylindrical coordinates on $\mathbb{R}^3$ consider the one-form
$$\omega = h(r)\, dr + (1 - h(r))\, dz,$$
where $h(r)$ is a smooth monotone nondecreasing function with $h(r) = 0$ for $r \leq 0$, $0 < h(r) < 1$ for $0 < r < 1$, and $h(r) = 1$ for $r \geq 1$. Then $d\omega = -h'dr \wedge dz$ and thus $\omega \wedge d\omega = 0$. The resulting foliation of the cylinder $D^2 \times \mathbb{R}$ gives a foliation of the solid torus $D^2 \times S^1$. The Reeb foliation of S^3 ([Rb 1], 1952) is obtained from two copies of $D^2 \times S^1$ by gluing them together along their common boundary $S^1 \times S^1$. The 2-torus appears thus as the unique closed ($\equiv$ compact without boundary) leaf of this foliation.

A theorem of Novikov ([No], 1964) states that every foliation of codimension one on a closed 3-manifold with finite fundamental group has a closed leaf. In fact Novikov proved that every such foliation has a Reeb component, i.e. a subset which is a union of leaves and which is diffeomorphic to $D^2 \times S^1$ with the foliation described above. The closed boundary leaf is thus necessarily a torus.

Note that the Reeb foliation on S^3 cannot be defined by a closed one-form ω. Since $H^1_{DR}(S^3) = 0$, such a form would be equal to df for some $f : S^3 \to \mathbb{R}$. At the critical points of f the form ω would then have to vanish. This argument applies to any codimension one foliation on a closed manifold with finite fundamental group.

For a codimension one foliation on a closed manifold M, the existence of a transversal line field shows that the Euler characteristic $\chi(M)$ vanishes. A theorem of Thurston ([Th4], 1976) states that conversely every closed manifold M with $\chi(M) = 0$ admits a codimension one foliation. It is important to note that these results all concern smooth foliations. Haefliger has shown (1958) that on a closed simply connected manifold no real analytic codimension one foliation exists.

We return to a discussion of the integrability condition.

2.1 PROPOSITION. *Let M be a smooth $(n+1)$-manifold, and $\omega \in \Omega^1(M)$ a nonsingular one-form. The following conditions are equivalent:*

(i) $\omega \wedge d\omega = 0$;
(ii) $\omega[V, W] = 0$ for all vector fields with V, W with $i(V)\omega = 0$, $i(W)\omega = 0$;
(iii) $d\omega = \alpha \wedge \omega$ for some $\alpha \in \Omega^1(M)$.

Proof. (i)$\Rightarrow$(iii): First we prove the local existence of α. Let U be an open set trivializing the dual tangent bundle $T^*M|U$. The nonsingular one-form $\omega = \omega_1$ can then be completed to a basis $\omega_1, \ldots, \omega_{n+1}$ of $\Omega^1(U)$. Thus $d\omega = \sum_{j<k} b_{jk}\omega_j \wedge \omega_k$ and by assumption
$$\omega \wedge d\omega = \omega_1 \wedge (\sum_{j<k} b_{jk}\omega_j \wedge \omega_k) = 0.$$

The 3-forms $\omega_1 \wedge \omega_j \wedge \omega_k$ for $1 < j < k$ are linearly independent, so that $b_{jk} = 0$ for $1 < j < k$. It follows that on U
$$d\omega = \sum_{k=2}^{n+1} b_{1k}\omega_1 \wedge \omega_k = (-\sum_{k=2}^{n+1} b_{1k}\omega_k) \wedge \omega_1.$$

Thus with $\alpha = -\sum_{k=2}^{n+1} b_{1k}\omega_k$, we have $d\omega = \alpha \wedge \omega$ on U.

The global existence of α follows from a partition of unity argument. Let $\mathcal{U} = \{U_\gamma\}$ be an open covering of M by subsets U_γ trivializing $T^*M|U_\gamma$, and $\lambda = \{\lambda_\gamma\}$ a smooth partition of 1 subordinate to $\mathcal{U}$. On each U_γ we obtain $\alpha_\gamma \in \Omega^1(U_\gamma)$, such that $d\omega = \alpha_\gamma \wedge \omega$ holds on U_γ. Define $\alpha = \sum \lambda_\gamma \alpha_\gamma \in \Omega^1(M)$. Then

$$\alpha \wedge \omega = \sum \lambda_\gamma(\alpha_\gamma \wedge \omega) = \sum \lambda_\gamma d\omega|U_\gamma = \sum \lambda_\gamma d\omega = d\omega,$$

which proves (iii).

(iii)$\Rightarrow$(ii): For $i(V)\omega = 0$, $i(W)\omega = 0$ we have

$$d\omega(V, W) = -\omega[V, W] \quad \text{and} \quad (\alpha \wedge \omega)(V, W) = 0.$$

Thus $d\omega = \alpha \wedge \omega$ implies (ii).

(ii)$\Rightarrow$(i): Again we extend locally ω to a basis $\omega_1, \ldots, \omega_{n+1}$ of $\Omega^1(U)$, with $\omega = \omega_1$. Let $V_2, \ldots, V_{n+1}$ be the dual frame of $TM|U$. Then $d\omega = \sum_{j<k} b_{jk}\omega_j \wedge \omega_k$ and

$$b_{jk} = d\omega(V_j, V_k) = -\omega[V_j, V_k].$$

Thus by assumption $b_{jk} = 0$ for $1 < j < k$. It follows that

$$d\omega = \sum_{k=2}^{n+1} b_{1k}\omega \wedge \omega_k, \text{ and therefore } \omega \wedge d\omega = 0. \qquad \square$$

To what extent is the one-form α determined? Let α', α both be one-forms with $d\omega = \alpha' \wedge \omega = \alpha \wedge \omega$. Then $(\alpha' - \alpha) \wedge \omega = 0$, i.e. the one-forms $\alpha' - \alpha$ and ω are linearly dependent. It follows that $\alpha' - \alpha = f\omega$ for some function $f : M \to \mathbb{R}$. This describes the indeterminacy in α.

Next we give a geometric construction of such a form α.

2.2 PROPOSITION. *Let Z be a vector field on M satisfying $i(Z)\omega = 1$ (e.g. a vector field orthogonal to the hyperplanes defined by $\omega = 0$, with respect to a Riemannian metric g on M, and scaled so as to satisfy $i(Z)\omega = 1$). Then the form $\alpha = -\theta(Z)\omega$ is a proper choice, i.e. $d\omega = -\theta(Z)\omega \wedge \omega$.*

Proof. The 2-forms $-\theta(Z)\omega \wedge \omega$ and $d\omega$ both vanish for V, W with $i(V)\omega = 0$, $i(W)\omega = 0$. Namely

$$(-\theta(Z)\omega \wedge \omega)(V, W) = (-\theta(Z)\omega)(V)\omega(W) - (-\theta(Z)\omega)(W)\omega(V) = 0,$$

and

$$d\omega(V, W) = V\omega(W) - W\omega(V) - \omega[V, W] = 0.$$

To prove their equality, it suffices therefore to prove for any vector field V on M

$$i(V)i(Z)(-\theta(Z)\omega \wedge \omega) = i(V)i(Z)d\omega .$$

This follows by a direct verification. $\qquad \square$

The existence of the global one-form α leads to the following construction of Godbillon and Vey ([Go-V], 1971).

2.3 THEOREM. *Let M be a smooth manifold, $\omega \in \Omega^1(M)$ a nonsingular integrable one-form, and α a global one-form such that $d\omega = \alpha \wedge \omega$. Then*

(i) $d(\alpha \wedge d\alpha) = 0$,
(ii) *the De Rham cohomology class $[\alpha \wedge d\alpha] \in H^3_{DR}(M)$ is independent of the choice of α satisfying $d\omega = \alpha \wedge \omega$ (this class is called the Godbillon-Vey class of ω);*
(iii) *the De Rham cohomology class associated to ω and $h\omega$ with nonzero h is the same.*

Proof. (i) $d\omega = \alpha \wedge \omega$ implies by differentiation $d\alpha \wedge \omega = 0$. By the argument used in the proof of Proposition 2.1, this implies $d\alpha = \gamma \wedge \omega$ for some one-form γ. Therefore
$$d(\alpha \wedge d\alpha) = d\alpha \wedge d\alpha = (\gamma \wedge \omega) \wedge (\gamma \wedge \omega) = 0.$$

(ii) Let α, α' be one-forms such that $d\omega = \alpha \wedge \omega = \alpha' \wedge \omega$. Then $\alpha' = \alpha + f\omega$, so $\alpha' \wedge d\alpha' = \alpha \wedge d\alpha' + f\omega \wedge d\alpha'$. It follows that
$$\alpha' \wedge d\alpha' = \alpha \wedge d\alpha + \alpha \wedge d(f\omega),$$
since $\omega \wedge d\alpha' = 0$ by the same argument used for α. Further
$$d(\alpha \wedge f\omega) = d\alpha \wedge f\omega - \alpha \wedge d(f\omega) = -\alpha \wedge d(f\omega).$$
Thus
$$\alpha' \wedge d\alpha' = \alpha \wedge d\alpha - d(\alpha \wedge f\omega),$$
which proves (ii).

(iii) Let $\omega' = h\omega$ with nonzero h. Then
$$d\omega' = dh \wedge \omega + h d\omega = \frac{1}{h} dh \wedge h\omega + h\alpha \wedge \omega = (d\log|h| + \alpha) \wedge \omega',$$
so that $\alpha' = \alpha + d\log|h|$ can be used for the definition of the Godbillon-Vey class of ω'. But then
$$\alpha' \wedge d\alpha' = (\alpha + d\log|h|) \wedge d\alpha = \alpha \wedge d\alpha + d(\log|h| \wedge d\alpha),$$
which proves (iii). □

Consider the canonical one-form $d\theta$ on S^1. A submersion $f : M \to S^1$ defines a closed one-form $\omega = f^*d\theta$ on M. The corresponding foliation of M by the fibers of f is therefore an example of a foliation with trivial Godbillon-Vey class (in fact by construction even with trivial Godbillon-Vey form, since $\alpha = 0$ is a proper choice).

The first example of a nontrivial Godbillon-Vey class was given by Roussarie in 1971 (see [Go-V]). His idea was to consider the foliation of a Lie group G by the left cosets gH of a connected subgroup $H \subset G$. Let further Γ be a discrete subgroup of G with orbit space $\Gamma\backslash G$ (left action). The foliation of G by H defines then a foliation of $\Gamma\backslash G$ which is a compact space for co-compact Γ. An interesting case is $G = PSL(2,\mathbb{R})$, the quotient of $SL(2,\mathbb{R})$ by the subgroup $\pm id$. This is the group of complex automorphisms of the unit disc D. Let M_g be a closed Riemannian surface of genus $g > 1$ and fundamental group Γ_g. The universal covering $\tilde{M}_g \cong D$, and Γ_g appears via deck transformations as a subgroup of G. But $\mathrm{Aut}(D)$ acts simply transitively on the unit tangent bundle $T_1 D$, so that $G \cong \mathrm{Aut}(D) \cong T_1 D \cong T_1 \tilde{M}_g$, and $\Gamma_g \backslash G \cong \Gamma_g \backslash T_1 \tilde{M}_g \cong T_1 M_g$, the unit tangent bundle of M_g. A basis of the Lie algebra of G consists of the matrices $A = \begin{pmatrix} 1/2 & 0 \\ 0 & -1/2 \end{pmatrix}$, $B^+ = \begin{pmatrix} 0 & 1 \\ 0 & 0 \end{pmatrix}$, $B^- = \begin{pmatrix} 0 & 0 \\ 1 & 0 \end{pmatrix}$. The subalgebra with basis A, B^+ defines a connected subgroup $H \subset G$, and a codimension one foliation by left cosets of H. The corresponding codimension one foliation of $\Gamma_g \backslash G = T_1 M_g$ has then a nonzero Godbillon-Vey number.

To verify this fact, consider the dual basis μ, ν^+, ν^- of $\mathfrak{g}^*$. Using the Lie algebra relations $[A, B^+] = B^+$, $[A, B^-] = -B^-$, $[B^+, B^-] = 2A$ one verifies the identity
$$d\nu^- = \mu \wedge \nu^-.$$
Detailed calculations are carried out e.g. in [Bo 2, p. 62–64].

Thus the integrable one-form ν^- defines the foliation of G by H. Its Godbillon-Vey form is $\mu \wedge d\mu$. One finds that
$$\mu \wedge d\mu = -2\mu \wedge \nu^+ \wedge \nu^-,$$
which is a left-invariant volume form on G. Thus it defines a volume form on the quotient $\Gamma_g \backslash G$. It follows that its integral over the closed manifold $\Gamma_g \backslash G$ is a nonzero number. Up to the factor $4\pi^2$, this is in fact the Euler characteristic
$$\chi(M_g) = 2 - 2g \quad (g > 1).$$

As in the example just discussed, the Godbillon-Vey class is of special interest for integrable one-forms ω on closed oriented 3-manifolds. Its integral over the fundamental cycle defines the Godbillon-Vey number of ω. A lot of work on this invariant has been done since its definition by Godbillon and Vey in 1971. For S^3 Thurston showed that every real number is realized as a Godbillon-Vey number in this fashion (1972). The Godbillon-Vey number of the Reeb foliation can be shown to be zero. Unpublished work of Duminy (Lille 1982, preprint) establishes that the Godbillon-Vey class vanishes if all the leaves of a codimension one foliation are manifolds with nonexponential growth.

Note that the foliations considered have an orientation of the normal direction defined by the sign of ω. A transversal orientation of a foliation $\mathcal{F}$ is an orientation of the normal bundle. Two codimension one transversally oriented foliations

$(M, \mathcal{F})$ and $(M', \mathcal{F}')$ on closed oriented n-manifolds are foliated cobordant if there exists a codimension one transversally oriented foliation $(W, \mathcal{H})$ of a compact oriented $(n+1)$-manifold W such that $\partial W = (-M) \cup M'$, $\mathcal{H}$ is transverse to ∂W and the restrictions $\mathcal{H} \mid M$ and $\mathcal{H} \mid M'$ coincide with $\mathcal{F}$ and $\mathcal{F}'$, respectively. The foliated cobordism classes form an additive group $\mathcal{F}\Omega_{n,1}$. The foliations $(M, \mathcal{F})$ representing the zero of this group are those cobordant to the empty set (null cobordant).

It is a consequence of Stokes' formula that the Godbillon-Vey number of $(M^3, \mathcal{F})$ depends only on its foliated cobordism class, and that $GV: \mathcal{F}\Omega_{3,1} \to \mathbb{R}$ is a homomorphism. This homomorphism is surjective by the result of Thurston mentioned above. It is still an open question if this map is injective or not. For an excellent survey on the then current knowledge on the Godbillon-Vey invariant of codimension one foliations on closed 3-manifolds, we refer to the survey article by T. Tsuboi, A characterization of the Godbillon-Vey invariant, Sugaku Expositions 8 (1995), 165–182.

The formula $\alpha = -\theta(Z)\omega$ proved in Proposition 2.2 yields the following formula due to Thurston.

2.4 PROPOSITION. *Let Z be a vector field with $i(Z)\omega = 1$. Then the Godbillon-Vey class of the foliation defined by ω is represented by the form*

$$-\omega \wedge \theta(Z)\omega \wedge \theta(Z)^2 \omega.$$

Proof. $d\alpha = d(-\theta(Z)\omega) = -\theta(Z) d\omega = -\theta(Z)(-\theta(Z)\omega \wedge \omega) = \theta(Z)^2 \omega \wedge \omega$. This yields the desired formula for $\alpha \wedge d\alpha$. □

As an illustration consider the flow ψ_t generated by Z. If $\psi_t^* \omega = \omega$ for all t, then

$$\theta(Z)\omega = \frac{d}{dt}\bigg|_{t=0} \psi_t^* \omega = 0,$$

so the Godbillon-Vey form of ω vanishes. In fact under this hypothesis we have even necessarily $d\omega = 0$. Conversely $d\omega = 0$ implies $\theta(Z)\omega = di(Z)\omega = 0$. So this invariance property under the flow of Z happens precisely for closed ω.

Next we give a geometric interpretation of the Godbillon-Vey construction in terms of a connection and its curvature form. For this purpose let Q^* denote the bundle of multiples of the form ω (the dual normal bundle). Since ω is non-singular, $Q^* \subset T^*M$ is a line bundle in the cotangent bundle. The form ω is a trivializing section of Q^*. A connection in Q^* is therefore completely determined by the formulas

$$\nabla_X \omega = \alpha(X) \cdot \omega$$
$$\nabla_X (f\omega) = Xf \cdot \omega + f \cdot \nabla_X \omega$$

for vector fields X and functions f. It is immediate to verify that ∇ satisfies all properties required of a linear connection. Let R be the curvature tensor of ∇.

2.5 PROPOSITION. *For vector fields X, Y we have a bundle map $R(X,Y) : Q^* \to Q^*$, which is given by $R(X,Y)\omega = d\alpha(X,Y)\omega$.*

Proof. $R(X,Y)\omega = \nabla_X \nabla_Y \omega - \nabla_Y \nabla_X \omega - \nabla_{[X,Y]}\omega = X\alpha(Y)\omega +$
$\alpha(Y)\alpha(X)\omega - Y\alpha(X)\omega - \alpha(X)\alpha(Y)\omega - \alpha[X,Y]\omega = d\alpha(X,Y)\omega.$ □

Assuming $i(V)\omega = 0$ and $i(W)\omega = 0$, it follows that $d\alpha(V,W) = 0$. Namely $d\omega = \alpha \wedge \omega$ implies $0 = d\alpha \wedge \omega$, hence locally $d\alpha = \gamma \wedge \omega$. Thus $d\alpha(V, W) = \gamma(V)\omega(W) - \gamma(W)\omega(V) = 0$. This implies by the proposition above that $R(V,W) = 0$. In other words the curvature is zero along the leaves of the foliation defined by ω. Thus the bundle Q^* is flat along the leaves.

In view of the observations above, the Godbillon-Vey form $\alpha \wedge d\alpha$ is the exterior product of the connection form α and the corresponding curvature 2-form. This is in contrast to the Chern-Weil construction, where characteristic differential forms are constructed via invariant polynomials from the curvature form alone. Our discussion indicates how to generalize the Godbillon-Vey construction, using more general products of connection and curvature forms. This leads to the theory of characteristic classes of foliations (see e.g. [K-To 1] and the account in [K-To 2]).

A final remark on the topic of integrable one-forms. The Frobenius condition $\omega \wedge d\omega = 0$ is equivalent to the local solvability of $\omega = g\, df$ with nonzero g. Assume this equation to hold globally. Then with $\alpha = d\, \log |g|$ we have

$$d\omega = dg \wedge df = \frac{1}{g}\, dg \wedge g\, df = \alpha \wedge \omega.$$

Since $d\alpha = 0$, the Godbillon-Vey class is necessarily zero. In other words, for an integrable form ω with nontrivial Godbillon-Vey class, no global integrating factor g can exist.

In the presence of a Riemannian metric g on M, one can discuss further properties of leaves of $\mathcal{F}$. As an illustration, consider the following problem. Let $\mathcal{F}$ be given by a nonsingular integrable one-form ω. Is it possible to find a Riemannian metric g on M such that each leaf $\mathcal{L}$ of $\mathcal{F}$ is a minimal submanifold of (M, g)? A foliation is called taut if this is the case. We prove the following result of Sullivan ([Su 3], 1979).

2.6 THEOREM. *Let M^3 be a closed 3-manifold with finite fundamental group. A foliation $\mathcal{F}$ of codimension one given by a nonsingular one-form is not taut.*

As an example, the Reeb foliation of S^3 is not taut.

Proof. Let (M,g) be equipped with a metric such that the leaves of $\mathcal{F}$ are minimal. We can assume that after normalization we have $|\omega| = 1$. For the following we will use an orientation of M^3. If necessary, we pass to an oriented 2-fold covering $\tilde{M}$, and lift $\mathcal{F}$ to a foliation $\tilde{\mathcal{F}}$ on $\tilde{M}$. The leaves of $\tilde{\mathcal{F}}$ are then also minimal with respect to the lifted metric $\tilde{g}$. Let $\chi = *\omega$, where $*$ is the Hodge star operator on (M, g). Then $\omega \wedge \chi = g(\omega, \omega)\mu = \mu$, where μ is the volume form of the orientable manifold (M, g). By the theorem of Novikov quoted earlier there is a closed leaf $\mathcal{L}$. Then $\int_{\mathcal{L}} \chi > 0$, since $\chi \mid \mathcal{L}$ is a volume form.

Assuming for a moment $d\chi = 0$, then $\chi = d\gamma$ with $\gamma \in \Omega^1(M)$, since $[\chi] \in H^2_{DR}(M)$ is necessarily 0 with the hypothesis of the theorem. But then

$$\int_{\mathcal{L}} \chi = \int_{\mathcal{L}} d\gamma = 0,$$

since $\mathcal{L}$ is closed, a contradiction.

It remains to prove $d\chi = 0$. Since $d\chi = d(*\omega) = \pm * \delta\omega$, where δ is the formal adjoint of d, it suffices to show that $\delta\omega = 0$. Let E_1, E_2, E_3 be an orthonormal frame of (M, g), such that $\omega(E_1) = 0$, $\omega(E_2) = 0$, and $\omega(E_3) = 1$. Then

$$\delta\omega = -\sum_{A=1}^{3}(\nabla_{E_A}\omega)(E_A) = -\sum_{A}\{E_A\omega(E_A) - \omega(\nabla_{E_A}E_A)\},$$

where ∇ denotes the Levi Civita connection of (M, g). Note that

$$\omega(\nabla_{E_3}E_3) = g(E_3, \nabla_{E_3}E_3) = \frac{1}{2}E_3 g(E_3, E_3) = 0.$$

It follows that

$$\delta\omega = \omega(\nabla_{E_1}E_1) + \omega(\nabla_{E_2}E_2).$$

The mean curvature vector field τ of the leaf surfaces is the E_3-component of the vector field $\nabla_{E_1}E_1 + \nabla_{E_2}E_2$. For the reader unfamiliar with this concept, we refer to the definition (3.21) and formula (3.22) in a more general context in Chapter 3. See also the geometric interpretation given in the paragraph following (3.22). In any case, in the present context it follows that

$$\delta\omega = \omega(\tau).$$

If the surfaces are minimal, then $\tau = 0$, and the desired result $\delta\omega = 0$ follows. □

An interesting study concerns umbilical codimension one foliations on a closed oriented 3-manifold M. A transversely oriented codimension one foliation on M is umbilical with respect to some Riemannian metric on M, if every leaf is an umbilical surface. An observation by Brunella and Ghys [Bru-Gh] is that this property is equivalent to the existence of a transversal flow, which is transversely holomorphic, i.e. defined locally by submersions to the complex plane which differ by holomorphic maps. This observation leads to a classification by Brunella and Ghys of transversely oriented codimension one umbilical foliations on M^3. A consequence of their classification is that if M^3 admits such a foliation, then it is, up to diffeomorphism, the total space of a Seifert fibration or of a torus bundle over the circle.

We consider next a codimension one foliation $\mathcal{F}$ given by a closed nonsingular one-form ω. Let M^{n+1} be closed. The De Rham duality map assigns to ω a current $c \in C_n$ by
$$c(\chi) = \int_M \chi \wedge \omega \quad \text{for} \quad \chi \in \Omega^n(M).$$
$\Omega(M)$ is equipped with the C^∞-topology of uniform convergence in the coefficients of forms and all its derivatives, and the currents are the continuous linear forms on $\Omega(M) : C_r = \Omega^r(M)'$ (topological dual). The current c is closed, since by definition
$$(\partial c)(\alpha) = \pm c(d\alpha) = \int_M d\alpha \wedge \omega = \int_M d(\alpha \wedge \omega) = 0$$
for $\alpha \in \Omega^{n-1}(M)$. Let moreover $\chi \in \Omega^n(M)$ be such that χ restricted to the leaves is a volume form on the leaves (as for $\chi = *\omega$ in the proof of Theorem 2.6). Then
$$c(\chi) = \int_M \chi \wedge \omega > 0,$$
since $\chi \wedge \omega$ is a volume form on M. These two properties are the defining properties of a Ruelle-Sullivan current for $\mathcal{F}$. In the case at hand it is given by the De Rham (Poincaré) dual of ω. This current reflects the dynamics of $\mathcal{F}$ and plays the role of a generalized closed leaf of $\mathcal{F}$. In the case of the Reeb foliation on S^3, the closed leaf defines a Ruelle-Sullivan current which does not arise in the fashion described above, since the Reeb foliation is not given by a closed one-form. In the case of the Kronecker foliation of the torus T^2 by lines of irrational slope, the Ruelle-Sullivan current represents a one-homology class playing the role of a generalized closed leaf, in spite of the fact that the leaves are dense in T^2.

At this point we raise the question if any n-manifold can occur as a leaf of a codimension one foliation on a closed $(n+1)$-manifold M? Ghys has shown that for every $n \geq 3$ there exists a noncompact n-manifold which does not occur as a leaf of a codimension one foliation of a closed M^{n+1} (Topology 24 (1985), 67-73). The idea is that any such leaf must spiral in on some nontrivial limit set. This imposes some recurrence properties on the leaf. The construction is a manifold whose topology gets forever changed as one goes to infinity.

Another situation where codimension one foliations occur is in the discussion of manifolds of cohomogeneity one. Let $G = \text{Iso}(M, g)$ be the (compact) isometry group of a closed Riemannian manifold. One way to analyze (M, g) is to examine the geometry of the orbits of this action. The simplest examples are homogeneous spaces, where there is just one orbit. This is the case for the space forms of constant curvature. The next simple class are the spaces where all orbits are of codimension one. They are said to be of cohomogeneity one.

We conclude this chapter with the following general comment. Some topological theorems have foliation proofs. The idea is the same as in tomography: slice the space, study the slices (leaves), then rebuild the space from the slices.

An example is Reeb's theorem on the structure of a closed manifold M^{n+1} with a function having no critical point except a single minimum and a single maximum. The level hypersurfaces of f form a codimension one foliation of M^{n+1} with two exceptional leaves for the extremal points. Reeb's observation is that this is enough to construct a homeomorphism with the sphere S^{n+1}, where the extremal points are mapped to the north and south poles, while the other level hypersurfaces are mapped to latitudinal hyperspheres on S^{n+1}, parallel to the equatorial hypersphere. This is a typical idea in the spirit of tomography. Another example is the construction of a holomorphic diffeomorphism of a simply connected region in $\mathbb{C}$ with the unit disk, as in the Riemann mapping theorem. Thinking of the family of curves with an exceptional singular point corresponding to the concentric circles in the unit disk centered at the origin, it is clearly enough to construct such a foliation for any simply connected region in $\mathbb{C}$ to prove the Riemann mapping theorem.

Chapter 3
Holonomy, Second Fundamental Form, Mean Curvature

We begin by discussing a construction of flat bundles. Let $h : F \to F$ be a diffeomorphism of a smooth manifold F. The product foliation defined by the inverse images of the projection $\mathbb{R} \times F \to F$ is invariant under the action $(t,y)^n = (t+n, h^n(y))$ of h^n, $n \in \mathbb{Z}$. This means that the quotient $\mathbb{R} \times_{\mathbb{Z}} F$ carries a one-dimensional foliation transverse to the fibers of $M_h = \mathbb{R} \times_{\mathbb{Z}} F \to \mathbb{R}/\mathbb{Z} \cong S^1$. The effect of h on a point $x \in M_h$ is obtained as the endpoint of the unique horizontal lift of S^1 to M_h with initial point x. For the corresponding flow on M_h transverse to the fibers, the map h is the Poincaré map of the flow. The diffeomorphism h is called the monodromy of the fibration $M_h \to S^1$. If h preserves a particular geometric structure on F, then there is a natural family of such structures on the fibers of $M_h \to S^1$. The open Moebius band is an example of such a bundle. In this case $F = (-1, 1)$ and $h(y) = -y$. It is the total space of the fibration $\mathbb{R} \times_{\mathbb{Z}} \mathbb{R} \to S^1$ with $(x, y) \sim (x+n, (-1)^n y)$ being the equivalence relation defining the total space. The leaves are circles, which are 2-fold coverings of the central circle $y = 0$, except for the central circle itself. If $F = S^1$ and $h : S^1 \to S^1$ is the rotation through an angle α, then the resulting foliation on $\mathbb{R} \times_{\mathbb{Z}} S^1 = T^2$ is a linear foliation of the 2-torus. An interesting situation arises if $h^p = id$ for some positive integer p. This is the case if the angle $\alpha = 2\pi/p$. Then the monodromy of $M_h \to S^1$ is the cyclic group $\mathbb{Z}_p$.

This construction generalizes as follows. Let B be a manifold with fundamental group $\Gamma = \pi_1 B$. Consider a homomorphism $h : \Gamma \to \text{Diff}(F)$ into the group of diffeomorphisms of another manifold F. The group Γ acts on the universal covering space $\tilde{B}$ by deck transformations and on F via h, thus on the product $\tilde{B} \times F$ by

$$(\tilde{b}, y)^\gamma = (\tilde{b} \cdot \gamma, h(\gamma^{-1})y), \quad \gamma \in \Gamma.$$

The projection $\tilde{B} \times F \to F$ defines a foliation $\tilde{\mathcal{F}}$, which is preserved by Γ. Thus the orbit space $\tilde{B} \times_\Gamma F$, defined by the equivalence relation $(\tilde{b} \cdot \gamma, y) \sim (\tilde{b}, h(\gamma)y)$, inherits a foliation $\mathcal{F}$, transverse to the fibers of the projection $\tilde{B} \times_\Gamma F \to \tilde{B}/\Gamma = B$.

The leaves carry locally the geometry of B, while the normal bundle of the foliation is the tangent bundle along the fibers of $\tilde{B} \times_\Gamma F \to B$. It carries any geometric structure of F preserved by the action of $h(\Gamma)$. For example, if Γ acts by isometries of a Riemannian structure on F, then a canonical Riemannian structure results on the normal bundle of the foliation. In this sense F models the normal or transversal geometry of the foliation, while B models its tangential or leaf geometry.

The representation h of Γ by diffeomorphisms of F is called the holonomy of $\mathcal{F}$. It can be recovered from the structure of the flat bundle $\tilde{B} \times_\Gamma F \stackrel{\pi}{\to} B$ as follows. The restriction of the projection to each leaf is a covering map, and thus has the unique path lifting property. This defines for each path $\alpha : [0,1] \to B$ a unique diffeomorphism of fibers $\tau(\alpha) : \pi^{-1}(\alpha(0)) \to \pi^{-1}(\alpha(1))$, and thus for a closed path γ with basepoint b_0 a diffeomorphism $\tau(\gamma) : \pi^{-1}(b_0) \to \pi^{-1}(b_0)$ of the fiber over b_0. The covering homotopy property implies that $\tau(\gamma)$ depends only on the homotopy class of γ. Then after identifying $\pi^{-1}(b_0)$ with F, the map $\tau(\gamma)$ can be identified with $h(\gamma)$.

Interesting examples are obtained by again considering as in Chapter 2 the unit tangent bundle $T_1 M_g$ of a closed Riemannian surface M_g of genus $g > 1$ and fundamental group Γ_g. The universal covering $\tilde{M}_g$ is diffeomorphic to the unit disc D and $T_1 M_g = \Gamma_g \backslash T_1 D$. The tangent circle bundle $T_1 D$ is diffeomorphic to the product $D \times S^1$, with $S^1 = \partial D$, by sending each unit vector v to the limit point on S^1 of the geodesic ray emanating from v. This trivialization defines a foliation on $T_1 D$ by projecting onto S^1. A leaf of this foliation consists of unit tangent vectors with asymptotic geodesic rays. This foliation is invariant under Γ_g, and hence descends to a foliation of codimension one of $T_1 M_g$. These foliations are called Anosov foliations. They are transversal to the fibers of $T_1 M_g \to M_g$.

Next we consider the case of flat principal bundles. Let G be a Lie group with Lie algebra $\mathfrak{g}$, $P \to M$ a principal G-bundle on M, and ω a connection on P. Thus $\omega \in \Omega^1(P, \mathfrak{g})$, and the following properties hold: (i) $\omega(X^*) = X$ for $X \in \mathfrak{g}$ and the corresponding vector field X^* on P, (ii) $R_g^* \omega = \mathrm{Ad}(g^{-1})\omega$ for $g \in G$ and the corresponding right action R_g on P. The curvature of ω is given by

$$\Omega = d\omega + \frac{1}{2}[\omega, \omega] \in \Omega^2(P, \mathfrak{g}).$$

Consider the horizontal subbundle $H \subset TP$, defined by $\omega = 0$.

3.1 Lemma. *H is integrable if and only if the curvature vanishes.*

Proof. Note first that for arbitrary vector fields X, Y on P

$$[\omega, \omega](X, Y) = [\omega(X), \omega(Y)] - [\omega(Y), \omega(X)] = 2[\omega(X), \omega(Y)].$$

Further for horizontal vector fields X, Y we have $d\omega(X, Y) = -\omega[X, Y]$. If H is integrable, it follows that $\Omega = 0$. If conversely $\Omega = 0$, for horizontal X, Y we have then $d\omega(X, Y) = \Omega(X, Y) = 0$, and thus $\omega[X, Y] = 0$, which proves the horizontality of $[X, Y]$. Thus H is integrable. $\square$

A connection ω with $\Omega = 0$ is a flat connection. A bundle $P \to M$ with a flat connection is said to be flat. It carries a foliation transverse to the fibers, where the leaves are the integral manifolds of M. Its parallel transport defines the holonomy homomorphism $h : \Gamma = \pi_1 M \to G$ and

$$P \cong \tilde{M} \times_\Gamma G$$

where Γ acts by left translations on G. This is a special case of the situation described before.

We show how to construct this isomorphism from the holonomy homomorphism h of the flat connection in P. We define a map $\tilde{\varphi} : \tilde{M} \times G \to P$ by the following procedure. Let $\tilde{x}_0 \in \tilde{M}$ be a base point over $x_0 \in M$ and $u_0 \in P$ over $x_0 \in M$. For a path $\tau : [0,1] \to \tilde{M}$ from $\tilde{x}_0$ to $\tilde{x}$ in $\tilde{M}$ let $\sigma : [0,1] \to M$ be its projection in M joining x_0 to x, and $\tilde{\sigma} : [0,1] \to P$ its horizontal path lift in P with $\tilde{\sigma}(0) = u_0$, $\tilde{\sigma}(1) = u$. Any other path from $\tilde{x}_0$ to $\tilde{x}$ in $\tilde{M}$ is homotopic to the original path, hence leads to a path σ_1 homotopic to σ in M and to a horizontal lift $\tilde{\sigma}_1$ in P. As the connection is assumed to be flat, both σ_1 and σ lie in the same integral manifold through u_0, and the end point $\tilde{\sigma}_1(1)$ depends only on the homotopy class $[\sigma_1] = [\sigma] \in \pi_1(M)$. This shows that $\tilde{\varphi}(\tilde{x}, g) = u \cdot g$ is well defined. In particular, the trivial section $\varphi(\tilde{x}, e)$ is mapped to the path lift $\tilde{\sigma}(1)$.

One verifies from the construction that $\tilde{\varphi}$ satisfies the following properties:

$$\tilde{\varphi}(\tilde{x} \cdot \gamma, g) = \tilde{\varphi}(\tilde{x}, h(\gamma)g),$$
$$\tilde{\varphi}(\tilde{x}, gg') = \tilde{\varphi}(\tilde{x}, g) \cdot g',$$

where $\gamma \in \Gamma = \pi_1 M$ and $g, g' \in G$. The first of these properties implies that $\tilde{\varphi}$ is constant on Γ-orbits, and hence induces a map $\tilde{M} \times_\Gamma G \to P$. The second of these properties implies that this map is in fact a map of principal G-bundles (over the identity), and hence necessarily an isomorphism.

The corresponding structure theorem for a smooth (not necessarily principal) fibration $F \to E \to B$ is as follows. Let $H \subset TE$ be an integrable subbundle which is transverse to the fiber at every point. If F is compact, then there is a homomorphism $h : \Gamma = \pi_1 B \to \text{Diff}(F)$ and a diffeomorphism

$$E \cong \tilde{B} \times_\Gamma F.$$

The interest of the flat bundle construction stems from the fact that locally every foliation has this structure. The idea is to take a tubular neighborhood of a leaf $\mathcal{L}$. It has the structure of a flat bundle.

For the following discussion, we chose to consider the (transversal or normal) vector bundle Q, rather than its (transversal or normal) frame bundle. The key fact is the existence of the Bott connection in Q defined by

(3.2) $$\overset{\circ}{\nabla}_V s = \pi[V, Y_s] = \theta(V)s \quad \text{for } V \in \Gamma L \text{ and } s \in \Gamma Q,$$

where $Y_s \in \Gamma TM$ is any vector field projecting to s under $\pi : TM \to Q$. It is a partial connection along L (only defined for $V \in \Gamma L$), but otherwise satisfies the usual connection properties. First we observe that the RHS in this definition is independent of the choice of Y_s. Namely the difference of two such choices is a vector field $W \in \Gamma L$, and $[V, W] \in \Gamma L$ so that $\pi[V, W] = 0$.

The curvature $\overset{\circ}{R}(V, W) = \overset{\circ}{\nabla}_V \overset{\circ}{\nabla}_W - \overset{\circ}{\nabla}_W \overset{\circ}{\nabla}_V - \overset{\circ}{\nabla}_{[V,W]}$ for $V, W \in \Gamma L$ is zero, as a consequence of the Jacobi identity for the bracket of vector fields. This means that Q restricted to each leaf $\mathcal{L}$ is a flat vector bundle. The parallel transport in Q along a path in $\mathcal{L}$ is a linearized version of a more fundamental holonomy explained below. The vanishing of $\overset{\circ}{R}$ along L is equivalent to the property that the parallel transport in $Q \mid L$ depends only on the homotopy class of a path in a leaf.

This is how one defines the more fundamental holonomy for a foliation on M alluded to above. A transversal manifold T is a submanifold of M such that at each point $x \in T$ the tangent space is a direct complement to $L_x \subset T_x M$. A transversal manifold may not hit each leaf, or may intersect a leaf in a complicated set. In the situation of a foliation on a flat bundle described earlier, all fibers are transversal manifolds, and the intersection with each leaf is discrete. In the general case, there exist transversal manifolds through each point. The holonomy is then defined for each path α in a leaf $\mathcal{L}$ as a germ of diffeomorphisms $\tau(\alpha)$ on the germ of transversal manifolds at the initial point $\alpha(0)$ to the germ of transversal manifolds at $\alpha(1)$, obtained by sliding transversal manifolds along the path α (for a foliation defined by a submersion, the holonomy is trivial). In general, the map $\tau(\alpha)$ depends only on the homotopy class of the path α in the given leaf. For a loop γ in a leaf $\mathcal{L}$ with basepoint x_0, we have therefore a germ of diffeomorphisms $h(\gamma)$, mapping the germ of transversal manifolds at x_0 to itself. h is a representation of $\pi_1(\mathcal{L}, x_0)$ by diffeomorphisms on the germ of transversal manifolds at x_0. This is the generalization of Poincaré's first return map from flows to foliations, and is an important tool in the theory of foliations.

What we have described earlier is a linearized version of this more fundamental holonomy. The linearized version takes place in Q, playing formally the role of the tangent bundle of the transversal manifold. It is defined via the partial Bott connection in Q. In the case we are mainly interested in, the (genuine) holonomy is acting by isometric transformations on (germs of) transversal manifolds. Since a local isometry is completely determined by its one-jet at a point, nothing is lost in the linearization process.

An adapted connection in Q is a connection restricting along L to the partial Bott connection $\overset{\circ}{\nabla}$. To show that such connections exist, consider a Riemannian metric g on M. Then TM splits orthogonally as $TM = L \oplus L^\perp$. This means that there is a bundle map $\sigma : Q \overset{\cong}{\to} L^\perp \subset TM$ splitting the exact sequence $0 \to L \to TM \to Q \to 0$, i.e. satisfying $\pi \circ \sigma = identity$. The metric g on TM is then a direct sum

$$g = g_L \oplus g_{L^\perp}.$$

With $g_Q = \sigma^* g_{L^\perp}$, the splitting map $\sigma : (Q, g_Q) \to (L^\perp, g_{L^\perp})$ is a metric isomorphism. Let now ∇^M be the Levi Civita connection associated to the Riemannian metric g on M. Then for $s \in \Gamma Q$ and $Y_s = \sigma(s) \in \Gamma L^\perp$ the definition

(3.3) $$\nabla_X s = \begin{cases} \pi[X, Y_s] & \text{for } X \in \Gamma L, \\ \pi(\nabla_X^M Y_s) & \text{for } X \in \Gamma L^\perp, \end{cases}$$

yields an adapted connection ∇ in Q. Its curvature R_∇ coincides with $\overset{\circ}{R}$ for V, $W \in \Gamma L$, hence $R_\nabla(V, W) = 0$ for $V, W \in \Gamma L$.

A connection ∇ in Q defines a connection ∇^* in Q^* by the formula

$$(\nabla_X^* \omega)(s) = X\omega(s) - \omega(\nabla_X s)$$

for $X \in \Gamma TM$, $\omega \in \Gamma Q^*$ and $s \in \Gamma Q$. For the partial Bott connection $\overset{\circ}{\nabla}$ this yields formally

$$(\overset{\circ}{\nabla}_V^* \omega)(Y) = V\omega(Y) - \omega[V, Y]$$

for $V \in \Gamma L$ and $Y \in \Gamma TM$ with $\pi(Y) = s$. Thus

(3.4) $$\overset{\circ}{\nabla}_V^* \omega = \theta(V)\omega \quad \text{for } \omega \in \Gamma Q^* \subset \Omega^1(M).$$

An adapted connection ∇ in Q is holonomy invariant, provided

(3.5) $$(\theta(V)\nabla)_Y s = 0 \quad \text{for } V \in \Gamma L, \quad Y \in \Gamma TM \text{ and } s \in \Gamma Q.$$

Here by definition

$$(\theta(V)\nabla)_Y s = \theta(V)\nabla_Y s - \nabla_{\theta(V)Y} s - \nabla_Y \theta(V) s,$$

and $\theta(V)s = \overline{[V, Y_s]}$ for $Y_s \in \Gamma TM$ with $\overline{Y}_s = s$ as in Chapter 1. We say that a connection in Q is basic, if it is adapted and holonomy invariant. The existence of a basic connection is by no means certain. In fact, there is a topological obstruction for the existence of such connections. This was found independently by Kamber-Tondeur and Molino and is discussed e.g. in [K-To 1] and [Mo 2]. From the fifth chapter on we will concentrate on Riemannian foliations, where this existence problem is not an issue, and basic connections always exist. The following properties of basic connections are fundamental.

3.6 PROPOSITION. *Let ∇ be a basic connection in Q with curvature R_∇. Then for $V \in \Gamma L$ the following holds:*
  (i) $i(V)R_\nabla = 0$,
  (ii) $\theta(V)R_\nabla = 0$.

Proof. (i) Let $Y \in \Gamma TM$ and $s \in \Gamma Q$. Then
$$R_\nabla(V,Y)s = \nabla_V \nabla_Y s - \nabla_Y \nabla_V s - \nabla_{[V,Y]} s = \theta(V)\nabla_Y s - \nabla_Y \theta(V) s - \nabla_{\theta(V)Y} s$$
$$= (\theta(V)\nabla)_Y s = 0.$$

(ii) Let $Y, Z \in \Gamma TM$ and $s \in \Gamma Q$. Then
$$(\theta(V) R_\nabla)(Y,Z)s$$
$$= \theta(V) R_\nabla(Y,Z)s - R_\nabla(\theta(V)Y, Z)s - R_\nabla(Y, \theta(V)Z)s - R_\nabla(Y,Z)\theta(V)s$$
$$= \theta(V)\{\nabla_Y \nabla_Z s - \nabla_Z \nabla_Y s - \nabla_{[Y,Z]} s\}$$
$$- \{\nabla_{\theta(V)Y} \nabla_Z s - \nabla_Z \nabla_{\theta(V)Y} s - \nabla_{[\theta(V)Y,Z]} s\}$$
$$- \{\nabla_Y \nabla_{\theta(V)Z} s - \nabla_{\theta(V)Z} \nabla_Y s - \nabla_{[Y,\theta(V)Z]} s\}$$
$$- \{\nabla_Y \nabla_Z \theta(V) s - \nabla_Z \nabla_Y \theta(V) s - \nabla_{[Y,Z]} \theta(V) s\}$$
$$= \nabla_Y (\theta(V)\nabla_Z s) - \nabla_Z(\theta(V)\nabla_Y s) - \nabla_{\theta(V)[Y,Z]} s$$
$$+ \nabla_Z \nabla_{\theta(V)Y} s + \nabla_{[\theta(V)Y,Z]} s - \nabla_Y \nabla_{\theta(V)Z} s + \nabla_{[Y,\theta(V)Z]} s$$
$$- \nabla_Y \nabla_Z \theta(V) s + \nabla_Z \nabla_Y \theta(V) s$$
$$= -\nabla_{\theta(V)[Y,Z]} s + \nabla_{[\theta(V)Y,Z]} s + \nabla_{[Y,\theta(V)Z]} s$$
$$= (-\nabla_{[V,[Y,Z]]} + \nabla_{[[V,Y],Z]} + \nabla_{[Y,[V,Z]]}) s = 0. \qquad \square$$

The object of fundamental interest for the study of a foliation is the set $\mathfrak{A}_B$ of all basic connections in Q. The gauge transformations preserving the foliated structure of Q form a subgroup $\mathfrak{G}_\mathcal{F}$ of the usual gauge group $\mathfrak{G}$, consisting of all bundle automorphisms of Q over the identity on M. Then $\mathfrak{G}_\mathcal{F}$ acts on $\mathfrak{A}_B$ by pull-back. The orbit space $\mathfrak{A}_B / \mathfrak{G}_\mathcal{F}$ is the space of gauge equivalence classes of basic connections. Thus all curvature considerations should be viewed as being parametrized by $\mathfrak{A}_B/\mathfrak{G}_\mathcal{F}$, and ideally one might wish to calculate averages over this space. Important examples of functions on $\mathfrak{A}_B/\mathfrak{G}_\mathcal{F}$ are Wilson loop functions. They are obtained as traces of the holonomy of basic connections in Q around closed loops. Since these traces are invariant under gauge transformations in Q, they give rise to functions on $\mathfrak{A}_B/\mathfrak{G}_\mathcal{F}$. We can consider these functions as primary observables.

In practice one will need to make particular choices of basic connections and calculate in terms of these choices. For the case of Riemannian foliations mainly considered in these notes, there is a canonical choice of basic connection, namely (3.3).

We return to the consideration of a connection ∇ in Q. The following ideas were developed in [K-To 5]. There is a torsion $T_\nabla \in \Omega^2(M, Q)$ defined by

(3.7) $$T_\nabla(Y, Y') = \nabla_Y \pi(Y') - \nabla_{Y'} \pi(Y) - \pi[Y, Y']$$

for $Y, Y' \in \Gamma TM$.

3.8 PROPOSITION. *For any metric g on M, and the adapted connection ∇ on Q defined by (3.3), we have $T_\nabla = 0$.*

Proof. For $V \in \Gamma L$, $Y \in \Gamma TM$ we have $\pi(V) = 0$ and
$$T_\nabla(V, Y) = \nabla_V \pi(Y) - \pi[V, Y] = 0.$$
For $Y, Y' \in \Gamma L^\perp$ we have
$$T_\nabla(Y, Y') = \pi(\nabla_Y^M Y') - \pi(\nabla_{Y'}^M Y) - \pi[Y, Y'] = \pi(T_{\nabla^M}(Y, Y')) = 0,$$
where T_{∇^M} is the (vanishing) torsion of ∇^M. Finally the bilinearity and skew symmetry of T_∇ imply the desired result. $\square$

For any connection ∇ in Q we consider the exterior derivative $d_\nabla : \Omega^r(M, Q) \to \Omega^{r+1}(M, Q)$, $0 \leqq r \leqq n = \dim M$. For vector fields $Y_1, \ldots, Y_{r+1}$ on M and $\omega \in \Omega^r(M, Q)$ we have

$$(3.9) \quad (d_\nabla \omega)(Y_1, \ldots, Y_{r+1}) = \sum_{i=1}^{r+1} (-1)^{i+1} \nabla_{Y_i} \omega(Y_1, \ldots, \hat{Y_i}, \ldots, Y_{r+1})$$
$$+ \sum_{i<j} (-1)^{i+j} \omega([Y_i, Y_j], Y_1, \ldots, \hat{Y_i}, \ldots, \hat{Y_j}, \ldots, Y_{r+1})$$

and $d_\nabla^2 \omega = R_\nabla \wedge \omega$. If ∇ is an adapted connection in Q, then $R_\nabla = 0$ along L. This implies that along L we get a complex $(\Omega^\bullet(M, Q), d_\nabla)$ and a resulting De Rham cohomology of forms along the leaves with values in Q. The foliated cohomology has shown up e.g. in Vaisman [V 1], Kamber-Tondeur (1971) and again in [K-To 1], Heitsch [Hi], and many more recent places.

We can consider the bundle map $\pi : TM \to Q$ as an element of $\Omega^1(M, Q)$. Then

$$(3.10) \quad (d_\nabla \pi)(Y, Y') = \nabla_Y \pi(Y') - \nabla_{Y'} \pi(Y) - \pi[Y, Y'].$$

This is precisely the formula (3.7) defining the torsion $T_\nabla \in \Omega^2(M, Q)$. By (3.8) it follows that $d_\nabla \pi = 0$ for the connection ∇ defined by (3.3) on (M, g).

In that situation we further have

$$(3.11) \quad (\nabla_X \omega)(Y_1, \ldots, Y_r) = \nabla_X \omega(Y_1, \ldots, Y_r) - \sum_{i=1}^{r} \omega(Y_1, \ldots, \nabla_X^M Y_i, \ldots, Y_r).$$

By definition
$$(\nabla \omega)(X; Y_1, \ldots, Y_r) = (\nabla_X \omega)(Y_1, \ldots, Y_r).$$
Because ∇^M is torsion free, it turns out that

$$(3.12) \quad (d_\nabla \omega)(Y_1, \ldots, Y_{r+1}) = \sum_{i=1}^{r+1} (-1)^{i+1} (\nabla_{Y_i} \omega)(Y_1, \ldots, \hat{Y_i}, \ldots, Y_{r+1}).$$

For $r = 1$ this formula reads

(3.13) $$(d_\nabla \omega)(X, Y) = (\nabla_X \omega)(Y) - (\nabla_Y \omega)(X).$$

Applied to $\pi \in \Omega^1(M, Q)$, it shows again that $d_\nabla \pi = 0$.

We consider now on (M, g) the Q-valued bilinear form

(3.14) $$\alpha = -\nabla \pi$$

involving the connection ∇ on Q given by (3.3). More explicitly

$$\alpha(X, Y) = -(\nabla \pi)(X, Y) = -\nabla_X \pi(Y) + \pi(\nabla_X^M Y).$$

3.15 PROPOSITION. α *is symmetric.*

Proof. $\alpha(X, Y) - \alpha(Y, X) = -\nabla_X \pi(Y) + \nabla_Y \pi(X) + \pi[X, Y] = -T_\nabla(X, Y)$, which vanishes. □

In the sequel we will use the symbol α for the restriction of this form to L. Thus $\alpha : L \otimes L \to Q$ is given by

(3.16) $$\alpha(U, V) = \pi(\nabla_U^M V) \quad \text{for } U, V \in \Gamma L.$$

α is the second fundamental form of the leaves of $\mathcal{F}$ in (M, g). The equation $\alpha = 0$ (along L) holds if and only if each leaf of $\mathcal{F}$ is a totally geodesic submanifold of (M, g).

To give an interpretation of the second fundamental form, consider $Y \in \Gamma L^\perp$ which is an infinitesimal automorphism of $\mathcal{F}$. For $U, V \in \Gamma L$ we have then

$$(\theta(Y)g)(U, V) = Yg(U, V) - g(\theta(Y)U, V) - g(U, \theta(Y)V).$$

But $\theta(Y)U = [Y, U] = \nabla_Y^M U - \nabla_U^M Y$, and similarly for V. Since

$$Yg(U, V) - g(\nabla_Y^M U, V) - g(U, \nabla_Y^M V) = 0,$$

this yields
$$(\theta(Y)g)(U, V) = g(\nabla_U^M Y, V) + g(U, \nabla_V^M Y).$$

The LHS involves only the restriction $g_L = g|L$. The RHS involves the term

$$g(\nabla_U^M Y, V) = Ug(Y, V) - g(Y, \nabla_U^M V) = -g(Y, \alpha(U, V)),$$

and an equal term for $g(U, \nabla_V^M Y)$. Thus finally

(3.17) $$(\theta(Y)g_L)(U, V) = -2g(Y, \alpha(U, V)).$$

This identifies the second fundamental form as the Lie derivative (up to a constant factor) of the metric along the leaves under the evolution of the flow generated by a transversal infinitesimal automorphism Y of $\mathcal{F}$. E.g. for a totally geodesic foliation this flow is by (3.17) a flow of isometries.

Associated to $\alpha : L \otimes L \to Q$ there is a shape operator or Weingarten map. For $s \in \Gamma Q$ this is the symmetric bundle map $W(s) : L \to L$ given by

$$\tag{3.18} g_Q(\alpha(U,V), s) = g(W(s)U, V)$$

for $U, V \in \Gamma L$. Let $\pi^\perp : TM \to L$ denote the orthogonal projection corresponding to the decomposition $TM = L \oplus L^\perp$ ($L^\perp \cong Q$). Then (3.16)–(3.18) imply, with Y_s corresponding to s under the identification $L^\perp \cong Q$,

$$g(W(s)U, V) = g(\nabla^M_U V, Y_s) = U g(V, Y_s) - g(V, \nabla^M_U Y_s),$$

hence

$$\tag{3.19} W(s)U = -\pi^\perp(\nabla^M_U Y_s).$$

Thus the characteristic polynomial of W is a geometric invariant associated to $\mathcal{F}$. Of particular interest is $\operatorname{Tr} W(s)$. It is linear in s, hence $\operatorname{Tr} W \in \Gamma Q^*$. We can extend $\operatorname{Tr} W$ to a one-form $\kappa \in \Omega^1(M)$ by setting

$$\tag{3.20} \begin{aligned} \kappa(V) &= 0 \quad \text{for } V \in \Gamma L, \\ \kappa(s) &= \operatorname{Tr} W(s) \quad \text{for } s \in \Gamma Q, \end{aligned}$$

where we have used the identification $L^\perp \cong Q$. This is the mean curvature one-form of $\mathcal{F}$ on (M, g). There is a unique $\tau \in \Gamma L^\perp \cong \Gamma Q$, such that

$$\tag{3.21} \kappa(s) = \operatorname{Tr} W(s) = g_Q(\tau, s) \quad \text{for } s \in \Gamma Q.$$

This is the mean curvature vector field of $\mathcal{F}$.

To evaluate κ and τ at a point $x \in M$, let $e_1, \ldots, e_n$ be an orthonormal basis of $T_x M$ such that $e_1, \ldots, e_p \in L_x$ and $e_{p+1}, \ldots, e_n \in L_x^\perp$. Then

$$\kappa(s)_x = \operatorname{Tr} W(s)_x = \sum_{i=1}^p g(W(s)e_i, e_i) = \sum_{i=1}^p g_Q(\alpha(e_i, e_i), s).$$

Comparing with (3.21) yields

$$\tag{3.22} \tau_x = \sum_{i=1}^p \alpha(e_i, e_i) \in L_x^\perp \cong Q_x.$$

Note that we have suppressed the usual factor $1/p$.

The significance of the mean curvature vector field is that it is the negative gradient vector field of the leaf volume functional for variations of the infinitesimal leaf volume under variations in normal directions. This is the content of the classical first variation formula.

The presence of the metric g on the oriented manifold M defines a star operator on $\Omega^{\cdot}(M)$. It extends to

$$* : \Omega^r(M,Q) \to \Omega^{n-r}(M,Q), \quad r \geq 0.$$

Note that the metric g on M defines the fiberwise metric g_Q on Q, so that $g(\omega, \omega')$ is well-defined for $\omega, \omega' \in \Omega^r(M,Q)$. Moreover the bilinear map $g_Q : Q \times Q \to \mathbb{R}$ gives rise to the exterior product $\omega \wedge_{g_Q} \omega' \equiv g_Q(\omega \wedge \omega')$. If μ denotes the volume form on M, then the $*$-operator above is characterized by the identity

$$g_Q(\omega \wedge *\omega') = g(\omega, \omega')\mu$$

on forms $\omega, \omega' \in \Omega^r(M, Q)$.

For any connection ∇ in Q and its associated exterior differentiation d_∇ in $\Omega^{\cdot}(M,Q)$, there is a codifferential $\delta_\nabla : \Omega^{r+1}(M,Q) \to \Omega^r(M,Q)$. It is given by

$$\delta_\nabla \omega = (-1)^{nr+1} * d_\nabla * \omega.$$

The evaluation formula for an orthonormal frame $E_1, \ldots, E_n$ is as follows:

$$\delta_\nabla \omega = -\sum_{A=1}^n i_{E_A} \nabla_{E_A} \omega.$$

For this operator to be formally the adjoint of d_∇, more compatibility conditions for ∇ and g are required. We leave this question aside for the moment, and evaluate the operator δ_∇ on $\pi \in \Omega^1(M,Q)$. Then

$$\delta_\nabla \pi = -\sum_{A=1}^n (\nabla_{E_A} \pi)(E_A).$$

Assume that $E_1, \ldots, E_p$ is a frame of L and $E_{p+1}, \ldots, E_n$ a frame of $L^\perp$. Then $\pi E_i = 0$ for $i = 1, \ldots, p$, and $\pi E_\alpha = E_\alpha$ for $\alpha = p+1, \ldots, n$. It follows that

$$\delta_\nabla \pi = -\{\sum_{\alpha=p+1}^n \nabla_{E_\alpha} E_\alpha - \sum_{A=1}^n \pi(\nabla^M_{E_A} E_A)\}.$$

Assuming that ∇ is given by (3.3), we conclude that

$$\delta_\nabla \pi = \sum_{i=1}^p \pi(\nabla^M_{E_i} E_i).$$

Comparing with (3.22), this proves the following fact.

3.23 PROPOSITION ([K-To 5]). *Let $\mathcal{F}$ be a foliation on (M,g) and ∇ the connection defined on Q via (3.3). Then the mean curvature vector field $\tau \in \Omega^0(M,Q)$ is given by the formula $\delta_\nabla \pi = \tau$.*

A foliation is harmonic, provided $\delta_\nabla \pi = 0$ (the condition $d_\nabla \pi = 0$ always holds). This means that each leaf is minimal (the mean curvature vanishes).

A case of particular interest occurs when $\tau \in V(\mathcal{F})$, i.e. $[V,\tau] \in \Gamma L$ for all $V \in \Gamma L$, or τ is projectable (see Chapter 1). In the presence of a metric, this condition can be expressed by

$$g([V,\tau], X) = 0$$

for $V \in \Gamma L$, $X \in \Gamma L^\perp$. In terms of the Bott connection $\overset{\circ}{\nabla}$, this says that $\overset{\circ}{\nabla}_V \tau = 0$ for $V \in \Gamma L$. In other words, τ is a section of $L^\perp$, which is Bott-parallel along leaves.

Consider the case of a transversally orientable foliation $\mathcal{F}$ of codimension one as in Chapter 2. It is defined by a completely integrable nonsingular one-form ω. On (M,g) there is a unit normal vector field $N \in \Gamma L^\perp$, and it is no restriction to assume that $\omega(N) = 1$.

3.25 PROPOSITION ([Walc 8]). *Let $\mathcal{F}$ be a foliation as above. The mean curvature vector field τ of $\mathcal{F}$ on (M,g) is projectable, if and only if*

$$\Theta(\tau)\omega = f\omega.$$

If this holds, then $f = g_Q(\nabla_N \tau, N)$.

Proof. We have with the notations above

$$(\theta(\tau)\omega)(N) = \tau\omega(N) - \omega([\tau, N]) = -\omega([\tau, N]).$$

Since the connection ∇^M in TM is torsion free, this yields

$$(\theta(\tau)\omega)(N) = \omega(\nabla_N^M \tau - \nabla_\tau^M N).$$

Note that $\omega(N) = 1$ means that for any normal vector field Y, $\omega(Y) = g(Y, N)$. It follows that

$$(\theta(\tau)\omega)(N) = g(\nabla_N^M \tau, N) - g(\nabla_\tau^M N, N).$$

But $2g(\nabla_\tau^M N, N) = \tau g(N,N) = 0$, so

(3.25) $$(\theta(\tau)\omega)(N) = g(\nabla_N^M \tau, N) = g_Q(\nabla_N \tau, N).$$

Assuming τ projectable, then we get for $V \in \Gamma L$

$$(\theta(\tau)w)(V) = \tau\omega(V) - \omega([\tau, V]) = 0,$$

thus together with (3.25), indeed $\theta(\tau)\omega = f\omega$ with $f = g_Q(\nabla_N \tau, N)$. If conversely $\theta(\tau)\omega$ is a multiple of ω, then for any $V \in \Gamma L$ necessarily $(\theta(\tau)\omega)(V) = 0$. But

$$(\theta(\tau)\omega)(V) = -\omega[\tau, V],$$

so that $[\tau, N] \in \ker \omega = L$. □

An example where this condition holds is the case of the foliation of (M, g) by equidimensional orbits of an isometric group action.

The mean curvature function of a codimension one foliation is defined by

$$h = g(\tau, N) = g_Q(\tau, N),$$

again suppressing the usual factor $1/p$. Note that

$$Nh = Ng(\tau, N) = g(\nabla_N^M \tau, N) = g_Q(\nabla_N \tau, N),$$

since $g(\tau, \nabla_N^M N) = 0$. This function was studied by Reeb. He observed the following fact.

3.26 THEOREM [Rb 1]. *Let h be the mean curvature function of a transversally oriented foliation of codimension one on the closed oriented Riemannian manifold (M^{n+1}, g). Then $\int_M h\mu = 0$, where μ is the volume form of M.*

Proof. We calculate div N in terms of an orthonormal frame $E_1, \ldots, E_{n+1}$ of TM^{n+1} with $E_{n+1} = N$. Then $E_1, \ldots, E_n$ are necessarily a frame of L. Note that

$$\operatorname{div} N = \sum_{i=1}^n g(\nabla_{E_i}^M N, E_i),$$

since the missing term $g(\nabla_N^M N, N) = \frac{1}{2}Ng(N, N)$ vanishes. Evaluating

$$\kappa(N) = g(\tau, N) = h$$

using (3.22) yields

$$\kappa(N) = \sum_{i=1}^n g(\alpha(E_i, E_i), N) = \sum_{i=1}^n g(\nabla_{E_i}^M N, E_i).$$

Comparing these formulas proves

(3.27) $$\operatorname{div} N = h,$$

and hence the desired result by the divergence theorem. □

3.28 COROLLARY. *Let the situation be as in Theorem 3.26, and assume moreover that τ is projectable. Then $\mathcal{F}$ has at least one minimal leaf.*

Proof. If τ is projectable, then $\tau_x = 0$ implies $\tau = 0$ for all points of the leaf through x. Thus it suffices to show that h vanishes at one point of M. This follows from Theorem 3.26. □

As an example, an orbit foliation of (M, g) by hypersurfaces for an isometric group action must have at least one minimal orbit.

As a simple illustration of the holonomy concept, we consider the case of a transversally orientable foliation $\mathcal{F}$ of codimension one defined by a nonsingular one-form ω. We consider a loop γ at x in the leaf $\mathcal{L}$ through $x \in M$. The tubular neighborhood of $\mathcal{L}$ restricted to the loop γ leads then to a 2-dimensional surface D, with two boundary components, of which one is γ. To describe the other boundary component, consider a point y of a transversal T through x. The path lift $\tilde{\gamma}$ of γ in the leaf through y leads to an endpoint $y' \in T$. The holonomy transformation $h(\gamma)$ is precisely given by $y' = h(\gamma)y$. The transversal path $\alpha_{y'y}$ in T completes $\tilde{\gamma}$ to a loop. The 2-dimensional region D is then bounded by $\tilde{\gamma} \cup \alpha \cup -\gamma$. Thus

$$\int_{\alpha_{y'y}} \omega = \int_{\partial D} \omega = \int_D d\omega .$$

The first equality is based on the fact that $\int_\gamma \omega = 0$ and $\int_{\tilde{\gamma}} \omega = 0$, since both γ and $\tilde{\gamma}$ are paths in leaves of the foliation defined by ω. Note that for a tangent vector Z to T we have $\omega(Z) \neq 0$. It follows that the LHS is 0 if and only if $y' = y$ for all $y \in T$, i.e. $h(\gamma) = id$.

For the particular case of a foliation $\mathcal{F}$ defined by a closed one-form ω, this implies the following fact.

3.29 Theorem [Rb 1]. *Let $\mathcal{F}$ be a foliation of codimension one on M. If $\mathcal{F}$ is defined by a closed nonsingular one-form, then the holonomy of every leaf is trivial.*

Another observation valid for such foliations is as follows.

3.30 Proposition [Rb 1]. *Let M be closed and $\mathcal{F}$ a foliation of codimension one defined by a closed nonsingular one-form ω. Then there exists a transversal vector field, whose flow consists of diffeomorphisms preserving $\mathcal{F}$, i.e. mapping leaves into leaves.*

Proof. Let Y be a vector field satisfying $\omega(Y) = 1$. Then

$$\theta(Y)\omega = i(Y)d\omega + di(Y)\omega = 0.$$

It follows that the flow φ_t of Y satisfies $\varphi_t^* \omega = \omega$, which proves the desired result. □

3.31 Corollary. *In the situation of Proposition 3.30 all leaves of $\mathcal{F}$ are diffeomorphic (M is assumed connected).*

Proof. The flow φ_t maps a leaf $\mathcal{L}$ into a diffeomorphic leaf $\varphi_t(\mathcal{L})$. For a fixed $\mathcal{L}$ consider $U = \bigcup_t \varphi_t(\mathcal{L})$ for $t \in \mathbb{R}$. Then $U \subset M$ is open and closed, hence coincides with M for connected M. □

Note that the map $\mathcal{L} \times \mathbb{R} \to M$ defined by $(x, t) \to \varphi_t(x)$ gives a product structure to a tubular neighborhood of $\mathcal{L}$ in M. This proves again Reeb's Theorem 3.29 (but this argument involves the completeness of the flow φ_t).

For the linear foliations on the 2-torus Corollary 3.31 applies as well to the foliation with dense leaves (diffeomorphic to $\mathbb{R}$), as to those with rational slope (and circle leaves). The latter foliations are defined by submersions while those with dense leaves are not.

The following result is of interest in this context.

3.32 THEOREM (Tischler [Ti]). *Let M be a closed manifold with a closed nonsingular 1-form ω. Then there is a fibration $f : M \to S^1$ over the circle. Moreover let $\omega' = f^*d\theta$. Then the fibration can be chosen such that $\|\omega' - \omega\| < \epsilon$, where $\epsilon > 0$ is any prescribed number.*

Here $\| \ \|$ denotes the global L^2-norm on forms, defined by a Riemannian metric on M. In terms of the foliations $\mathcal{F}$ and $\mathcal{F}'$ defined by the closed one-forms ω and ω', the statement is that for the given $\mathcal{F}$ defined by a closed 1-form there exists arbitrarily close foliations $\mathcal{F}'$ which are given by fibrations $M \to S^1$.

Tischler's Theorem characterizes the closed manifolds fibering over S^1 as those admitting a foliation of codimension one of the type described. Before turning to the proof, we consider a particularly interesting situation. Assume that the typical fiber F itself carries a foliation preserved by a diffeomorphism $h : F \to F$. Then the construction at the beginning of this chapter yields a manifold $M_h \cong \mathbb{R} \times_{\mathbb{Z}} F$ fibering over the circle S^1 (and a further one-dimensional foliation transverse to the fiber). For the case of a torus fiber $F = T^2$, the resulting 3-manifold has been analyzed in many special cases. An interesting example is the case of a matrix $A \in SL(2, \mathbb{Z})$, e.g. $A = \begin{pmatrix} 1 & 1 \\ 1 & 2 \end{pmatrix}$. Since the linear map A preserves the integral lattice $\mathbb{Z}^2$, it induces a diffeomorphism A_0 of the torus $T^2 = \mathbb{R}^2/\mathbb{Z}^2$. If, as in the example above, $\operatorname{Tr} A > 2$, the characteristic polynomial has two real eigenvalues, and the corresponding eigenvectors give rise to two complementary one-dimensional foliations invariant under the induced diffeomorphism A_0 of T^2. The resulting type of 3-manifolds has often been considered. They are surface bundles over the circle, and in particular occur prominently in Thurston's work. There it is shown that such a manifold has a canonical partition into pieces which all have a simple geometric structure. This is a typical situation conjectured by Thurston to hold for all closed 3-manifolds.

Proof of Theorem 3.32. The guideline for this proof is the observation that the periods of the one-form $f^*d\theta$ for a smooth map $f : M \to S^1$ are integer multiples of 2π. This follows from the fact that the induced homomorphism $f_* :$

$H_1(M, \mathbb{Z}) \to \mathbb{Z}$ annihilates elements of finite order, and thus induces a homomorphism of $H_1(M, \mathbb{Z})/\text{torsion}$ to $\mathbb{Z}$, sending generating cycles c to $\int_c f^* d\theta = \int_{f_*c} d\theta$. Now we turn to any closed nonsingular 1-form ω on M. We consider the universal covering $p: \tilde{M} \to M$ with base points $\tilde{x}_0$ and $x_0 = p(\tilde{x}_0)$. Then for $\tilde{x} \in \tilde{M}$ the integral of $p^*\omega$ from $\tilde{x}_0$ to $\tilde{x}$ does not depend on the choice of the path γ from $\tilde{x}_0$ to $\tilde{x}$, since $d\omega = 0$. It follows that this integral defines a map $f_\omega : \tilde{M} \to \mathbb{R}$. An equivalent fact is that the map integrating ω over the loops γ of M at x_0 depends only on the homotopy class of $\gamma \in \pi_1(M, x_0)$. Thus there is an induced map $\text{per}_\omega : \pi_1(M, x_0) \to \mathbb{R}$ which is a homomorphism. Note that under the isomorphisms

$$\text{Hom}(\pi_1 M, \mathbb{R}) \cong \text{Hom}(H_1(M, \mathbb{R}), \mathbb{R}) \cong H^1_{DR}(M),$$

the period map per_ω corresponds to the De Rham class $[\omega]$. The image of per_ω in $\mathbb{R}$ is the group generated by the periods of ω. The case $[\omega] = 0$ is excluded by the compactness assumption on M, since a function g with $\omega = dg$ would give rise to singularities of ω at the critical points of the function g.

The map $f_\omega : \tilde{M} \to \mathbb{R}$ is equivariant with respect to per_ω, i.e.

$$f_\omega(\tilde{x} \cdot \gamma) = f_\omega(\tilde{x}) + \text{per}_\omega(\gamma),$$

and thus induces a map of quotients

$$f_\omega : M \to \mathbb{R}/\text{im}\,\text{per}_\omega.$$

There are two possibilities.

(i) Suppose that some (say positive) constant multiple ω' of ω has rational periods, i.e. for all cycles c in $H_1(M, \mathbb{Z})$ the values $\omega'(c)$ are rational numbers. Then some integer multiple of ω' will have integer periods, and the corresponding period group is infinite cyclic. Replacing ω by such a form produces then a fibration $f : M \to \mathbb{R}/\mathbb{Z} = S^1$.

(ii) If no multiple of ω has rational periods, we can approximate ω by such a form ω'. More precisely, we show that for $\epsilon > 0$ there is a closed nonsingular one-form ω' with rationally related periods, and such that the global norm of $\omega' - \omega$ is less than ϵ.

Let $\omega = \omega_0 + dg$ be the De Rham-Hodge decomposition of ω with respect to a Riemannian metric. ω_0 denotes the harmonic representative of $[\omega]$ and g is a function. There is no term $\delta\alpha$, $\alpha \in \Omega^2(M)$, since $d\omega = d\delta\alpha = 0$, and thus $\langle d\delta\alpha, \alpha \rangle = \langle \delta\alpha, \delta\alpha \rangle = 0$ for the global scalar product, hence $\delta\alpha = 0$. The space of harmonic one-forms $\mathcal{H}^1$ is isomorphic to $H^1_{DR}(M)$, in which the rational points $H^1(M, \mathbb{Q})$ are dense. These are represented by harmonic forms with rational periods. Thus for any $\epsilon > 0$ we can find $\omega'_0 \in \mathcal{H}^1$ with rational periods, and such that $\|\omega'_0 - \omega_0\| < \epsilon$. The form $\omega' = \omega'_0 + dg$ for sufficiently small ϵ remains nonsingular, and thus has the desired property. $\square$

Note that the modification of the given closed one-form with irrational periods in the proof of Theorem 3.32 changes the cohomology class $[\omega]$. This cannot be achieved by an isotopy of M, which acts trivially on De Rham cohomology.

Chapter 4
Basic Forms, Spectral Sequence, Characteristic Form

Basic forms

Let $\mathcal{F}$ be an arbitrary foliation on a manifold M. A differential form $\omega \in \Omega^r(M)$ is basic, if
$$i(V)\omega = 0, \; \theta(V)\omega = 0 \quad \text{for all} \quad V \in \Gamma L.$$
In a distinguished chart $(x_1, \ldots, x_p; y_1, \ldots, y_q)$ of $\mathcal{F}$ this means that
$$\omega = \sum_{\alpha_1 < \cdots < \alpha_r} \omega_{\alpha_1 \cdots \alpha_r} dy_{\alpha_1} \wedge \cdots \wedge dy_{\alpha_r},$$
where the functions $\omega_{\alpha_1 \cdots \alpha_r}$ are independent of x, i.e. $\frac{\partial}{\partial x_i} \omega_{\alpha_1 \cdots \alpha_r} = 0$. The exterior derivative preserves basic forms, since $\theta(V)d\omega = d\theta(V)\omega = 0$, $i(V)d\omega = \theta(V)\omega - di(V)\omega = 0$ for ω basic. Thus the set $\Omega_B^{\cdot} \equiv \Omega_B^{\cdot}(\mathcal{F})$ of all basic forms constitutes a subcomplex
$$d: \Omega_B^r \to \Omega_B^{r+1}$$
of the De Rham complex $\Omega^{\cdot}(M)$. We also denote $d|\Omega_B = d_B$. Its cohomology
$$H_B^{\cdot} \equiv H_B^{\cdot}(\mathcal{F}) = H(\Omega_B^{\cdot}(\mathcal{F}), d_B)$$
is the basic cohomology of $\mathcal{F}$. It plays the role of the De Rham cohomology of the leaf space of the foliation.

For the case of codimension one, there are just two groups $H_B^0(\mathcal{F})$ and $H_B^1(\mathcal{F})$. In the general case we have the following facts.

Let $r = 0$ (and M connected). The functions in $\Omega_B^0(\mathcal{F})$ are constant along the leaves. The 0-cycles $Z_B^0(\mathcal{F})$ are functions $f: M \to \mathbb{R}$ which are locally constant, hence constant. Thus
$$H_B^0(\mathcal{F}) \cong \mathbb{R}.$$

For $r = 1$ the situation is as follows.

4.1 PROPOSITION. *The inclusion $\Omega_B^{\cdot}(\mathcal{F}) \to \Omega^{\cdot}(M)$ induces an injective map*
$$H_B^1(\mathcal{F}) \to H_{DR}^1(M).$$

Proof. Let $\omega \in \Omega^1_B(\mathcal{F})$ such that $\omega = df$ for $f : M \to \mathbb{R}$. For $V \in \Gamma L$ we have

$$Vf = i(V)df = i(V)\omega = 0.$$

It follows that $f \in \Omega^0_B(\mathcal{F})$ and $[\omega] = 0 \in H^1_B(\mathcal{F})$. $\square$

The groups $H^r_B(\mathcal{F})$ are defined for $0 \leq r \leq q$, but may be infinite-dimensional for $r \geq 2$. For the Riemannian foliations considered in the next chapter, they turn out to be finite-dimensional, provided M is a closed manifold.

Spectral sequence of $\mathcal{F}$

This is the spectral sequence determined by the following multiplicative filtration of the De Rham complex $\Omega^{\cdot} = \Omega^{\cdot}(M)$ [K-To 1,2,10].

(4.2) $\quad F^r \Omega^m = \{\omega \in \Omega^m | i(V_1) \cdots i(V_{m-r+1})\omega = 0 \text{ for } V_1, \ldots, V_{m-r+1} \in \Gamma L\}.$

It is a convenient tool to formulate many properties of foliations. F^r is a decreasing filtration by differential ideals. Clearly

$$F^0 \Omega^m = \Omega^m \quad \text{and} \quad F^{m+1} \Omega^m = 0.$$

Further

(4.3) $\quad F^r \Omega^{p+r} = \Omega^{p+r} \quad (p = \dim \mathcal{F}),$

since $(p+r) - r + 1 = p + 1$, and every $(p+r)$-form evaluated on $p+1$ vector fields tangent to $\mathcal{F}$ vanishes. Note that $F^{r+1} \Omega^{p+r}$ consists of all $(p+r)$-forms evaluating to zero on $(p+r) - (r+1) + 1 = p$ vector fields tangent to $\mathcal{F}$. Thus by definition

(4.4) $\quad F^{r+1} \Omega^{p+r} \equiv \text{``}\mathcal{F}\text{-trivial''} \ (p+r)\text{-forms}$

(forms annihilated by p vector fields tangent to $\mathcal{F}$).

An equivalent description of the filtration (4.2) is as forms $\omega \in \Omega^m$, which are locally sums of products

(4.5) $\quad \alpha \wedge \beta \text{ with } \alpha \in \Lambda^r Q^*, r \leq m$

and β of degree $m - r$. Clearly for such a form $i(V_1) \cdots i(V_{m-r+1})\omega = 0$, and conversely every $\omega \in F^r \Omega^m$ has locally a representation as a sum of forms as in (4.5). Since $\Lambda^{q+1} Q^* = 0$, this shows that

(4.6) $\quad F^{q+1} \Omega^m = 0 \quad (q = \text{codim } \mathcal{F}).$

A metric g on M gives rise to the following bigrading of $\Omega = \Omega(M)$

(4.7) $\quad \Omega^{u,v} = \Gamma(\Lambda^u Q^* \wedge \Lambda^v L^*),$

where $Q = L^\perp$. Then the filtration (4.2) is given by

$$(4.8) \qquad F^r\Omega = \oplus_{u \geq r}\Omega^{u,\cdot}.$$

The nonzero range of the groups $\Omega^{u,v}$ is $0 \leq u \leq q$ and $0 \leq v \leq p$.

In local coordinates $(x_1, \ldots, x_p; y_1, \ldots, y_q)$ the forms of bidegree (u, v) are given by

$$\sum \omega_{i_1 \cdots i_v; \alpha_1 \cdots \alpha_u} dx_{i_1} \wedge \cdots \wedge dx_{i_v} \wedge dy_{\alpha_1} \wedge \cdots \wedge dy_{\alpha_u}.$$

The exterior differential d is the sum of bihomogeneous differentials

$$(4.9) \qquad d = d_{0,1} + d_{1,0} + d_{2,-1},$$

where

$$d_{0,1} : \Omega^{u,v} \to \Omega^{u,v+1},$$
$$d_{1,0} : \Omega^{u,v} \to \Omega^{u+1,v},$$
$$d_{2,-1} : \Omega^{u,v} \to \Omega^{u+2,v-1}.$$

These maps are given by the following formulas on $\omega \in \Omega^{u,v}$. For $Y_1, \ldots, Y_u \in \Gamma L^\perp$ and $V_1, \ldots, V_{v+1} \in \Gamma L$

$$(d_{0,1}\omega)(Y_1, \ldots, Y_u; V_1, \ldots, V_{v+1})$$
$$= \sum_{i=1}^{v+1}(-1)^{u+i+1} V_i \omega(Y_1, \ldots, Y_u; V_1, \ldots, \hat{V}_i, \ldots, V_{v+1})$$
$$(4.10) \quad + \sum_{1 \leq i < j \leq v+1}(-1)^{i+j+u}\omega(Y_1, \ldots, Y_u; [V_i, V_j], V_1, \ldots, \hat{V}_i, \ldots, \hat{V}_j, \ldots, V_{v+1})$$
$$+ \sum_{\alpha=1}^{u}\sum_{j=1}^{v+1}(-1)^{u+\alpha+j}\omega(\pi[Y_\alpha, V_j], Y_1, \ldots, \hat{Y}_\alpha, \ldots, Y_u; V_1, \ldots, \hat{V}_j, \ldots, V_{v+1}).$$

For $Y_1, \ldots, Y_{u+1} \in \Gamma L^\perp$ and $V_1, \ldots, V_v \in \Gamma L$
$$(4.11)$$
$$(d_{1,0}\omega)(Y_1, \ldots, Y_{u+1}; V_1, \ldots, V_v)$$
$$= \sum_{\alpha=1}^{u+1}(-1)^{\alpha+1} Y_\alpha \omega(Y_1, \ldots, \hat{Y}_\alpha, \ldots, Y_{u+1}; V_1, \ldots, V_v)$$
$$+ \sum_{1 \leq \alpha < \beta \leq u+1}(-1)^{\alpha+\beta}\omega(\pi[Y_\alpha, Y_\beta], Y_1, \ldots, \hat{Y}_\alpha, \ldots, \hat{Y}_\beta, \ldots, Y_{u+1}; V_1, \ldots, V_v)$$
$$+ \sum_{\alpha=1}^{u+1}\sum_{j=1}^{v}(-1)^{\alpha+j}\omega(Y_1, \ldots, \hat{Y}_\alpha, \ldots, Y_{u+1}; \pi^\perp[Y_\alpha, V_j], V_1, \ldots, \hat{V}_j, \ldots, V_v).$$

Finally for $Y_1, \ldots, Y_{u+2} \in \Gamma L^\perp$ and $V_1, \ldots, V_{v-1}$

(4.12)
$(d_{2,-1}\omega)(Y_1, \ldots, Y_{u+2}; V_1, \ldots, V_{v-1})$
$$= \sum_{1 \leq \alpha < \beta \leq u+2} (-1)^{u+\alpha+\beta} \omega(Y_1, \ldots, \hat{Y}_\alpha, \ldots, \hat{Y}_\beta, \ldots, Y_{u+2}; \pi^\perp[Y_\alpha, Y_\beta], V_1, \ldots, V_{v-1})$$

The property $d^2 = 0$ implies the relations

$$d_{0,1}^2 = 0, \qquad d_{2,-1}^2 = 0,$$
$$d_{0,1}d_{1,0} + d_{1,0}d_{0,1} = 0,$$
$$d_{1,0}d_{2,-1} + d_{2,-1}d_{1,0} = 0,$$
$$d_{1,0}^2 + d_{2,-1}d_{0,1} + d_{0,1}d_{2,-1} = 0.$$

The initial term $E_0^{r,s}$ of the spectral sequence is defined as the quotient in the exact sequence

(4.13) $\qquad 0 \to F^{r+1}\Omega^{r+s} \to F^r\Omega^{r+s} \to G^r\Omega^{r+s} \equiv E_0^{r,s} \to 0.$

In the presence of a metric g this sequence is split, and an element $\bar{\omega} \in E_0^{r,s}$ is represented by $\omega \in \Omega^{r,s}$. The differential d_0 of bidegree (0,1) induced by $d_{0,1}$ in

$$E_0^{r,s} \cong \mathrm{Hom}\,(\Lambda^s L, \Lambda^r Q^*)$$

corresponds to the Chevalley-Eilenberg differential d_C on the RHS. Its homology

$$E_1^{r,s} = H^s_{d_C}(\mathrm{Hom}(\Lambda^\cdot L, \Lambda^r Q^*))$$

gives for $s = 0$
$$E_1^{r,0} = \Gamma((\Lambda^r Q^*)^L) \cong \Omega_B^r(\mathcal{F}).$$

The differential d_1 in E_1 of bidegree (1,0) induces on $\Omega_B(\mathcal{F})$ precisely the differential d_B, so that
$$E_2^{r,0} \cong H_B^r(\mathcal{F}).$$

The basic cohomology of $\mathcal{F}$ is thus the basis of the E_2-term of the spectral sequence associated to $\mathcal{F}$. For the case of a fibration, this corresponds to the position of the De Rham cohomology of the basis in the spectral sequence associated to the fibration.

Infinitesimal automorphisms

An infinitesimal automorphism $Y \in V(\mathcal{F})$ acts on $s \in \Gamma Q$ by

(4.14) $\qquad \theta(Y)s = \pi[Y, Y_s],$

where $Y_s \in \Gamma TM$ is such that $\pi(Y_s) = s$. Because $Y \in V(\mathcal{F})$, the RHS does not depend on the choice of Y_s. For the particular case of $Y = V \in \Gamma L$, this reduces to the action $\theta(V)s$ considered in (3.2).

Let more generally ω be any covariant tensor of degree r on Q. Then an infinitesimal automorphism Y acts on ω by

$$(4.15) \qquad (\theta(Y)\omega)(s_1,\ldots,s_r) = Y\omega(s_1,\ldots,s_r) - \sum_{i=1}^{r} \omega(s_1,\ldots,\theta(Y)s_i,\ldots,s_r).$$

Let now $\omega \in \Omega_B^r(\mathcal{F})$ and $Y \in V(\mathcal{F})$. Then the formulas

$$i(V)\theta(Y)\omega = \theta(Y)i(V)\omega + i[V,Y]\omega$$
$$\theta(V)\theta(Y)\omega = \theta(Y)\theta(V)\omega + \theta[V,Y]\omega$$

show that $\theta(Y)\omega$ is again a basic form. Similarly $i(Y)\omega$ is also a basic form. Thus we have $V(\mathcal{F})$ acting on $\Omega_B(\mathcal{F})$ via $i(Y)$ and $\theta(Y)$ with the usual relations. This is just one aspect of the fact that infinitesimal automorphisms are compatible with the filtration (4.2).

Characteristic form

Assume L to be oriented. The foliation $\mathcal{F}$ is then said to be tangentially oriented. The characteristic form $\chi_\mathcal{F}$ of $\mathcal{F}$ on (M,g) is defined as follows. It is a p-form on M, which for arbitrary $Y_1,\ldots,Y_p \in \Gamma TM$ is given by

$$(4.16) \qquad \chi_\mathcal{F}(Y_1,\ldots,Y_p) = \det(g(Y_i, E_j)_{ij}).$$

Note that $\chi_\mathcal{F}$ evaluated on a local oriented orthonormal frame E_i ($i = 1,\ldots,p$) of L gives the value 1 (i.e. it is the canonical volume form associated to g_L), and $i(Y)\chi_\mathcal{F} = 0$ for $Y \in \Gamma L^\perp$.

For every vector field $Y \in \Gamma TM$ we define an operator $A_Y : TM \to TM$ by the formula

$$(4.17) \qquad A_Y X = -\nabla_X^M Y, \quad X \in TM.$$

The following fact holds.

4.18 THEOREM. *Let $\mathcal{F}$ be tangentially oriented on (M,g), and $\chi_\mathcal{F}$ the characteristic form of $\mathcal{F}$. Then for $Y \in \Gamma TM$*

$$(4.19) \qquad \theta(Y)\chi_\mathcal{F} \equiv -\operatorname{Tr}_L A_Y \chi_\mathcal{F} \mod F^1\Omega^p(M).$$

The congruence modulo the $\mathcal{F}$-trivial p-forms $F^1\Omega^p$ means by (4.2) that these forms have the same evaluations on L.

Proof. Let $E_1, \ldots, E_p$ be a (local) orthonormal frame of L. Then

$$(\theta(Y)\chi_{\mathcal{F}})(E_1, \ldots, E_p) = Y\chi_{\mathcal{F}}(E_1, \ldots, E_p) - \sum_{i=1}^{p} \chi_{\mathcal{F}}(E_1, \ldots, \theta(Y)E_i, \ldots, E_p)$$

$$= -\sum_i g(\theta(Y)E_i, E_i)\chi_{\mathcal{F}}(E_1, \ldots, E_p)$$

$$= -\sum_i g([Y, E_i], E_i).$$

On the other hand,

$$\operatorname{Tr}_L A_Y = \sum_{i=1}^{p} g(A_Y E_i, E_i) = -\sum_{i=1}^{p} g(\nabla^M_{E_i} Y, E_i)$$

$$= -\sum_{i=1}^{p} g(\nabla^M_Y E_i + [E_i, Y], E_i) = -\sum_{i=1}^{p} g([E_i, Y], E_i),$$

where we have used $g(\nabla^M_Y E_i, E_i) = \frac{1}{2} Y g(E_i, E_i) = 0$. Comparison proves the desired identity. $\square$

We consider two special cases.

4.20 COROLLARY. *Let the situation be as in 4.18, and let $Y \in \Gamma L^\perp$. Then*

(4.21) $$\theta(Y)\chi_{\mathcal{F}} \equiv -\kappa(Y) \cdot \chi_{\mathcal{F}} \mod F^1 \Omega^p,$$

where κ is the mean curvature one-form of $\mathcal{F}$ (Rummler [Ru 1]).

This formula plays the role of a nonintegrated version of the first variation formula for the volume of the leaves.

Proof. For $Y \in \Gamma L^\perp$ we have

$$\operatorname{Tr}_L A_Y = \sum_{i=1}^{p} g(A_Y E_i, E_i) = -\sum_i g(\nabla^M_{E_i} Y, E_i)$$

$$= -\sum_i g(\pi^\perp(\nabla^M_{E_i} Y), E_i) = \operatorname{Tr} W(Y) = \kappa(Y). \quad \square$$

4.22 COROLLARY. *Let the situation be as in 4.18, and let $V \in \Gamma L$. Then*

(4.23) $$\theta(V)\chi_{\mathcal{F}} \equiv \operatorname{div}_L V \chi_{\mathcal{F}} \mod F^1 \Omega^p$$

(tangential divergence formula).

Proof. For $V \in \Gamma L$

$$\operatorname{Tr}_L A_V = -\sum_{i=1}^{p} g(\nabla^M_{E_i} V, E_i) = -\operatorname{div}_L V. \qquad \square$$

The following fact is also worth noting. For all $Y \in \Gamma TM$.

(4.24) $$\nabla^M_Y \chi_{\mathcal{F}} \equiv 0 \mod F^1 \Omega^p(M).$$

Proof. For an orthonormal frame $E_1, \ldots, E_p$ of L we have

$$(\nabla^M_Y \chi_{\mathcal{F}})(E_1, \ldots, E_p) = Y \chi_{\mathcal{F}}(E_1, \ldots, E_p) - \sum_{i=1}^{p} \chi_{\mathcal{F}}(E_1, \ldots, \nabla^M_Y E_i, \ldots, E_p).$$

Only the projections $g(\nabla^M_Y E_i, E_i) E_i$ contributes, and $g(\nabla^M_Y E_i, E_i) = \frac{1}{2} Y g(E_i, E_i) = 0$. This proves the desired result. $\qquad \square$

For the particular choice $Y = \tau$ we have by (3.21)

$$\kappa(\tau) = g_Q(\tau, \tau) = |\tau|^2,$$

and by (4.21) we find

(4.25) $$(\theta(\tau) \chi_{\mathcal{F}}) \equiv -|\tau|^2 \chi_{\mathcal{F}} \mod F^1 \Omega^p.$$

τ is the direction of steepest change for $\chi_{\mathcal{F}}$ under the transversal flow φ_t of τ. The vanishing of τ means roughly the invariance of $\chi_{\mathcal{F}}$ under transversal flows, the precise meaning being given by (4.25).

Formula (4.21) can be restated equivalently as

(4.26) $$d\chi_{\mathcal{F}} + \kappa \wedge \chi_{\mathcal{F}} = \varphi_0 \in F^2 \Omega^{p+1}.$$

Under which conditions can one conclude that $\varphi_0 = 0$? We prove first the following result.

4.27 Proposition. *Let $\mathcal{F}$ be a tangentially oriented foliation on (M, g), and $\chi_{\mathcal{F}}$ the characteristic form of L. Assume $L^\perp$ to be involutive. Then*

(4.28) $$\theta(Y) \chi_{\mathcal{F}} + \kappa(Y) \chi_{\mathcal{F}} = 0 \quad \text{for } Y \in \Gamma L^\perp.$$

Proof. Let $\alpha = \theta(Y)\chi_\mathcal{F} + \kappa(Y)\chi_\mathcal{F} \in \Omega^p(M)$. We know by (4.21) that $\alpha|L = 0$. It suffices to show that $i(X)\alpha = 0$ for $X \in \Gamma L^\perp$. Note that

$$(4.29) \qquad i(X)\theta(Y) = \theta(Y)i(X) - i[Y, X].$$

Thus

$$i(X)\alpha = i(X)(\theta(Y)\chi_\mathcal{F} + \kappa(Y)\chi_\mathcal{F})$$
$$= \theta(Y)i(X)\chi_\mathcal{F} - i[Y, X]\chi_\mathcal{F} + \kappa(Y)i(X)\chi_\mathcal{F}.$$

For involutive $L^\perp$ we have $[Y, X] \in L^\perp$, and $i(X)\alpha = 0$ follows. □

4.30 PROPOSITION. *Let $\mathcal{F}$ be as in Proposition 4.27 with involutive $L^\perp$. Then the following conditions are equivalent:*

(i) $\kappa = 0$, *i.e.* $\mathcal{F}$ *is harmonic;*
(ii) $\theta(Y)\chi_\mathcal{F} = 0$ *for* $Y \in \Gamma L^\perp$;
(iii) $d\chi_\mathcal{F} = 0$.

Proof. (i)⇔(ii) follows by (4.28). (iii)⇒(ii) follows since $i(Y)\chi_\mathcal{F} = 0$ for $Y \in \Gamma L^\perp$. It remains to show (ii)⇒(iii). Thus we assume $i(Y)d\chi_\mathcal{F} = 0$, and we need to show that $\beta = i(V)d\chi_\mathcal{F} = 0$ for $V \in \Gamma L$. We observe first that

$$(4.31) \qquad \beta \mid L = 0.$$

Since β is a p-form, it suffices to prove that β evaluates to zero on a local orthonormal frame E_i $(i = 1, \ldots, p)$ of L. But

$$\beta(E_1, \ldots, E_p) = d\chi_\mathcal{F}(V, E_1, \ldots, E_p) = 0,$$

since V is a linear combination of the E_i's. This proves (4.31).

To show $\beta = 0$, it suffices now to prove that moreover $i(Y)\beta = 0$ for $Y \in \Gamma L^\perp$. But

$$i(Y)i(V)d\chi_\mathcal{F} = -i(V)i(Y)d\chi_\mathcal{F},$$

and $i(Y)d\chi_\mathcal{F} = \theta(Y)\chi_\mathcal{F} = 0$ by assumption. □

Some properties of harmonic foliations

Consider a transversally oriented foliation $\mathcal{F}$ of codimension q on a Riemannian manifold (M, g), admitting a holonomy invariant transverse volume form $\nu \in \Gamma \Lambda^q Q^*$. Since $i(V)\nu = 0$ for $V \in \Gamma L$, the holonomy invariance condition $\theta(V)\nu = 0$ for $V \in \Gamma L$ shows that $\nu \in \Omega^q_B(\mathcal{F})$. It follows that $d\nu = 0$. It is of interest to examine the cohomology class $[\nu] \in H^q_B(\mathcal{F})$, which plays the role of the orientation class for the leaf space of $\mathcal{F}$. Of particular importance are conditions implying the nontriviality of this class.

Before doing this, we consider the star operator $*: \Omega^r(M) \to \Omega^{n-r}(M)$ associated to g on the oriented Riemannian manifold (M,g) with volume form μ. For local orthonormal frames $E_1, \ldots, E_p \in \Gamma L$ and $E_{p+1}, \ldots, E_n \in \Gamma L^\perp$ the transverse volume form ν is dual to the q-vector $E_{p+1} \wedge \cdots \wedge E_n$. Then with the choice of compatible orientations on L, $L^\perp$ and TM we have $*\nu = \chi_\mathcal{F}$, the characteristic form of $\mathcal{F}$. With this convention

$$\nu \wedge \chi_\mathcal{F} = \nu \wedge *\nu = \mu.$$

4.32 Theorem. *Let $\mathcal{F}$ be a transversally oriented harmonic foliation of codimension q, with invariant transversal volume ν on a closed oriented Riemannian manifold (M,g). Then $[\nu] \neq 0$ in $H_B^q(\mathcal{F})$.*

Proof. Assume $\nu = d\alpha$ with $\alpha \in \Omega_B^{q-1}$, and let $\chi_\mathcal{F} = *\nu$ be the characteristic form of $\mathcal{F}$. Then

$$d(\alpha \wedge \chi_\mathcal{F}) = d\alpha \wedge \chi_\mathcal{F} + (-1)^{q-1} \alpha \wedge d\chi_\mathcal{F}.$$

Since $\alpha \in \Omega_B^{q-1}$, it follows that $\alpha \in F^{q-1}\Omega^{q-1}$. The harmonicity assumption on $\mathcal{F}$ implies by (4.26) that $d\chi_\mathcal{F} \in F^2\Omega^{p+1}$. It follows that the second term on the RHS is of filtration degree $q - 1 + 2 = q + 1$, and hence vanishes by (4.6). Thus

$$d(\alpha \wedge \chi_\mathcal{F}) = d\alpha \wedge \chi_\mathcal{F} = \nu \wedge \chi_\mathcal{F} = \mu = \text{volume form of } g.$$

This contradicts the fact that $[\mu] \neq 0$ in $H^n(M)$. $\square$

Before discussing another application, recall from Chapter 2 that $\mathcal{F}$ is taut if there exists a metric on M such that all the leaves of $\mathcal{F}$ are minimal submanifolds. Consider now a (transversally) symplectic foliation [Du][K-To 3,10][Sco]. For such a foliation $\mathcal{F}$ of even codimension $q = 2m$ on M^n the defining property is the existence of a basic and closed 2-form $\omega \in \Omega_B^2(\mathcal{F})$ such that ω^m is a nowhere zero q-form. Note that ω^k for $k = 1, \ldots, m$, is closed and satisfies for $V \in \Gamma L$

$$i(V)\omega^k = \sum \pm \omega \wedge \cdots \wedge i(V)\omega \wedge \cdots \wedge \omega = 0$$
$$\theta(V)\omega^k = \sum \omega \wedge \cdots \wedge \theta(V)\omega \wedge \cdots \wedge \omega = 0.$$

Thus the forms $\omega^k \in \Omega_B^{2k}(\mathcal{F})$ give rise to basic cohomology classes $[\omega^k] \in H_B^{2k}(\mathcal{F})$.

4.33 Theorem. *Let $\mathcal{F}$ be a taut and (transversally) symplectic foliation of codimension $q = 2m$ on a closed oriented manifold M^n. Then the cohomology classes $[\omega^k] \in H_B^{2k}(\mathcal{F})$ for $k = 1, \ldots, m$ are all nontrivial.*

Proof. Let g be a metric such that all leaves are minimal. The q-form $\tilde{\nu} = \omega^m$ is basic, thus it is a nowhere zero multiple $\lambda \nu$ of the transversal volume ν associated to the induced transversal metric (the volume ν is not necessarily holonomy invariant). This follows from the fact that these forms are both sections of the line bundle $\Lambda^q Q^*$. Now consider

$$\int_M \tilde{\nu} \wedge \chi_\mathcal{F} = \int_M \lambda \nu \wedge \chi_\mathcal{F} = \int_M \lambda \mu,$$

where μ is the volume form of M. Since $\int_M \mu \neq 0$, and λ has a fixed sign, it follows that $\int_M \tilde{\nu} \wedge \chi_\mathcal{F} \neq 0$. A relation $\tilde{\nu} = d\alpha$ with $\alpha \in \Omega_B^{q-1}(\mathcal{F})$ would then lead to a contradiction as in the proof of Theorem 4.32. Thus $[\tilde{\nu}] = [\omega^m] \neq 0 \in H_B^q(\mathcal{F})$.

Assume now that $[\omega^k] = 0 \in H_B^{2k}(\mathcal{F})$ for some $1 \leq k < m$. Then $\omega^k = d\alpha$ with $\alpha \in \Omega_B^{2k-1}(\mathcal{F})$ and

$$d(\omega^{m-k} \wedge \alpha) = \omega^{m-k} \wedge d\alpha = \omega^{m-k} \wedge \omega^k = \omega^m,$$

which contradicts what we just proved. □

Next we wish to state and prove a transversal divergence theorem, for a harmonic foliation with a holonomy invariant transversal volume ν. Let $Y \in V(\mathcal{F})$ be an infinitesimal automorphism of $\mathcal{F}$. We define the transversal divergence $\text{div}_B Y$ by

(4.34) $$\theta(Y)\nu = \text{div}_B Y \nu.$$

Here we use the fact that $\theta(Y)\nu \in \Omega_B^q(\mathcal{F}) \subset \Gamma \Lambda^q Q^*$. Observe that $\text{div}_B Y \in \Omega_B^0(\mathcal{F})$ and in fact depends only on $\bar{Y} = \pi(Y)$. Thus we denote it also by $\text{div}_B \bar{Y}$. We prove the following result [K-To-T].

4.35 TRANSVERSAL DIVERGENCE THEOREM. *Let $\mathcal{F}$ be a transversally oriented harmonic foliation with holonomy invariant transversal volume ν on a closed oriented Riemannian manifold (M, g). Let Y be an infinitesimal automorphism of $\mathcal{F}$ and $\bar{Y} = \pi(Y)$. Then*

$$\int_M \text{div}_B \bar{Y} \mu = 0.$$

Proof. Let $\chi_\mathcal{F} = *\nu$ be the characteristic form of $\mathcal{F}$, and assume M to be oriented by $\mu = \nu \wedge \chi_\mathcal{F}$. Then

$$\text{div}_B \bar{Y} \mu = (\text{div}_B \bar{Y} \nu) \wedge \chi_\mathcal{F} = \theta(Y)\nu \wedge \chi_\mathcal{F} = (di(Y)\nu) \wedge \chi_\mathcal{F}$$
$$= d(i(Y)\nu \wedge \chi_\mathcal{F}) + (-1)^q i(Y)\nu \wedge d\chi_\mathcal{F}.$$

We have $\nu \in F^q$ and $i(Y)\nu \in F^{q-1}$. Since $\mathcal{F}$ is harmonic, $d\chi_\mathcal{F} \in F^2$ by (4.21). It follows that $i(Y)\nu \wedge d\chi_\mathcal{F}$ is of filtration degree $q - 1 + 2 = q + 1$, and hence vanishes. Stokes' theorem implies now the desired result. □

The value of this integral for the case of a Riemannian (but not necessarily harmonic) foliation is given in the next chapter (see Theorem 5.24).

Chapter 5
Transversal Riemannian Geometry

A Riemannian metric g_Q on the normal bundle Q of a foliation $\mathcal{F}$ is holonomy invariant, if

(5.1) $$\theta(V)g_Q = 0 \quad \text{for all} \quad V \in \Gamma L.$$

Here we have by definition for $s, s' \in \Gamma Q$

$$(\theta(V)g_Q)(s, s') = V g_Q(s, s') - g_Q(\theta(V)s, s') - g_Q(s, \theta(V)s').$$

A Riemannian foliation is a foliation $\mathcal{F}$ with a holonomy invariant transversal metric g_Q. The study of these foliations was initiated by Reinhart in 1959 [Re 2].

In terms of the definition of a foliation by a foliated atlas, the requirement is that the local transition functions $\tau_{\beta\alpha}$ relating submersions $f_\alpha : U_\alpha \to \mathbb{R}^q$ and $f_\beta : U_\beta \to \mathbb{R}^q$ are local isometries of suitable Riemannian metrics on $\mathbb{R}^q$. The local submersions f_α define then by pull-back a well-defined Riemannian metric g_Q on the bundle Q. These conditions can be summarized by saying that the Haefliger cocycle defining $\mathcal{F}$ corresponds to a Riemannian groupoid.

A Riemannian metric g on M is bundle-like for $\mathcal{F}$, if the induced metric on the normal bundle is holonomy invariant.

The transverse orientability of $\mathcal{F}$ is defined by the orientability of Q. For a holonomy invariant transversal metric g_Q there is a corresponding transversal volume form $\nu \in \Gamma \Lambda^q Q^* \subset \Omega^q(M)$. It is holonomy invariant, i.e.

(5.2) $$\theta(V)\nu = 0 \quad \text{for all} \quad V \in \Gamma L,$$

where

$$(\theta(V)\nu)(s_1, \ldots, s_q) = V\nu(s_1, \ldots, s_q) - \sum_{\alpha=1}^{q} \nu(s_1, \ldots, \theta(V)s_\alpha, \ldots, s_q)$$

for $s_1, \ldots, s_q \in \Gamma Q$. For $q = 1$ the concept of a transversal and holonomy invariant volume coincides with the concept of a holonomy invariant transversal metric.

A simple example of a Riemannian foliation is given by a nonsingular Killing vector field V on (M, g). This means that $\theta(V)g = 0$ or equivalently

(5.3) $$V g(Y, Y') = g([V, Y], Y') + g(Y, [V, Y'])$$

for any vector fields $Y, Y' \in \Gamma TM$. Let $\mathcal{F}$ be the foliation of M by the orbits of V. Then V is a nontrivial section of $L \subset TM$. The complement $L^\perp$ is preserved by the flow and for the induced metric g_Q on $L^\perp$ we have $\theta(V)g_Q = 0$. The holonomy invariance in this case is precisely the invariance under the flow generated by V.

More generally consider a Lie group G, acting by isometries on (M, g). If the orbits of the G-action have all the same dimension, this gives rise to a Riemannian foliation. The point is that the sections of L are linear combinations of Killing vector fields arising from the group action, so that the previous arguments apply. This situation occurs in particular for actions of compact groups, since any metric on M can be averaged to an invariant metric under the action.

We return now to the general situation of a Riemannian foliation on (M, g). The Lie derivative for any metric g_Q on Q is given by

(5.4) $\qquad (\theta(V)g_Q)(s, s') = V g_Q(s, s') - g_Q(\theta(V)s, s') - g_Q(s, \theta(V)s')$

for $V \in \Gamma L$ and $s, s' \in \Gamma Q$. More generally for any covariant r-tensor ω on Q we have

(5.5) $\qquad (\theta(V)\omega)(s_1, \ldots, s_r) = V\omega(s_1, \ldots, s_r) - \sum_{\alpha=1}^{r} \omega(s_1, \ldots, \theta(V)s_\alpha, \ldots, s_r)$

for $V \in \Gamma L$ and $s_1, \ldots, s_r \in \Gamma Q$. Condition (5.1) can be restated as

(5.6) $\qquad V g_Q(s, s') = g_Q(\pi[V, Z_s], s') + g_Q(s, \pi[V, Z_{s'}])$

for $V \in \Gamma L$, sections $s, s' \in \Gamma Q$ and $Z_s = \sigma(s)$, $Z_{s'} = \sigma(s') \in \Gamma L^\perp$. We compare this with the condition that the connection ∇ defined by (3.3) is a metric connection in the bundle Q equipped with the induced metric g_Q. This condition reads for $s, s' \in \Gamma Q$

(5.7) $\qquad Y g_Q(s, s') = g_Q(\nabla_Y s, s') + g_Q(s, \nabla_Y s'),$

but now for all $Y \in \Gamma TM$ (not only $V \in \Gamma L$), and thus implies (5.6).

5.8 THEOREM. *Let $\mathcal{F}$ be a foliation on (M, g), g_Q the induced metric on Q, and ∇ the connection on Q defined by (3.3). Then $\mathcal{F}$ is Riemannian and g bundle-like, if and only if ∇ is a metric connection.*

Proof. It suffices to verify that for $\mathcal{F}$ Riemannian the condition (5.7) holds for g_Q and ∇ as in the theorem. It suffices to verify this for $Z \in \Gamma L^\perp$. But we have for $Z_s = \sigma(s)$, $Z_{s'} = \sigma(s')$

$$Z g_Q(s, s') = Z g(Z_s, Z'_s) = g(\nabla^M_Z Z_s, Z'_s) + g(Z_s, \nabla^M_Z Z'_s)$$
$$= g_Q(\pi(\nabla^M_Z Z_s), s') + g_Q(s, \pi(\nabla^M_Z Z'_s)) = g_Q(\nabla_Z s, s') + g_Q(s, \nabla_Z s').$$
$\square$

5.9 THEOREM. *Let g_Q be a holonomy invariant metric in the normal bundle Q of $\mathcal{F}$. Then there is a unique metric and torsion free connection in Q.*

Proof. The existence follows by constructing ∇ via a bundle-like metric g. It remains to prove the uniqueness. Let ∇ be a metric and torsion free connection in Q. Then

$$\begin{aligned}(5.10)\quad 2g_Q(\nabla_Y s, s') &= Yg_Q(s,s') + Z_s g_Q(\pi(Y), s') - Z_{s'} g_Q(\pi(Y), s) \\ &\quad + g_Q(\pi[Y, Z_s], s') + g_Q(\pi[Z_{s'}, Y], s) - g_Q(\pi[Z_s, Z_{s'}], \pi(Y))\end{aligned}$$

for $Y \in \Gamma TM$, $s, s' \in \Gamma Q$, $Z_s, Z_{s'} \in \Gamma TM$ with $\pi(Z_s) = s$, $\pi(Z_{s'}) = s'$. This formula is proved by expanding the first three terms on the RHS using (5.7), and then using the vanishing of the torsion.

Note that (5.10) is independent of the choices of Z_s and Z_s for a holonomy invariant g_Q. (5.10) implies the uniqueness of ∇. $\square$

The unique metric and torsion free connection ∇ in the normal bundle of a Riemannian foliation $\mathcal{F}$ is the transversal Levi Civita connection of $\mathcal{F}$. It is worth repeating that the (holonomy invariant) transversal metric g_Q completely determines ∇. Formula (3.3) shows that the covariant derivative in the transversal directions corresponds under the local Riemannian submersion, to the effect of the Levi Civita connections on the Riemannian manifolds modeling the foliation. The transition functions being isometries, the pull-backs are invariantly defined. In particular all curvature data associated to ∇ have an invariant meaning.

The following fact is not surprising.

5.11 THEOREM. *The metric torsion free connection ∇ in the normal bundle of a Riemannian foliation is holonomy invariant.*

Proof. This is a consequence of (5.10). $\square$

5.12 COROLLARY. *Let R_∇ be the curvature of ∇. Then for $V \in \Gamma L$*

(i) $i(V)R_\nabla = 0$,
(ii) $\theta(V)R_\nabla = 0$.

Proof. This follows from Proposition 3.6. $\square$

We return to the situation of an arbitrary foliation on (M, g), and the metric g_Q induced on Q. Identifying $(Q, g_Q) \cong (L^\perp, g|L^\perp)$, we have then for $V \in \Gamma L$ and $X, Y \in \Gamma L^\perp$

$$\begin{aligned}(5.13)\quad (\theta(V)g_Q)(X,Y) &= Vg_Q(X,Y) - g_Q(\pi[V,X], Y) - g_Q(X, \pi[V,Y]) \\ &= Vg(X,Y) - g([V,X], Y) - g(X, [V,Y]).\end{aligned}$$

Note that the vanishing of this symmetric bilinear form follows already from the vanishing of the corresponding quadratic form on unit vectors. Using the torsion free connection ∇^M, (5.13) can be rewritten as

$$(5.14)\quad (\theta(V)g_Q)(X,Y) = g(\nabla_X^M V, Y) + g(X, \nabla_Y^M V).$$

Observe that

(5.15) $$g(\nabla^M_X V, Y) = -g(V, \nabla^M_X Y),$$

and similarly for $g(X, \nabla^M_Y V)$. It follows from (5.14) that

(5.16) $(\theta(V) g_Q)(X, Y) = -g(V, \nabla^M_X Y + \nabla^M_Y X) = -2g(V, \nabla^M_X Y) + g(V, [X, Y]).$

These formulas due to Reinhart are summarized as follows.

5.17 THEOREM. *Let $\mathcal{F}$ be a foliation on (M, g). Then the following conditions are equivalent:*

(i) *$\mathcal{F}$ is Riemannian and g bundle-like;*
(ii) *$g(\nabla^M_X V, Y) + g(X, \nabla^M_Y V) = 0$ for $V \in \Gamma L$ and $X, Y \in \Gamma L^\perp$;*
(iii) *$g(\nabla^M_X Y + \nabla^M_Y X, V) = 0$ for $V \in \Gamma L$ and $X, Y \in \Gamma L^\perp$;*
(iv) *$2g(\nabla^M_X Y, V) = g([X, Y], V)$ for $V \in \Gamma L$ and $X, Y \in \Gamma L^\perp$.*

Proof. (i)⇔(ii) follows from (5.14). (i)⇔(iii)⇔(iv) follows from (5.16). □

We comment on property (iii) in Theorem 5.17. This identity states that the appropriately defined second fundamental form of the normal bundle $L^\perp$ vanishes. Thus this bundle can be viewed as a totally geodesic subbundle of TM.

An important characterization of Riemannian foliations on (M, g) is the property that a geodesic orthogonal to the leaves of $\mathcal{F}$ at one point is orthogonal to the leaves of $\mathcal{F}$ at any of its points. It suffices to prove this for a Riemannian submersion $M \to B$. If γ is a geodesic in B and $\tilde{\gamma}$ a horizontal lift of γ in M, then $\tilde{\gamma}$ realizes (locally) a minimum of the distance between its points, hence is a geodesic. Conversely if geodesics γ of (M, g) orthogonal to a leaf of $\mathcal{F}$ at one point $\gamma(0)$ are orthogonal to the leaves of $\mathcal{F}$ at all points $\gamma(t)$, then the metric g is necessarily bundle-like for $\mathcal{F}$ (see Molino [Mo 10, Proposition 6.1]). This was one of the initial observations of Reinhart in starting this subject in 1959.

The motivating example might have been the special case of Riemannian submersions, but in fact this was developed somewhat later by O'Neill ([ON], 1966) and Gray ([Gr], 1967). In any case such submersions are abundant in Riemannian geometry. A striking example is the structure theorem of Cheeger and Gromoll for noncompact nonnegatively curved manifolds M. Such a manifold contains a (not necessarily unique) totally convex and totally geodesic submanifold S without boundary, $0 \leq \dim S \leq \dim M$, such that M is diffeomorphic to the total space of the normal bundle of the soul S in M. Perelman has proved this so called Soul Theorem in 1994 [Prl] as the consequence of the existence of a Riemannian submersion $M \to S$.

An illustrative and very important example of a manifold with a Riemannian foliation is a flat bundle $M = \tilde{B} \times_\Gamma F \to B$ associated to a homomorphism $h : \Gamma = \pi_1 B \to \text{Diff}(F)$ as explained at the beginning of Chapter 3. The foliation

$\mathcal{F}$ transversal to the fibers is orthogonal to the fibers with respect to the metric g induced by the product metric $p^*g_B \oplus g_F$ on $\tilde{B} \times F$. Here g_B denotes a metric on B and p^*g_B its pull-back to the universal covering $\tilde{B}$, while g_F is a metric on F. The foliation by the fibers has g_B as the induced metric in the normal bundle and is Riemannian. But note that $\mathcal{F}$ is not necessarily Riemannian. In fact, its normal bundle is the tangent bundle $T(f)$ along the fibers of $f: M \to B$, and its holonomy maps in $T(f)$ need not be isometries. For the diffeomorphism $T^2 \to T^2$ induced by $A = \begin{pmatrix} 1 & 1 \\ 1 & 2 \end{pmatrix} \in SL(2,\mathbb{Z})$ e.g., the foliation transverse to the fibers of the resulting torus fibration $T_A = \mathbb{R} \times_\mathbb{Z} T^2 \to S^1$ has nonisometric holonomy, hence cannot be Riemannian. Another example is the Roussarie foliation on the unit tangent bundle of a Riemannian surface M_g ($g > 1$), considered in Chapter 2. It is a codimension one foliation transverse to the circle fibers in $T_1 M_g \cong D^2 \times_\Gamma S^1$, $\Gamma = \pi_1(M_g)$. This foliation is definitely not Riemannian. This follows from the nontriviality of its Godbillon-Vey class. Namely a (transversally oriented) Riemannian foliation is in particular a $SL(q)$-foliation. This implies the vanishing of its Godbillon-Vey class. This class is constructed from a 1-form α satisfying $d\nu = \alpha \wedge \nu$ for the transversal volume form ν as the De Rham cohomology class $[\alpha \wedge (d\alpha)^q]$. Theorem 2.3 is the case corresponding to $q = 1$. For a holonomy invariant ν we have $d\nu = 0$ and $\alpha = 0$ is a suitable choice, hence the De Rham class in question vanishes. If however $\tilde{B} \times_\Gamma F \to B$ arises from a representation of Γ by isometries of a Riemannian metric g_F on F, then this will turn $\mathcal{F}$ into a Riemannian foliation.

Consider again an arbitrary foliation on (M,g), with induced metric g_Q on Q. For an infinitesimal automorphism $Y \in V(\mathcal{F})$ we have then

$$(5.18) \quad (\theta(Y)g_Q)(X,X') = Y g_Q(X,X') - g_Q(\pi[Y,X],X') - g_Q(X,\pi[Y,X']),$$

where $X, X' \in \Gamma L^\perp$. Therefore

$$(\theta(Y)g_Q)(X,X') = Y g(X,X') - g([Y,X],X') - g(X,[Y,X'])$$
$$= g(\nabla_X^M Y, X') + g(X, \nabla_{X'}^M Y).$$

Let now $Y = \pi^\perp(Y) + \pi(Y) = V + \bar{Y}$. Then $\nabla_X^M Y = \nabla_X^M V + \nabla_X^M \bar{Y}$. It follows that

$$(5.19) \quad (\theta(Y)g_Q)(X,X') = (\theta(V)g_Q)(X,X') + g_Q(\nabla_X \bar{Y}, X') + g_Q(X, \nabla_{X'}\bar{Y}).$$

For a Riemannian foliation the first term on the RHS vanishes, and one has the following result.

5.20 PROPOSITION. *Let $\mathcal{F}$ be a Riemannian foliation. For $Y \in V(\mathcal{F})$ and $X, X' \in \Gamma Q$ we have*

$$(5.21) \quad (\theta(Y)g_Q)(X,X') = g_Q(\nabla_X \bar{Y}, X') + g_Q(X, \nabla_{X'}\bar{Y}).$$

Note that the RHS involves only $\bar{Y} = \pi(Y)$.

An infinitesimal automorphism Y is transversally metric, if $\theta(Y)g_Q = 0$. If this holds, $\bar{Y} = \pi(Y)$ is called a transversal Killing field (Molino [Mo7,8]). For the point foliation with $L = 0$ this is the usual definition of a Killing vector field.

In terms of the basic one-form ω given by $\omega(X) = g_Q(\bar{Y}, X)$ for $X \in \Gamma L^\perp$, formula (5.21) can be written in the form

(5.22) $\qquad (\theta(Y)g_Q)(X, X') = (\nabla_X \omega)(X') + (\nabla_{X'}\omega)(X).$

Since by definition

$$(\nabla_X \omega)(X') = X\omega(X') - \omega(\nabla_X X') = X g_Q(\bar{Y}, X) - g_Q(\bar{Y}, \nabla_X X') = g_Q(\nabla_X \bar{Y}, X'),$$

this proves the equivalence of (5.21) and (5.22). Thus $\frac{1}{2}\theta(Y)g_Q$ is the symmetric part of the 2 linear form $\nabla \omega$ on Q.

A transversally oriented Riemannian foliation has a canonical holonomy invariant transversal volume ν. We state the following fact.

5.23 THEOREM ([K-To 8]). *Let $\mathcal{F}$ be a transversally oriented and taut Riemannian foliation of codimension q on a closed oriented manifold M^n. Then $H_B^q(\mathcal{F}) \neq 0$.*

Proof. Recall that $\mathcal{F}$ is taut if there exists a metric on M such that all the leaves of $\mathcal{F}$ are minimal submanifolds. For a Riemannian foliation, this can be achieved with a bundle-like metric [K-To 5, Corollary 2.31]. The corresponding transversal volume ν is then holonomy invariant, and Theorem 4.32 applies to prove $[\nu] \neq 0$. It follows that $H_B^q(\mathcal{F}) \neq 0$. $\square$

Next we calculate the value of the integral of $\text{div}_B \bar{Y}$ for $Y \in V(\mathcal{F})$ in case of a Riemannian foliation.

5.24 THEOREM. *Let $\mathcal{F}$ be a transversally oriented Riemannian foliation on a closed oriented Riemannian manifold (M, g). Let $Y \in V(\mathcal{F})$. Then*

$$\int_M \text{div}_B \bar{Y} \mu = \int_M g_Q(\tau, \bar{Y})\mu \equiv \langle \tau, \bar{Y} \rangle$$

(the global scalar product of the sections τ and $\bar{Y}$ of Q).

Proof. With the notations as in the proof of Theorem 4.35 we have

$$\text{div}_B \bar{Y} \mu = d(i(Y)\nu \wedge \chi_\mathcal{F}) + (-1)^q i(Y)\nu \wedge d\chi_\mathcal{F}.$$

We replace $d\chi_\mathcal{F}$ using (4.26). Since $\varphi_0 \in F^2 \Omega^{p+1}$, and $i(Y)\nu \in F^{q-1}\Omega^{q-1}$, it follows that the term $i(Y)\nu \wedge \varphi_0$ is of filtration degree $q+1$, and hence vanishes. Thus

$$\text{div}_B \bar{Y} \mu = d(i(Y)\nu \wedge \chi_\mathcal{F}) + \kappa \wedge i(Y)\nu \wedge \chi_\mathcal{F}.$$

Since $\kappa \wedge \nu \in \Gamma \Lambda^{q+1} Q^*$, and hence vanishes, we have further
$$0 = i(Y)\kappa\nu - \kappa \wedge i(Y)\nu.$$
Thus
$$\operatorname{div}_B \bar{Y}\mu = d(i(Y)\nu \wedge \chi_{\mathcal{F}}) + i(Y)\kappa \cdot \mu.$$
Since $\kappa(Y) = g_Q(\tau, Y)$, the desired result follows now by Stokes' Theorem. □

For another proof of this Transversal Divergence Theorem see also Chapter 7 (the comments preceding formula (7.15)).

The fundamental Gray and O'Neill tensors T and A

For arbitrary vector fields E and F on the foliated Riemannian manifold (M, g) we define as in [Gr][ON]

(5.25) $$T_E F = \pi \left(\nabla^M_{\pi^\perp E} \pi^\perp F \right) + \pi^\perp \left(\nabla^M_{\pi^\perp E} \pi F \right).$$

Clearly $T_E = T_{\pi^\perp E}$.

In the following formulas $U, V, W \in \Gamma L$ and $X, Y, Z \in \Gamma L^\perp$. We will adhere to this convention throughout the rest of this chapter. Then

(5.26a) $$T_X U = 0, \quad T_X Y = 0;$$

(5.26b) $$T_U V = \pi \left(\nabla^M_U V \right), \quad T_U X = \pi^\perp \left(\nabla^M_U X \right);$$

(5.26c) $$T_U V = T_V U;$$

(5.26d) $T_U V$ is alternating, in particular $g(T_U V, X) = -g(V, T_U X)$.

(5.26b) shows that $T_U V$ is the second fundamental form of the leaves of $\mathcal{F}$, and thus is symmetric, as claimed in (5.26c). If $T_U V = 0$ for all $U, V \in \Gamma L$, then $T = 0$. The vanishing of T is thus equivalent to the property that all leaves of $\mathcal{F}$ are totally geodesic submanifolds of (M, g). Such a foliation is called totally geodesic.

Reversing the role of L and $L^\perp$ in (5.25), we define

(5.27) $$A_E F = \pi^\perp(\nabla^M_{\pi E} \pi F) + \pi(\nabla^M_{\pi E} \pi^\perp F).$$

Clearly $A_E = A_{\pi E}$.

These two tensors were introduced by O'Neill [ON] and Gray [Gr], and turn out to be very useful computational tools. The following properties of A are easily verified.

(5.28a) $$A_U X = 0, \quad A_U V = 0;$$

(5.28b) $$A_X U = \pi(\nabla^M_X U), \quad A_X Y = \pi^\perp(\nabla^M_X Y);$$

(5.28c) $$A_X Y = -A_Y X;$$

(5.28d) A_X is alternating, in particular $g(A_X Y, U) = -g(Y, A_X U)$.

As an illustration, we prove (5.28c). It suffices to prove $A_X X = 0$, and we can even limit ourselves to projectable $X \in \Gamma L^\perp$. Then

$$g(U, A_X X) = g(U, \nabla_X^M X) = -g(\nabla_X^M U, X)$$
$$= -g(\nabla_U^M X, X) - g([X, U], X)$$
$$= -\frac{1}{2} U g(X, X) = 0.$$

The last equality follows from the fact that $g(X, X)$ is a basic function for projectable X. But since $A_X X \in \Gamma L$, this vanishing property for all $U \in \Gamma L$ implies $A_X X = 0$.

5.29 Proposition (O'Neill [ON]). *Let $X, Y \in \Gamma L^\perp$. Then*

(5.30) $$A_X Y = \frac{1}{2} \pi^\perp [X, Y].$$

Proof. This follows from

$$\pi^\perp [X, Y] = \pi^\perp (\nabla_X^M Y - \nabla_Y^M X) = A_X Y - A_Y X = 2 A_X Y. \quad \square$$

(5.30) shows that $L^\perp$ is an integrable bundle if and only if $A_X Y = 0$ for all $X, Y \in \Gamma L^\perp$. But then $A = 0$ by (5.28a), (5.28b) and (5.28d). The vanishing of A is thus equivalent to the integrability of $L^\perp$. A is called the integrability tensor of $\mathcal{F}$.

We can describe the relations between the connection ∇^M on (M, g) and the induced connection $\hat{\nabla}_U V = \pi^\perp (\nabla_U^M V)$ along the leaves as follows:

(5.31a) $$\nabla_U^M V = \hat{\nabla}_U V + T_U V;$$

(5.31b) $$\nabla_U^M X = T_U X + \pi(\nabla_U^M X);$$

(5.31c) $$\nabla_X^M U = \pi^\perp(\nabla_X^M U) + A_X U;$$

(5.31d) $$\nabla_X^M Y = A_X Y + \pi(\nabla_X^M Y).$$

Covariant derivatives of T and A

We begin with the following facts proved in [ON].

5.32 Lemma.

$$(\nabla_V^M A)_W = -A_{T_V W}, \quad (\nabla_X^M T)_Y = -T_{A_X Y},$$
$$(\nabla_X^M A)_W = -A_{A_X W}, \quad (\nabla_V^M T)_Y = -T_{T_V Y}.$$

Proof. Let $E \in \Gamma TM$. Then
$$(\nabla^M_V A)_W E = \nabla^M_V (A_W E) - A_{\nabla^M_V W} E - A_W(\nabla^M_V E)$$
$$= -A_{\nabla^M_V W} E = -A_{\pi(\nabla^M_V W)} E = -A_{T_V W} E.$$

The other formulas are proved similarly. $\square$

Similar proofs establish the following facts.

5.33 LEMMA.
$$g((\nabla^M_U A)_X V, W) = g(T_U V, A_X W) - g(T_U W, A_X V).$$

5.34 LEMMA.

(i) $g((\nabla^M_E A)_X Y, V)$ *is skew-symmetric in* X, Y;
(ii) $g((\nabla^M_E T)_V W, X)$ *is symmetric in* V, W.

5.35 LEMMA. *For the cyclic sum* $\mathfrak{S}$ *over* X, Y, $Z \in \Gamma L^\perp$

(5.36)
$$\mathfrak{S}\, g((\nabla^M_Z A)_X Y, V) = \mathfrak{S}\, g(A_X Y, T_V Z).$$

Curvature identities

Aside from the connections ∇^M on TM and $\hat{\nabla}$ along the leaves, we have the connection ∇ in $L^\perp \cong Q$ defined by (3.3). To these connections ∇^M, $\hat{\nabla}$ and ∇, there are associated curvature tensors R^M, $\hat{R}$ and R^∇, respectively.

5.37 THEOREM ([Gr], [ON]). *With the conventions above, we have the following identities:*

(5.37a) $\quad g(R^M(U,V)W, W') = g(\hat{R}(U,V)W, W')$
$$+ g(T_U W, T_V W') - g(T_V W, T_U W');$$

(5.37b) $\quad g(R^M(U,V)W, X) = -g((\nabla^M_V T)_U W, X) + g((\nabla^M_U T)_V W, X);$

(5.37c) $\quad g(R^M(X,U)Y, V) = -g((\nabla^M_X T)_U V, Y) + g(T_U X, T_V Y)$
$$- g((\nabla^M_U A)_X Y, V) - g(A_X U, A_Y V);$$

(5.37d) $\quad g(R^M(U,V)X, Y) = -g((\nabla^M_U A)_X Y, V) + g((\nabla^M_V A)_X Y, U)$
$$- g(A_X U, A_Y V) + g(A_X V, A_Y U)$$
$$+ g(T_U X, T_V Y) - g(T_V X, T_U Y);$$

(5.37e) $\quad g(R^M(X,Y)Z, U) = -g((\nabla^M_Z A)_X Y, U) - g(A_X Y, T_U Z)$
$$+ g(A_Y Z, T_U X) + g(A_Z X, T_U Y);$$

(5.37f) $\quad g(R^M(X,Y)Z, Z') = g((R^\nabla(X,Y)Z, Z') + 2g(A_X Y, A_Z Z')$
$$- g(A_Y Z, A_X Z') + g(A_X Z, A_Y Z').$$

Proof. (5.37a) is the equation of Gauss along leaves (viewed as submanifolds of M). (5.37b) is the Codazzi equation along leaves. We omit the proof of the other identities. $\square$

5.38 COROLLARY. *Let K^M, $\hat{K}$ and K^∇ be the sectional curvatures of g, $g_L = g|L$ and $g_Q = g|L^\perp$, respectively. Then for $|X| = 1$, $|U| = 1$, $|X \wedge Y| = 1$, $|U \wedge V| = 1$ we have*

(5.38a) $\qquad K^M(U,V) = \hat{K}(U,V) + |T_U V|^2 - g(T_U U, T_V V);$

(5.38b) $\qquad K^M(X,U) = g((\nabla_X^M T)_U U, X) - |T_U X|^2 + |A_X U|^2;$

(5.38c) $\qquad K^M(X,Y) = K^\nabla(X,Y) - 3|A_X Y|^2.$

Proof. This follows from (5.37a), (5.37c) and (5.37f). $\square$

A consequence of (5.38c) is that if $K^M > 0$, then a fortiori $K^\nabla > 0$.

Let $\mathcal{F}$ be a totally geodesic (and Riemannian) foliation. Then $T = 0$. It follows from (5.38b) that the mixed sectional curvature $K^M(X,U) = |A_X U|^2 \geqq 0$.

Relations between A, T, $\nabla^M A$ and $\nabla^M T$

$\qquad (\nabla_{E_1}^M T)_{E_2}$, $(\nabla_{E_1}^M A)_{E_2}$ are alternating;

$\qquad g((\nabla_E^M T)_U V, X)$ is symmetric in U, V;

$\qquad g((\nabla_E^M A)_X Y, U)$ is alternating in X, Y;

$\qquad (\nabla_X^M T)_Y = -T_{A_X Y}; \qquad (\nabla_U^M T)_X = -T_{T_U X};$

$\qquad (\nabla_U^M A)_V = -A_{T_U V}; \qquad (\nabla_X^M A)_U = -A_{A_X U};$

$\qquad g((\nabla_X^M T)_U V, W) = g(A_X V, T_U W) - g(A_X W, T_U V);$

$\qquad g((\nabla_U^M A)_X V, W) = g(T_U V, A_X W) - g(T_U W, A_X V);$

$\qquad g((\nabla_X^M T)_U Y, Z) = g(A_X Y, T_U Z) - g(A_X Z, T_U Y);$

$\qquad g((\nabla_U^M A)_X Y, Z) = g(A_X Z, T_U Y) - g(A_X Y, T_U Z);$

$\qquad g((\nabla_U^M T)_V W, W') = g(T_U W, T_V W') - g(T_U W', T_V W);$

$\qquad g((\nabla_X^M A)_Y U, V) = g(A_X U, A_Y V) - g(A_X V, A_Y U);$

$\qquad g((\nabla_U^M T)_V X, Y) = g(T_U X, T_V Y) - g(T_V X, T_U Y);$

$\qquad g((\nabla_X^M A)_Y Z, Z') = g(A_X Z, A_Y Z') - g(A_X Z, A_X Z);$

$\qquad g((\nabla_X^M A)_Y Z, U) + g((\nabla_Y^M A)_Z X, U) + g((\nabla_X^M A)_X Y, U)$
$\qquad = g(A_X Y, T_U Z) + g(A_Y Z, T_U X) + g(A_Z X, T_U Y);$

$\qquad g((\nabla_U^M A)_X Y, V) + g((\nabla_V^M A)_X Y, U) = g((\nabla_Y^M T)_U V, X) - g((\nabla_X^M T)_U V, Y).$

Transversally symmetric Riemannian foliations

These are Riemannian foliations whose transversal geometry is locally modeled on a Riemannian symmetric space. They were discussed in [To-V 4]. The first result is a characterization of transversal symmetry by a condition on the canonical Levi Civita connection ∇ of the normal bundle (Theorem 5.39). For a totally geodesic foliation $\mathcal{F}$ this characterization can be sharpened in the analytic case (Theorem 5.44), using the results of [To-V 1], [To-V 2], [To-V 3].

Next we examine the influence of the geometry of the ambient space M on the properties discussed above. A typical illustration is the following. For a space of constant curvature the total geodesic property for the leaves of $\mathcal{F}$ implies the transversal symmetry of $\mathcal{F}$ (Theorem 5.45). Related results are Theorem 5.46 and Corollary 5.48.

Conversely, the existence of a transversally symmetric foliation has strong implications for the geometry of the space (M, g). A typical result is that the transversal symmetry of the foliation defined by a Killing vector field of unit length on a complete, simply connected (M, g) implies that (M, g) is a naturally reductive space (Theorem 5.49).

The results just mentioned are contained in [To-V 4]. The first main fact is as follows.

THEOREM 5.39. *Let $\mathcal{F}$ be a Riemannian foliation on (M, g), and g a bundle-like metric. The following conditions are equivalent:*

(i) *$\mathcal{F}$ is transversally symmetric;*
(ii) *the local geodesic symmetries (geodesic reflections) on the model space are isometries;*
(iii) *$\nabla_X R^\nabla_{XYXY} = 0$ for all $X, Y \in \Gamma L^\perp$;*
(iv) *$\nabla^M_X R^M_{XYXY} + 2R^M_{XA_XYXY} = -6g((\nabla^M_X A)_X Y, A_X Y)$ for all $X, Y \in \Gamma L^\perp$.*

All these conditions are purely local, and they are automatically satisfied for a Riemannian foliation of codimension one. Hence a Riemannian foliation of codimension one is always transversally symmetric.

We have used above the notation

$$R^M_{VWXY} = -g(R^M_{VW}X, Y), \quad R^\nabla_{VWXY} = -g_Q(R^\nabla_{VW}X, Y),$$

the latter being defined for $X, Y \in \Gamma L^\perp$. In the arguments below we make extensive use of the fact that ∇ is a basic connection. This is expressed by the property (i) in Corollary 5.12. As a consequence, it suffices to evaluate $R^\nabla \in \Omega^2(M, \text{End}(Q))$ on $X, Y \in \Gamma L^\perp$. Since locally $L^\perp$ can be framed by projectable normal vector fields, denoted $\Gamma Q^L \subset \Gamma Q \cong \Gamma L^\perp$, it is often enough to consider $R^\nabla(X, Y)$ for projectable vector fields X, Y. For given $\bar{X} \in \Gamma Q^L$ there is a unique projectable vector field $X \in \Gamma L^\perp$ with $\pi(X) = \bar{X}$ under the projection $\pi : TM \to Q$. We will identify X and $\bar{X}$.

Proof of Theorem 5.39. The proof is based on the relationship

(5.40) $$f_* R^\nabla(X,Y)Z = R^N(f_*X, f_*Y)f_*Z$$

between R^∇ and the curvature R^N of the local model in a distinguished chart, where $\mathcal{F}$ is defined via the local submersion f and $X, Y, Z \in \Gamma Q^L$. This is a consequence of the definition (3.3) of ∇.

Now it is classical that the local symmetry of the model space is characterized by (ii) or equivalently by

$$\nabla^N_{\bar{X}} R^N_{\bar{X}\bar{Y}\bar{X}\bar{Y}} = 0$$

for vector fields $\bar{X}, \bar{Y}$ in the model space, and where

$$R^N_{\bar{X}\bar{Y}\bar{X}\bar{Y}} = -g(R^N_{\bar{X}\bar{Y}}\bar{X}, \bar{Y}).$$

For $X, Y \in \Gamma L^\perp$ which are f-related to $\bar{X}, \bar{Y}$, we have then

(5.41) $$f_* \nabla_X R^\nabla_{XYXY} = \nabla^N_{\bar{X}} R^N_{\bar{X}\bar{Y}\bar{X}\bar{Y}}.$$

This follows from the fact that $\nabla^M_X Y$ in each argument of R^∇ can be replaced in view of (5.12), (i) by $\pi(\nabla^M_X Y) = \nabla_X Y$.

It remains to prove the equivalence (iii)⇔(iv). First we note that by (5.37f) we have for $X, Y \in \Gamma Q^L$

(5.42) $$R^M_{XYXY} = R^\nabla_{XYXY} - 3g(A_X Y, A_X Y).$$

By [Re 9, p. 156] we may assume $\nabla^M_X X = 0$. Then

$$\nabla^M_X R^M_{XYXY} = X(R^M_{XYXY}) - 2R^M_{X\nabla^M_X YXY}$$

and similarly

$$\nabla_X R^\nabla_{XYXY} = X(R^\nabla_{XYXY}) - 2R^\nabla_{X\nabla_X YXY}.$$

Further, it follows then by (5.42) that

$$\nabla^M_X R^M_{XYXY} - \nabla_X R^\nabla_{XYXY} = -3Xg(A_X Y, A_X Y) - 2R^M_{X\nabla^M_X YXY} + 2R^\nabla_{X\nabla_X YXY}.$$

Now using again (5.37f) yields

$$R^M_{X\nabla_X YXY} - R^\nabla_{X\nabla_X YXY} = -3g(A_X \nabla_X Y, A_X Y)$$

and thus

(5.43) $$\begin{aligned}\nabla^M_X R^M_{XYXY} - \nabla_X R^\nabla_{XYXY} + 2R^M_{XA_X YXY} \\ = -6g(\nabla^M_X(A_X Y), A_X Y) + 6g(A_X \nabla_X Y, A_X Y) \\ = -6g((\nabla^M_X A)_X Y, A_X Y).\end{aligned}$$

In the last equality we have used

$$g(A_X \nabla_X^M Y, A_X Y) = g(A_X \nabla_X Y, A_X Y),$$

which is a consequence of definition (5.27). The identity (5.43) establishes the equivalence of (iii) and (iv), and completes the proof of Theorem 5.39. □

Next we assume the foliation $\mathcal{F}$ to be in addition totally geodesic, i.e. all leaves are totally geodesic submanifolds. The (local) reflection $\varphi_\mathcal{L}$ in each leaf $\mathcal{L}$ (or relative to $\mathcal{L}$) is defined as the local geodesic symmetry for normal geodesics to $\mathcal{L}$ in a sufficiently small tubular neighborhood of $\mathcal{L}$. For $m \in \mathcal{L}$ and p on a sufficiently short normal geodesic γ emanating from m, and parametrized by arc length, i.e. $p = \exp_m(rX) = \gamma(r)$ for some unit vector $X \in \Gamma L_m^\perp$, we have $\varphi_\mathcal{L}(p) = \exp_m(-rX) = \gamma(-r)$. (For more details about reflections see e.g. [To-V 1].)

For a Riemannian foliation it is immediate that the reflection $\varphi_\mathcal{L}$ sends leaves into leaves, and corresponds to a geodesic symmetry on the (local) model space for the transversal geometry at the point corresponding to the leaf $\mathcal{L}$.

It is well-known that when all the reflections are isometries, then the leaves $\mathcal{L}$ are necessarily totally geodesic. In [To-V 2] there is a discussion of conditions to impose on the reflections in a totally geodesic submanifold, so as to guarantee that they are isometries. They involve the shape operator $T_p(m) : T_m G_p \to T_m G_p$ of the geodesic sphere $G_p \subset M$ with center $p = \gamma(r)$ and radius r. We have then $L_m \subset T_m G_p$, the inclusion being an identity only for $q = 1$. In [To-V 3] similar conditions are discussed using the Ricci operator $\tilde{Q}_p(m) : T_m G_p \to T_m G_p$ of G_p.

The characterizations for transversal symmetry in Theorem 5.39 can then be sharpened. The following result is a corrected version of Theorem 3 of [To-V 4, p. 311].

5.44 THEOREM. *Let $\mathcal{F}$ be a totally geodesic and Riemannian foliation on (M, g) of codimension $q > 1$, and g a bundle-like metric. Assume all data to be analytic. Consider the following conditions:*

(i) *$\mathcal{F}$ is transversally symmetric and $R_{XYXV}^M = 0$ for all $X, Y \in \Gamma L^\perp$ and all $V \in \Gamma L$;*
(ii) *$\nabla_X^\nabla R_{XYXY}^M = 0$ for all $X, Y \in \Gamma L^\perp$ and $R_{XYXV}^M = 0$ for all $V \in \Gamma L$;*
(iii) *$R_{XYXV}^M = 0$ and $\nabla_X^M R_{XYXY}^M = 0$ for all $X, Y \in \Gamma L^\perp$ and $V \in \Gamma L$;*
(iv) *the reflections φ in the leaves are isometries;*
(v) *$\varphi_*(m) \circ T_p(m) = T_{\varphi(p)}(m) \circ \varphi_*(m)$ for all $m \in M$, all unit $X \in L_m^\perp$, and all $p = \exp_m(rX)$ for all sufficiently small r;*
(vi) *same condition as in (v), but applied only to normal vectors $Y \in L_m^\perp \cap T_m G_p$;*
(vii) *$\varphi_*(m) \circ \tilde{Q}_p(m) = \tilde{Q}_{\varphi(p)}(m) \circ \varphi_*(m)$ for all $m \in M$, all unit $X \in L_m^\perp$, and all $p = \exp_m(rX)$ for all sufficiently small r, if $\dim M > 3$ and $2 \dim M = 2n \neq 3(n - q + 1)$.*

Then (i), (ii) *and* (iii) *are equivalent;* (iv)–(vii) *are equivalent and imply* (i), (ii) *and* (iii).

Proof. The equivalence of (i) and (ii) was proved above. Further, the equivalence of (iv) and (v) has been proved in [To-V 2], and the equivalence of (iv) and (vii) in [To-V 3]. These proofs use the analyticity assumption made.

The implication (vi)⇒(iii) follows at once from the detailed computations in [To-V 2].

Now we prove the implication (iii)⇒(ii). For totally geodesic $\mathcal{F}$ we have to put $T = 0$ (see (3.25) for the definition of T) in (5.37e), and this yields

$$R^M_{XYXV} = g((\nabla^M_X A)_X Y, V).$$

By assumption this term vanishes for all $V \in \Gamma L$. In particular

$$g((\nabla^M_X A)_X Y, A_X Y) = 0$$

for $X, Y \in \Gamma L^\perp$. Thus condition (iv) of Theorem 5.39 is satisfied. This completes the proof of the theorem. □

Consequences of constant ambient curvature

In this section we apply the previous considerations to a manifold (M, g) of constant sectional curvature, i.e. to real space forms.

As observed in [To-V 1], in a space of constant curvature the reflections in totally geodesic submanifolds are isometries. Hence from this and Theorem 5.44 we obtain the following fact.

5.45 THEOREM. *Let $\mathcal{F}$ be a Riemannian foliation on a space (M, g) of constant curvature, and g a bundle-like metric. If $\mathcal{F}$ is totally geodesic, it is necessarily transversally symmetric.*

More generally we have the following result.

5.46 THEOREM. *Let $\mathcal{F}$ be a Riemannian foliation on a space (M, g) of constant curvature, and g a bundle-like metric. Then $\mathcal{F}$ is transversally symmetric if and only if*

(5.47) $$g(A_X Y, T_{A_X Y} X) = 0 \quad \text{for all} \quad X, Y \in \Gamma L^\perp.$$

Note that for totally geodesic $\mathcal{F}$, condition (5.47) is satisfied, and thus Theorem 5.45 follows from Theorem 5.46.

Proof of Theorem 5.46. By assumption $\nabla^M R^M = 0$ and for $X, Y \in \Gamma L^\perp$ we have

$$R^M_{XY} X = c\{g(X,X)Y - g(X,Y)X\},$$

where c is the constant curvature of (M, g). Thus for $V \in \Gamma L$ it follows $R^M_{XYXV} = 0$. In particular

$$R^M_{XYXA_XY} = 0.$$

By (5.37e) we have on the other hand

$$R^M_{XYXA_XY} = g((\nabla^M_X A)_X Y, A_X Y) + 2g(A_X Y, T_{A_X Y} X).$$

Then (5.43) implies

$$-\nabla_X R^\nabla_{XYXY} = -6g((\nabla^M_X A)_X Y, A_X Y) = 12g(A_X Y, T_{A_X Y} X).$$

This, together with Theorem 5.39, completes the proof. $\square$

For a Riemannian foliation with integrable $L^\perp$, we have $A = 0$ and hence (5.47) implies the following fact.

COROLLARY 5.48. *Let $\mathcal{F}$ be a Riemannian foliation on a space of constant curvature, and g a bundle-like metric. If $L^\perp$ is integrable, then $\mathcal{F}$ is transversally symmetric.*

Effect on the ambient metric

In this section we treat some aspect of the following question. How does the existence of a transversally symmetric foliation influence the geometry of the ambient space?

Since we restrict our attention to foliations of dimension one, the reader may wish to first consult Chapter 6. Thus we consider a Riemannian flow (a Riemannian foliation with one-dimensional leaves), and we assume moreover $\mathcal{F}$ to be generated by the flow lines of a Killing vector field ξ of unit length. The leaves are then necessarily geodesics. We prove the following result, correcting Theorem 10 of [To-V 4, p. 314].

5.49 THEOREM. *Let $\mathcal{F}$ be the Riemannian flow defined by a unit Killing vector field ξ on (M, g). If $\mathcal{F}$ is transversally symmetric and $R_{XYXV} = 0$ for all X, $Y \in \Gamma L^\perp$ and all $V \in \Gamma L$, the space (M, g) is locally homogeneous. If moreover (M, g) is complete and simply connected, it is a naturally reductive homogeneous space.*

Proof. By F. Tricerri and L. Vanhecke, Lecture Note Series London Math. Soc. 83, Cambridge Univ. Press, 1983 one has to prove the existence of a (1,2)-tensor field T (unrelated to O'Neill's tensor earlier in this chapter), such that for the new connection $\bar{\nabla} = \nabla^M - T$ we have the equations of Ambrose and Singer

(5.50) $$\bar{\nabla} g = 0, \bar{\nabla} T = 0, \bar{\nabla} R^M = 0,$$

and

(5.51) $$T_X X = 0$$

for all tangent vector fields X. (5.50) guarantees the local homogeneity, while (5.51) guarantees that the homogeneous structure T is of natural reductive type. A complete and simply connected manifold with such a T is a natural reductive homogeneous space.

To prove the existence of such a T, let A^* be the tensor field defined by

(5.52) $$A^*_\xi \xi = 0, \quad A^*_X \xi = A_\xi X, \quad A^*_\xi X = -A_X \xi, \quad A^*_X Y = 0$$

for $X, Y \in \Gamma L^\perp$. Then let

(5.53) $$T = A - A^*$$

and define $\bar{\nabla}$ by

$$\bar{\nabla} = \nabla^M - T.$$

Note that the properties of the O'Neill tensor A (see (5.28)) imply

(5.54)
$$\begin{aligned} & A_X \text{ is alternating,} \\ & A_X Y = -A_Y X, \\ & A_\xi X = A_\xi \xi = 0, \\ & A_X \xi = \pi \nabla^M_X \xi, \\ & A_X Y = \pi^\perp \nabla^M_X Y \end{aligned}$$

and hence $A_X Y \sim \xi$. Hence, from (5.52), (5.53) and (5.54) we get

(5.55) $$T_\xi \xi = 0, \quad T_X \xi = A_X \xi, \quad T_\xi X = -A_X \xi, \quad T_X Y = A_X Y.$$

Now, from this we see at once that (5.51) is satisfied. Further, $\bar{\nabla} g = 0$ is equivalent to

$$g(T_X Y, Z) + g(T_X Z, Y) = 0$$

and we see easily that this condition is also satisfied.

Further, when $\mathcal{F}$ is transversally symmetric, ξ is Killing and

$$R^M_{XYZ\xi} = 0,$$

then, using also the properties of the O'Neill tensor, a lengthy but straightforward computation shows that $\bar{\nabla} R^M = 0$ and $\bar{\nabla} T = 0$, which completes the proof. □

$\mathcal{F}$-Jacobi fields

In the remainder of this chapter we discuss Jacobi fields along geodesics orthogonal to the leaves, and an associated Riccati equation for the shape operator of the leaves along such geodesics. This material was presented in [Kih], [Kih-To] and [To-V 7].

In the sequel we will make use of the metric connection ∇^L on L defined by

(5.56) $$\nabla^L_E V = \pi^\perp \nabla^M_E V$$

for $E \in \Gamma TM$, $V \in \Gamma L$. For a unit speed geodesic γ orthogonal to the leaves of $\mathcal{F}$, we consider $V \in \Gamma L_\gamma$, i.e. vector fields along γ which are tangential to $\mathcal{F}$.

DEFINITION. $V \in \Gamma L_\gamma$ is an $\mathcal{F}$-Jacobi vector field along γ, if

(5.57) $$\pi^\perp \{ \ddot{V} + R^M(V, \dot{\gamma})\dot{\gamma} \} = 0.$$

Here $\ddot{V}$ denotes as usual $\left(\frac{\nabla^M}{dt}\right)^2 V = \nabla^M_{\dot{\gamma}} \nabla^M_{\dot{\gamma}} V$ along γ. The curvature R^M of ∇^M gives rise to the operator $\bar{R}_{\dot{\gamma}} : L_\gamma \to L_\gamma$ by

(5.58) $$\bar{R}_{\dot{\gamma}} V = \pi^\perp R^M(V, \dot{\gamma})\dot{\gamma}.$$

Let A be the integrability tensor defined by (5.27). In particular for $X \in L^\perp$ we have $A_X : L \to L^\perp$, $A_X : L^\perp \to L$ and $A_X^2 | L : L \to L$. Then (5.57) can be equivalently expressed by

(5.59) $$\left(\frac{\nabla^L}{dt}\right)^2 V + \bar{R}_{\dot{\gamma}} V + (A_{\dot{\gamma}})^2 V = 0.$$

This follows immediately from $A_{\dot{\gamma}} V = \pi \nabla^M_{\dot{\gamma}} V$ and

$$A_{\dot{\gamma}} A_{\dot{\gamma}} V = A_{\dot{\gamma}} \pi \nabla^M_{\dot{\gamma}} V = \pi^\perp \nabla^M_{\dot{\gamma}} \pi \nabla^M_{\dot{\gamma}} V = \pi^\perp \nabla^M_{\dot{\gamma}} (\nabla^M_{\dot{\gamma}} V - \pi^\perp \nabla^M_{\dot{\gamma}} V)$$
$$= \pi^\perp \nabla^M_{\dot{\gamma}} \nabla^M_{\dot{\gamma}} V - \nabla^L_{\dot{\gamma}} \nabla^L_{\dot{\gamma}} V.$$

Clearly tangential Jacobi fields along γ in the usual sense are $\mathcal{F}$-Jacobi fields along γ. What about the converse? We consider first the shape operator $S_{\dot{\gamma}} : L_\gamma \to L_\gamma$ of $\mathcal{F}$, defined for $V \in L_\gamma$ by

(5.60) $$S_{\dot{\gamma}} V = \pi^\perp \nabla^M_V \dot{\gamma},$$

where $\dot{\gamma}$ is extended to a normal vector field in the neighborhood of $\gamma(t)$. Thus $S_{\dot{\gamma}} = -W(\dot{\gamma})$ in terms of the notation in Chapter 3. We prove the following auxiliary fact. Let V be an $\mathcal{F}$-Jacobi field along γ satisfying the initial conditions

$$V(0) = v, \quad \left(\frac{\nabla^L}{dt} V\right)(0) = S_m v$$

at $\gamma(0) = m$. Then we have necessarily

(5.61) $$\pi^\perp [V, \dot\gamma] = 0,$$

and as a consequence

(5.62) $$S_{\dot\gamma} V = \pi^\perp \nabla^M_{\dot\gamma} V.$$

The bracket in (5.61) is well-defined as a vector field along γ. The same interpretation has to be given to several expressions in the calculation to follow. Since V is $\mathcal{F}$-Jacobi along the geodesic γ, we have

$$\pi^\perp (\nabla^M_{\dot\gamma} \nabla^M_{\dot\gamma} V + R^M(V, \dot\gamma)\dot\gamma)$$
$$= \pi^\perp (\nabla^M_{\dot\gamma} \nabla^M_V \dot\gamma + \nabla^M_{\dot\gamma}[\dot\gamma, V] - \nabla^M_{[V,\dot\gamma]}\dot\gamma - \nabla^M_{\dot\gamma} \nabla^M_V \dot\gamma)$$
$$= 0.$$

We set $U = \pi^\perp[\dot\gamma, V] \in \Gamma L_\gamma$, $X = \pi[\dot\gamma, V] \in \Gamma L_\gamma^\perp$. Then the identity above yields

$$\pi^\perp \nabla^M_{\dot\gamma}(U + X) + \pi^\perp \nabla^M_{U+X}\dot\gamma = 0.$$

Observe that

$$\pi^\perp \nabla^M_{\dot\gamma} X + \pi^\perp \nabla^M_X \dot\gamma = A_{\dot\gamma} X + A_X \dot\gamma = 0,$$

since A is skew-symmetric on orthogonal vectors to $\mathcal{F}$ (5.28c). Then

$$\pi^\perp \nabla^M_{\dot\gamma} U + \pi^\perp \nabla^M_U \dot\gamma = 0.$$

Equivalently,

$$\frac{\nabla^L}{dt} U + S_{\dot\gamma} U = 0.$$

Further

$$U(0) = \pi^\perp[\dot\gamma, V](0) = (\pi^\perp \nabla^M_{\dot\gamma} V - \pi^\perp \nabla^M_V \dot\gamma)(0)$$
$$= \left(\frac{\nabla^L}{dt} V\right)(0) - (S_{\dot\gamma} V)(0) = 0$$

by our assumption on V. Since U solves a linear first order ODE with trivial initial condition, necessarily $U = 0$. This establishes (5.61).

5.63 THEOREM. *Let γ be a unit speed geodesic orthogonal to $\mathcal{F}$. Then the following holds:*

(i) *an ordinary Jacobi vector field V along γ is tangential to $\mathcal{F}$ if and only if it satisfies the initial conditions at $m = \gamma(0)$:*

(5.64) $$V(0) = v \in L_m, \ \left(\frac{\nabla^M}{dt} V\right)(0) = S_m v + A_{\dot\gamma(0)} v;$$

(ii) *an $\mathcal{F}$-Jacobi vector field $V \in \Gamma L_\gamma$ is a tangential ordinary Jacobi vector field if and only if it satisfies the initial conditions at $m = \gamma(0)$:*

$$V(0) = v \in L_m, \quad \left(\frac{\nabla^L}{dt}V\right)(0) = S_m v. \tag{5.65}$$

The main point of the arguments in the following proof is the fact that given γ, the choice of $v \in L_m$, $m = \gamma(0)$ determines a unique tangential Jacobi vector field V along γ satisfying (5.64).

Proof. (i) Let $V \in \Gamma L_\gamma$ be an ordinary Jacobi vector field with $V(0) = v \in L_m$. Then

$$\frac{\nabla^M}{dt}V = \pi^\perp \nabla^M_{\dot\gamma} V + \pi \nabla^M_{\dot\gamma} V = \pi^\perp \nabla^M_{\dot\gamma} V + A_{\dot\gamma} V.$$

Further, since V is Jacobi along γ and $V \perp \dot\gamma$, we can use (5.62). Hence

$$\frac{\nabla^M}{dt}V = S_{\dot\gamma} V + A_{\dot\gamma} V,$$

and (5.64) follows. Conversely let $\bar V \in \Gamma TM_\gamma$ be an ordinary Jacobi vector field satisfying (5.64). The initial condition $\bar V(0) = v \in L_m$ defines a tangential Jacobi vector field V along γ by variations of γ through orthogonal geodesics (if $\alpha(s)$ is a curve in the leaf through m with $\alpha(0) = m$, $\dot\alpha(0) = v$, the geodesic γ_s with $\gamma_s(0) = \alpha(s)$ has the initial velocity given by the unique horizontal lift of $\dot\gamma(0) \in L_m^\perp$ to $L_{\alpha(s)}^\perp$). Since $\bar V$, V are both Jacobi fields, and satisfy the same initial conditions (5.64), it follows that $\bar V = V$ and $\bar V$ is necessarily tangential.

(ii) Let $V \in \Gamma L_\gamma$ be an ordinary Jacobi vector field with $V(0) = v \in L_m$. Then (5.64) holds. The ordinary Jacobi equation implies the $\mathcal{F}$-Jacobi equation. Moreover (5.64) implies (5.65). Conversely let $\bar V \in \Gamma L_\gamma$ be the $\mathcal{F}$-Jacobi vector field satisfying (5.65). Then $\bar V$ coincides with the Jacobi vector field $V \in \Gamma L_\gamma$ satisfying (5.64). □

Riccati equation

For the following discussion let $\{e_i\}_{i=1,\ldots,p}$ be an orthonormal basis of L_m at $m = \gamma(0)$. By parallel translation along γ with respect to the metric connection ∇^L we obtain an orthonormal frame field $\{E_i\}$ of L along γ. Further, let Y_i ($i = 1, \ldots, p$) be the $\mathcal{F}$-Jacobi vector fields along γ satisfying the initial conditions

$$Y_i(0) = e_i, \quad \left(\frac{\nabla^L}{dt} Y_i\right)(0) = S_m e_i. \tag{5.66}$$

This gives rise to a linear operator $D : L_\gamma \to L_\gamma$ given by

$$Y_i = D E_i. \tag{5.67}$$

Clearly
$$\frac{\nabla^L}{dt}Y_i = \left(\frac{\nabla^L}{dt}D\right)E_i,$$

and
$$\left(\frac{\nabla^L}{dt}\right)^2 Y_i = \left(\left(\frac{\nabla^L}{dt}\right)^2 D\right)E_i.$$

Then from (5.59) we obtain

(5.68) $$\left(\frac{\nabla^L}{dt}\right)^2 D + (\bar{R}_{\dot\gamma} + A_{\dot\gamma}^2)D = 0.$$

Thus D is the $\mathcal{F}$-Jacobi tensor (endomorphism) field along γ satisfying the initial conditions

(5.69) $$D(0) = I, \qquad \left(\frac{\nabla^L}{dt}D\right)(0) = S_m.$$

5.70 Theorem. *Let γ be a unit speed geodesic orthogonal to $\mathcal{F}$, and $D : L_\gamma \to L_\gamma$ the endomorphism field defined by (5.66), (5.67). Then*

(5.71) $$S_{\dot\gamma} = \frac{\nabla^L}{dt}D \cdot D^{-1}.$$

Proof. By construction
$$S_{\dot\gamma}Y_i = \pi^\perp \nabla^M_{Y_i}\dot\gamma.$$

By (5.62) it follows then that
$$S_{\dot\gamma}Y_i = \frac{\nabla^L}{dt}Y_i,$$

or
$$S_{\dot\gamma}DE_i = \frac{\nabla^L}{dt}DE_i,$$

i.e. $S_{\dot\gamma}D = \frac{\nabla^L}{dt}D$. Since D is invertible, Theorem 5.70 follows. □

The next fact was established in [Kih-To] by a direct calculation.

5.72 Theorem. *Let γ be a unit speed geodesic orthogonal to $\mathcal{F}$. Then the shape operator $S_{\dot\gamma} : L_\gamma \to L_\gamma$ of $\mathcal{F}$ along γ satisfies the Riccati equation*

(5.73) $$\frac{\nabla^L}{dt}S_{\dot\gamma} + S_{\dot\gamma}^2 + \bar{R}_{\dot\gamma} + A_{\dot\gamma}^2 = 0.$$

Proof. By (5.71) we have

$$(5.74) \qquad S_{\dot\gamma} D = \frac{\nabla^L}{dt} D.$$

Differentiating covariantly along γ, we get

$$\left(\frac{\nabla^L}{dt} S_{\dot\gamma}\right) D + S_{\dot\gamma}\left(\frac{\nabla^L}{dt} D\right) = \left(\frac{\nabla^L}{dt}\right)^2 D.$$

Using (5.74) and (5.68), this implies

$$\left(\frac{\nabla^L}{dt} S_{\dot\gamma}\right) D + S_{\dot\gamma}(S_{\dot\gamma} D) + (\bar{R}_{\dot\gamma} + A_{\dot\gamma}^2) D = 0.$$

Since D is invertible, the desired result follows. $\square$

Remark. This Riccati equation can also be obtained from formula (5.37c) for $X = Y = \dot\gamma$. Observing that $T_U \dot\gamma = S_{\dot\gamma} U$, the Riccati equation readily follows.

Returning to an $\mathcal{F}$-Jacobi vector field V along γ as described in Theorem 5.63, we observe that

$$\dot V \equiv \frac{\nabla^M}{dt} V = \pi^\perp \frac{\nabla^M}{dt} V + \pi \frac{\nabla^M}{dt} V = S_{\dot\gamma} V + A_{\dot\gamma} V.$$

It follows that

$$\ddot V \equiv \left(\frac{\nabla^M}{dt}\right)^2 V = (\dot S_{\dot\gamma} + \dot A_{\dot\gamma}) V + (S_{\dot\gamma} + A_{\dot\gamma}) \dot V$$
$$= [(\dot S_{\dot\gamma} + \dot A_{\dot\gamma}) + (S_{\dot\gamma} + A_{\dot\gamma})^2] V.$$

Since

$$\pi^\perp \ddot V + \bar{R}_{\dot\gamma} V = 0,$$

and all this holds for a frame field of L_γ, it follows that on L

$$(5.75) \qquad \pi^\perp(\dot S_{\dot\gamma} + \dot A_{\dot\gamma}) + \pi^\perp(S_{\dot\gamma} + A_{\dot\gamma})^2 + \bar{R}_{\dot\gamma} = 0.$$

We have established the following fact.

5.76 THEOREM. *Let γ be a unit speed geodesic orthogonal to $\mathcal{F}$. Then (5.75) holds on L.*

We wish to show that equation (5.75) conversely implies (5.73). For this it suffices to show that on L

$$\pi^\perp \dot A_{\dot\gamma} + \pi^\perp(S_{\dot\gamma} A_{\dot\gamma} + A_{\dot\gamma} S_{\dot\gamma}) = 0.$$

Let $U, V \in L$. It suffices to show

$$g((\nabla^M_{\dot\gamma} A_{\dot\gamma}) U + (S_{\dot\gamma} A_{\dot\gamma} + A_{\dot\gamma} S_{\dot\gamma}) U, V) = 0,$$

or equivalently

$$g((\nabla^M_{\dot\gamma} A)_{\dot\gamma} U, V) + g(A_{\dot\gamma} U, S_{\dot\gamma} V) - g(S_{\dot\gamma} U, A_{\dot\gamma} V) = 0.$$

The first term vanishes by an earlier formula (see the list following (5.38)). The second term vanishes since $A_{\dot\gamma}U \in L^\perp$, while $S_{\dot\gamma}V \in L$. The third term vanishes similarly. This completes the proof of the equivalence of equations (5.73) and (5.75).

The Wronskian

We now introduce a notion naturally associated to the defining equation for $\mathcal{F}$-Jacobi vector fields.

5.77 DEFINITION. *If D, E are fields of endomorphisms of L along γ, the Wronskian is the field of endomorphisms along γ given by*

$$(5.78) \qquad W(D,E) = \frac{\nabla^L}{dt}{}^t D \circ E - {}^t D \circ \frac{\nabla^L}{dt} E.$$

If $D, E \in \Gamma \operatorname{End} L_\gamma$ are both solutions of the differential equation along γ

$$(5.79) \qquad \left(\frac{\nabla^L}{dt}\right)^2 F + \bar{R}_{\dot\gamma} F + A_{\dot\gamma}^2 F = 0,$$

then we have the following fact.

5.80 THEOREM. *Let γ be a unit speed geodesic orthogonal to $\mathcal{F}$ and D, E endomorphism fields of L along γ satisfying (5.79). Then $W(D,E)$ is (covariantly) constant along γ.*

Proof. We have to show that

$$\frac{\nabla^L}{dt} W(D,E) = 0.$$

We note that while $A_{\dot\gamma}$ has skew-symmetric aspects, the operator $A_{\dot\gamma}^2 : L \to L$ is symmetric, and so is $\bar{R}_{\dot\gamma}$. It follows that

$$\begin{aligned}
\frac{\nabla^L}{dt} W(D,E) &= \left(\frac{\nabla^L}{dt}\right)^2 {}^t D \circ E + \frac{\nabla^L}{dt}{}^t D \circ \frac{\nabla^L}{dt} E \\
&\quad - \frac{\nabla^L}{dt}{}^t D \circ \frac{\nabla^L}{dt} E - {}^t D \circ \left(\frac{\nabla^L}{dt}\right)^2 E \\
&= -{}^t D \circ {}^t(\bar{R}_{\dot\gamma} + A_{\dot\gamma}^2) \circ E + {}^t D \circ (\bar{R}_{\dot\gamma} + A_{\dot\gamma}^2) \circ E \\
&= 0,
\end{aligned}$$

since ${}^t(\bar{R}_{\dot\gamma} + A_{\dot\gamma}^2) = \bar{R}_{\dot\gamma} + A_{\dot\gamma}^2$. □

5.81 COROLLARY. *Let D be an endomorphism field of L along γ, satisfying the initial conditions at $m = \gamma(0)$:*

$$D(0) = I, \quad \left(\frac{\nabla^L}{dt} D\right)(0) = S_m.$$

Then $W(D,D) = 0$.

Proof. This follows from Theorem 5.80 and

$$W(D,D)(0) = \left(\frac{\nabla^L}{dt}{}^t D\right)(0) - \left(\frac{\nabla^L}{dt}D\right)(0) = {}^t S_m - S_m = 0. \qquad \square$$

Applications

(a) Riemannian foliations $\mathcal{F}$ with bundle like g and involutive normal bundle $L^\perp$. The last property is expressed by $A = 0$, and thus in particular $A_{\dot\gamma} = 0$. Equations (5.68)(5.73) yield

(5.82) $$\left(\frac{\nabla^L}{dt}\right)^2 D + \bar R_{\dot\gamma} D = 0$$

for the endomorphism field D of L along γ defined by (5.67), and

(5.83) $$\frac{\nabla^L}{dt} S_{\dot\gamma} + S_{\dot\gamma}^2 + \bar R_{\dot\gamma} = 0$$

for the field of shape operators $S_{\dot\gamma}$ along γ.

If (M, g) is of constant curvature c, then $\bar R_{\dot\gamma} = cI : L_\gamma \to L_\gamma$. Equations (5.82), (5.83) then read

(5.84) $$\left(\frac{\nabla^L}{dt}\right)^2 D + cD = 0,$$

(5.85) $$\frac{\nabla^L}{dt} S_{\dot\gamma} + S_{\dot\gamma}^2 + cI = 0.$$

Note that $\mathcal{F}$ can only be totally geodesic if $c = 0$. Even if A is not assumed to vanish, the Riccati equation shows that $\mathcal{F}$ can only be totally geodesic provided $c \geqq 0$ (taking traces yields then $c = \frac{1}{p}|A_{\dot\gamma}|^2$, where p is the dimension of the leaves of $\mathcal{F}$). Returning to the case of foliations with involutive normal bundle, we observe that for $c > 0$ the explicit solution of equation (5.84) along the geodesic $\gamma(r)$ is given by

$$D(r) = \cos(\sqrt{c}\, r)I + \frac{\sin(\sqrt{c}\, r)}{\sqrt{c}} S(0).$$

Then $S = \dot D D^{-1}$ turns out to be diagonal, and of the form

$$S(r) = \begin{pmatrix} \dfrac{\kappa_1 - \sqrt{c}\tan(\sqrt{c}\, r)}{1 + \dfrac{\tan(\sqrt{c}\, r)}{\sqrt{c}}\kappa_1} & & & & 0 \\ & \ddots & & & \\ & & \bullet & & \\ & & & \bullet & \\ 0 & & & & \bullet \end{pmatrix},$$

where $\kappa_1, \ldots, \kappa_p$ are the principal curvatures of $S(0)$.

Harmonic foliations are characterized by $\operatorname{Tr} S = 0$. Taking the trace in (5.83) yields

$$(5.86) \qquad |S|^2 + \sum_{i=1}^{p} g(R^M(E_i, \dot{\gamma})\dot{\gamma}, E_i) = 0.$$

This implies the following fact ([K-To 11, 2.27]).

5.87 PROPOSITION. *Let $\mathcal{F}$ be a Riemannian foliation with bundle-like g on (M, g). Assume $L^\perp$ to be involutive. If the sectional curvature $K^M \geqq 0$, then the harmonicity of $\mathcal{F}$ implies that $\mathcal{F}$ is totally geodesic.*

Proof. (5.86) implies $|S|^2 = 0$, hence $S = 0$. □

Remark. Riemannian foliations with involutive normal bundle and totally geodesic leaves are locally Riemannian products. The proof is an application of De Rham's holonomy theorem, together with the fact that in this case the decomposition $TM \cong L \oplus L^\perp$ is preserved under parallel transport.

The condition $A = 0$ holds in particular for the case of foliations of codimension $q = 1$. An example is the following conclusion.

5.88 PROPOSITION. *If $\mathcal{F}$ is of codimension one and harmonic on (M, g) with nonnegative Ricci curvature, then $\mathcal{F}$ is totally geodesic.*

Global arguments for this conclusion on a closed M were given in [Os 1] and [K-To 11] (see also [To 2, Theorem 7.50]). While the argument to follow is local in nature, and thus applies equally well off the singular set of a Riemannian foliation, the global arguments in [Os 1] and [K-To 11] imply the result as well as the Riemannian property of $\mathcal{F}$, while in the present context $\mathcal{F}$ is assumed to be Riemannian to start with.

Proof. For $q = 1$ equation (5.86) implies

$$(5.89) \qquad |S|^2 + \operatorname{Ric}(\dot{\gamma}, \dot{\gamma}) = 0.$$

Thus for nonnegative Ricci curvature $|S|^2 = 0$, hence $S = 0$. □

As (5.89) moreover shows, the positivity of the Ricci operator at even a single point of M is incompatible with the existence of a foliation satisfying our assumptions.

(b) Riemannian foliations $\mathcal{F}$ with bundle-like g on (M, g) with strictly negative sectional curvatures for all 2-planes at a single point of (M, g).

5.90 PROPOSITION. *With these assumptions $\mathcal{F}$ cannot be totally geodesic.*

Proof. Assume $S = 0$. It follows from (5.73) that $B = \bar{R}_{\dot\gamma} + A_{\dot\gamma}^2 = 0$. But

$$g(BE_i, E_i) = g(\bar{R}_{\dot\gamma} E_i, E_i) - |A_{\dot\gamma} E_i|^2.$$

Let now $\gamma(t)$ be the point at which $K^M < 0$ on all 2-planes. Then at this point

$$g(\bar{R}_{\dot\gamma} E_i, E_i) = g(R^M(E_i, \dot\gamma)\dot\gamma, E_i) = K^M(\dot\gamma, E_i) < 0.$$

This contradicts

$$K^M(\dot\gamma, E_i) - |A_{\dot\gamma} E_i|^2 = 0. \qquad \square$$

As in [Kih-To], it is convenient to consider the partial Ricci curvature form Ric^L defined by

$$\mathrm{Ric}^L(x, y) = \sum_{i=1}^{p} g(R^M(e_i, x)y, e_i)$$

for $x, y \in L_m^\perp$, and an orthonormal basis $\{e_i\}_{i=1,\ldots,p}$ of L_m. The quadratic form associated to the bilinear symmetric form Ric^L on $L^\perp$ is then given by

$$\mathrm{Ric}^L(x, x) = \sum_{i=1}^{p} K^M(x, e_i).$$

Taking $\operatorname{Tr} B$ over L in the preceding argument, the conclusion is as follows.

5.91 PROPOSITION. *If $\mathcal{F}$ is a Riemannian foliation with bundle-like g on (M, g), and $\mathrm{Ric}^L < 0$ at a single point of M, then $\mathcal{F}$ cannot be totally geodesic.*

In the case of codimension $q = 1$ we have $\mathrm{Ric}^L = \mathrm{Ric}^M$, and the condition above concerns the ordinary Ricci operator at a point of M.

(c) Mean curvature conditions. Consider $\operatorname{Tr} S$. Note that S depends on the choice of a normal vector. Thus along a geodesic γ orthogonal to $\mathcal{F}$, the mean curvature function $h = \operatorname{Tr} S_{\dot\gamma}$ is well defined. Let

$$w = \frac{1}{p} h,$$

and consider the operator $S_{\dot\gamma} - wI : L \to L$. Then

$$|S_{\dot\gamma} - wI|^2 = \sum_{i=1}^{p} g((S_{\dot\gamma} - wI)E_i, (S_{\dot\gamma} - wI)E_i)$$
$$= |S_{\dot\gamma}|^2 - 2wh + pw^2 = |S_{\dot\gamma}|^2 - pw^2,$$

or

$$|S_{\dot\gamma}|^2 = pw^2 + |S_{\dot\gamma} - wI|^2.$$

Taking traces over L in the Riccati equation (5.73) implies

(5.92) $$p(\dot{w} + w^2) + |S_{\dot{\gamma}} - wI|^2 + \mathrm{Ric}^L(\dot{\gamma}, \dot{\gamma}) - |A_{\dot{\gamma}}|^2 = 0.$$

Here we have used

$$\mathrm{Tr}(A_{\dot{\gamma}}^2) = \sum_{i=1}^{p} g(A_{\dot{\gamma}}^2 E_i, E_i) = -\sum_{i=1}^{p} g(A_{\dot{\gamma}} E_i, A_{\dot{\gamma}} E_i) = -\sum_{i=1}^{p} |A_{\dot{\gamma}} E_i|^2 = -|A_{\dot{\gamma}}|^2,$$

while

$$\mathrm{Tr}(S_{\dot{\gamma}}^2) = \sum_{i=1}^{p} g(S_{\dot{\gamma}}^2 E_i, E_i) = \sum_{i=1}^{p} g(S_{\dot{\gamma}} E_i, S_{\dot{\gamma}} E_i) = \sum_{i=1}^{p} |S_{\dot{\gamma}} E_i|^2 = |S_{\dot{\gamma}}|^2.$$

To illustrate the method used in [Kih-To], we prove the following fact as an application of (5.92) under the additional completeness condition for the manifold (M, g). We refer to [Kih-To] and [To-V 7] for further applications.

5.93 PROPOSITION. *Let $\mathcal{F}$ be a foliation with bundle-like metric g on a complete Riemannian manifold (M, g). If for each geodesic γ orthogonal to $\mathcal{F}$ we have $\mathrm{Ric}^L(\dot{\gamma}, \dot{\gamma}) \geqq |A_{\dot{\gamma}}|^2$, then $\mathcal{F}$ is totally geodesic.*

Proof. (5.92) implies
$$\dot{w} + w^2 + c + r = 0,$$
with $c = \frac{1}{p}|S_{\dot{\gamma}} - wI|^2$, $r = \frac{1}{p}(\mathrm{Ric}^L(\dot{\gamma}, \dot{\gamma}) - |A_{\dot{\gamma}}|^2)$. Since $c \geqq 0$, $r \geqq 0$ it is clear that the solution decreases not less rapidly than the solution of $\dot{w} + w^2 = 0$ with the same initial condition. But that solution goes to $-\infty$ in finite time, contrary to the completeness assumption. This implies $w = 0$. Thus $c + r = 0$. Since $c \geqq 0$ and $r \geqq 0$, this implies $c = 0$. This means that $S_{\dot{\gamma}} = wI$, or, since $w = 0$, $S_{\dot{\gamma}} = 0$. □

As pointed out in [Kih-To], the inequality in the preceding proposition is in fact sharp. The hypotheses are in particular realized for a foliation of codimension $q = 1$ on (M, g) with $\mathrm{Ric}^M \geq 0$.

Chapter 6
Flows

In this chapter we discuss the case of tangentially oriented 1-dimensional foliations, in which many of the previously discussed concepts take a particularly simple form.

Let V be a vector field on M^{n+1}. For the orbits (integral curves) of V to define a foliation $\mathcal{F}$ on M, one has to assume that V is nonsingular. An example is provided by a circle action $S^1 \times M \to M$ without fixed points.

Let g be a Riemannian metric on M. Normalizing lengths shows that a foliation $\mathcal{F}$ given by a nonsingular vector field V is also given by a unit vector field T with respect to g. The dual 1-form $\chi \in \Omega^1(M)$ defined by

$$(6.1) \qquad \chi(Y) = g(T, Y) \text{ for } Y \in \Gamma TM$$

is precisely the characteristic form of $\mathcal{F}$. The induced metric g_L is related to χ by

$$(6.2) \qquad g_L(\lambda T, \lambda T) = \lambda^2, \ \chi(\lambda T) = \lambda.$$

For the mean curvature vector field $\tau \in \Gamma L^\perp$, we find by (3.22)

$$\tau = \pi(\nabla_T^M T).$$

But $g(T, \nabla_T^M T) = \frac{1}{2} T g(T, T) = 0$, so that $\nabla_T^M T$ already is orthogonal to the leaves. It follows that

$$(6.3) \qquad \tau = \nabla_T^M T.$$

For the dual mean curvature 1-form we find

$$(6.4) \qquad \kappa = \theta(T)\chi.$$

Proof. Let $X \in \Gamma L^\perp$. Then

(6.5) $$\kappa(X) = g(\nabla_T^M T, X) = -g(T, \nabla_T^M X).$$

On the other hand

$$\begin{aligned}(\theta(T)\chi)(X) &= T\chi(X) - \chi(\theta(T)X) = -\chi([T, X]) = -g(T, [T, X]) \\ &= g(T, \nabla_X^M T - \nabla_T^M X) = -g(T, \nabla_T^M X) .\end{aligned}$$

Comparing this with (6.5) shows

$$\kappa(X) = (\theta(T)\chi)(X).$$

To prove (6.4) it suffices to verify that

$$(\theta(T)\chi)(T) = 0.$$

But indeed

$$(\theta(T)\chi)(T) = T\chi(T) - \chi(\theta(T)T) = 0 \qquad \square$$

From this discussion, and the results in Chapter 4, we obtain the following characterization of harmonic or geodesic flows.

6.6 PROPOSITION. *For a flow $\mathcal{F}$ defined by a nonsingular vector field V (with normalized $T = 1/|V| \cdot V$) on (M, g), the following conditions are equivalent:*

(i) *$\mathcal{F}$ is harmonic;*
(ii) *the orbits of V are geodesics;*
(iii) *$\theta(T)\chi = 0$;*
(iv) *$\nabla_T^M T = 0$;*
(v) *g_L is invariant under flows of vector fields orthogonal to V;*
(vi) *$d\chi \in F^2\Omega^2(M)$.*

If $L^\perp$ is involutive, then these conditions are further equivalent to the conditions:

(vii) *$d\chi = 0$;*
(viii) *$\theta(X)\chi = 0$ for $X \in \Gamma L^\perp$.*

6 FLOWS

Proof. This follows from (3.17), (4.21), and Proposition 4.30. □

A flow is geodesible, if there exists a Riemannian metric such that the leaves of $\mathcal{F}$ are geodesics. Note that for a flow this concept is equivalent to tautness. Note further that the conditions (iii) to (viii) in Proposition 6.6 already involve a Riemannian metric. Sullivan [SU 2] gave a purely topological property characteristic for such flows (see Theorems 6.10 and 6.17 below for other properties characterizing such Riemannian flows). It is based on the following result of Gluck [Gl 2] and Sullivan [Su 2].

6.7 PROPOSITION. *Let $\mathcal{F}$ be a flow given by the nonsingular vector field V on M^{n+1}. Then the following conditions are equivalent:*

(i) *there exists a Riemannian metric on M, making the orbits of V geodesics, and V of unit length;*
(ii) *there exists a 1-form $\chi \in \Omega^1(M)$, such that $\chi(V) = 1$, and $\theta(V)\chi = 0$;*
(iii) *there exists a 1-form $\chi \in \Omega^1(M)$, such that $\chi(V) = 1$, and $i(V)d\chi = 0$;*
(iv) *there exists an n-plane bundle $E \subset TM$, complementary to L, such that $[V, X] \in \Gamma E$ for all $X \in \Gamma E$.*

Proof. (i) ⇒ (ii): We can assume that $V = T$ with the notations at the beginning of this chapter. Define χ by (6.1). Since the orbits of T are geodesics, we have $\nabla^M_T T = 0$. Thus by Proposition 6.6, part (iii), we have $\theta(T)\chi = 0$.

(ii) ⇒ (iii): We have
$$0 = \theta(V)\chi = i(V)d\chi + di(V)\chi = i(V)d\chi.$$

(iii) ⇒ (iv): Let $E = \ker \chi$, and $X \in \Gamma E$. Then
$$\chi[V, X] = -d\chi(V, X) + V\chi(X) - X\chi(V) = 0,$$
and $[V, X] \in \Gamma E$.

(iv) ⇒ (i): Let g be a metric, such that $V = T$ is of unit length, and E the orthogonal complement of T, with an arbitrary choice of metric on E. It suffices to show that $\kappa(X) = 0$ for $X \in \Gamma E$. But
$$\kappa(X) = g(\nabla^M_T T, X) = -g(T, \nabla^M_T X) = -g(T, \nabla^M_X T + [T, X]).$$
Since by assumption $[T, X] \in \Gamma E$, it follows that
$$\kappa(X) = -g(T, \nabla^M_X T) = -\frac{1}{2} X g(T, T) = 0. \qquad \square$$

6.8 PROPOSITION [Su 2]. *Let V be a nonsingular vector field on M. Then the following conditions are equivalent:*

(i) *the flow of V is geodesible;*
(ii) *there exists a 1-form $\chi \in \Omega^1(M)$ such that $\chi(V) > 0$ and $i(V)d\chi = 0$.*

Proof. (i) $\Rightarrow$ (ii): Let g be a metric such that the orbits of V are geodesics. Normalize V to a unit vector field. Then (i) $\Rightarrow$ (iii) in Proposition 6.7 implies the desired result.

(ii)$\Rightarrow$(i): Let χ be given as indicated. Then $V' = \frac{1}{\chi(V)}V$ satisfies property (iii) in Proposition 6.7. The implication (iii) $\Rightarrow$ (i) in Proposition 6.7 implies that the orbits of V' are geodesics (and V' is of unit length). But the orbits of V and V' are the same. □

Before turning to Riemannian flows, we wish to mention at least the definition of Anosov flows, which exhibit a totally distinct behavior. They are nonsingular flows which have attracted an immense amount of interest, but fail to be Riemannian. A nonsingular vector field V on a closed manifold $M^{n+1}(n \geq 2)$ is Anosov, if the following conditions are satisfied:

(i) the tangent bundle of M has a direct sum decomposition $TM = E^s \oplus E^u \oplus E^0$ into vector bundles, preserved by the flow $\varphi_t = \exp tV$;

(ii) the bundle E^0 is a line bundle, and is spanned by V;

(iii) the flow φ_t acting on E^s is exponentially contracting (s stands for stable), and the flow φ_t acting on E^u is exponentially dilating (u stands for unstable).

In terms of a Riemannian metric g on M, the exponentially contracting and dilating condition means that there exist constants $c \geq 1, \alpha > 0$ such that

$$|(\varphi_t)_*X| \leq ce^{-\alpha t}|X| \quad \text{for} \quad X \in E^s, t \geq 0;$$
$$|(\varphi_t)_*X| \geq c^{-1}e^{\alpha t}|X| \quad \text{for} \quad X \in E^u, t \geq 0.$$

But the notion is independent of the particular metric. It is an important observation that the geodesic flow on the unit tangent bundle of a closed Riemannian surface of constant negative curvature is an Anosov flow (D. V. Anosov's original observation on the subject). While this specific flow is not Riemannian, it does preserve a transversal volume. It can be shown conversely, that a closed smooth 3-manifold, with a transversal volume preserving Anosov flow with smooth contracting and dilating bundles, has precisely the structure of the unit tangent bundle of a closed Riemannian surface of constant negative curvature [Gh 6].

After this digression, we return to Riemannian flows. Let V be a nonsingular Killing vector field on (M, g). If we renormalize the metric by

(6.9)
$$\bar{g}|L = \frac{1}{|V|^2}g|L,$$
$$\bar{g}|L^\perp = g|L^\perp,$$

then V is still a Killing vector field for $(M, \bar{g})$, and moreover of unit length in the new metric.

A nonsingular Killing vector field clearly defines a Riemannian flow.

A flow $\mathcal{F}$ on M is isometric, if there exists a Riemannian metric g such that $\mathcal{F}$ is given by the orbits of a nonsingular Killing vector field with respect to g. Such a flow is necessarily Riemannian. The following fact was proved by Carrière.

6.10 THEOREM [Ca 1][Ca 2, Proposition 1, p. 49]. *Let $\mathcal{F}$ be a Riemannian flow on M. Then the following conditions are equivalent:*

(i) *$\mathcal{F}$ is isometric;*
(ii) *$\mathcal{F}$ is geodesible.*

Proof. (i) $\Rightarrow$ (ii): Let V be a Killing vector field of g, such that $\mathcal{F}$ is given by the orbits of V. Then the orthogonal complement E of V is preserved by the flow of V, and hence property (iv) of Proposition 6.7 is satisfied. (Renormalizing the metric by (6.9), V is a unit vector field, and the orbits are geodesics by the proof of (iv) $\Rightarrow$ (i) in Proposition 6.7.)

(ii) $\Rightarrow$ (i): Let $\mathcal{F}$ be a geodesible Riemannian flow. There exists a bundle-like metric and a unit vector field V, such that properties (i) and (iv) with $E = L^\perp$ of Proposition 6.7 hold. For such a metric V is a Killing vector field. $\square$

For the proof of the implication (ii) $\Rightarrow$ (i), we used the assumption that the given flow is Riemannian.

For these special Riemannian flows, the basic cohomology can be described in more detail [K-To 10]. Let V be a nonsingular Killing vector field on a closed Riemannian manifold M. By renormalizing the metric as in (6.9), if necessary, we can assume that $V = T$ is a unit vector field. The closure of $\exp tT$ in the compact isometry group of g is compact and abelian, hence a torus G. For the basic forms of the foliation $\mathcal{F}$ by the orbits of T we have then

$$\Omega_B(\mathcal{F}) \subset \Omega(M)^G,$$

where $\Omega(M)^G$ denotes the G-invariant forms on M. On the other hand the operator $i(T)$ on $\Omega(M)$ gives rise to a map of degree -1

(6.11) $$\Omega^{\cdot}(M)^G \xrightarrow{i(T)} \Omega_B^{\cdot -1}(\mathcal{F}).$$

Namely for $\omega' = i(T)\omega$, $\omega \in \Omega(M)^G$ we have

$$i(T)\omega' = i(T)^2\omega = 0, \quad \theta(T)\omega' = \theta(T)i(T)\omega = i(T)\theta(T)\omega = 0.$$

This map is compatible with differentials (up to sign), since $di(T) + i(T)d = \theta(T)$, which vanishes on $\Omega(M)^G$. We further show that this map is surjective. Let χ be the characteristic form. Consider for $\omega \in \Omega_B^{r-1}(\mathcal{F})$ the r-form $\chi \wedge \omega$. Then

$$\theta(T)(\chi \wedge \omega) = \theta(T)\chi \wedge \omega + \chi \wedge \theta(T)\omega = 0,$$

and $\chi \wedge \omega \in \Omega(M)^G$. Moreover

$$i(T)(\chi \wedge \omega) = i(T)\chi \, \omega \pm \chi \wedge i(T)\omega = \omega,$$

which proves the surjectivity of the map (6.11). As a consequence one has the following result.

6.12 PROPOSITION. *There is an exact sequence of complexes*

$$0 \to \Omega_B^{\cdot}(\mathcal{F}) \hookrightarrow \Omega^{\cdot}(M)^G \xrightarrow{i(T)} \Omega_B^{\cdot -1}(\mathcal{F}) \to 0.$$

Proof. It only remains to verify the exactness in the middle. For a basic form ω we have $i(T)\omega = 0$, which proves im $\subset$ ker $i(T)$. Let conversely $\omega \in \Omega(M)^G$ with $i(T)\omega = 0$. Since $\theta(T)\omega = 0$, $\omega \in \Omega_B(\mathcal{F})$. □

Since $H(\Omega^\cdot(M)^G) \xrightarrow{\cong} H(\Omega^\cdot(M))$, as for any compact Lie group acting on M, the following facts are a consequence of Proposition 6.12.

6.13 THEOREM. *Let T be a unit Killing vector field on the closed manifold (M^{n+1}, g).*

(i) *There is a long exact cohomology sequence*

$$(6.14) \quad \cdots \longrightarrow H_B^r(\mathcal{F}) \longrightarrow H_{DR}^r(M) \xrightarrow{i(T)_*} H_B^{r-1}(\mathcal{F}) \xrightarrow{\Delta} H_B^{r+1}(\mathcal{F}) \longrightarrow \cdots$$

with connecting homomorphism Δ.

(ii) *The groups $H_B^r(\mathcal{F})$ are all finite-dimensional for $0 \leq r \leq n$ (and 0 otherwise).*

(iii) *The Euler characteristic*

$$\chi_B(\mathcal{F}) = \sum_{r=0}^{n}(-1)^r \dim H_B^r(\mathcal{F})$$

is a well-defined integer.

As an illustration consider the case when M^{n+1} has the De Rham cohomology of a sphere S^{n+1}, $n = 2k$. If T is a unit Killing vector field, then one concludes recursively that $H_B^r(\mathcal{F}) \cong H_{DR}^r(P^k\mathbf{C})$ for $r = 0, \ldots, 2k$.

We note that the cohomology sequence in (6.14) ends with the exact sequence

$$(6.15) \quad 0 \to H_{DR}^{n+1}(M) \xrightarrow[\cong]{i(T)_*} H_B^n(\mathcal{F}) \to 0.$$

Thus for a closed oriented manifold, the volume form μ of M maps to a nontrivial element $i(T)\mu$ of $H_B^n(\mathcal{F}) \cong \mathbb{R}$. This is a special case of the situation described earlier in Theorem 5.23.

The boundary map Δ in (6.14) is given as follows. Let $\omega \in \Omega_B^{r-1}(\mathcal{F})$ be closed. Then $i(T)(\chi \wedge \omega) = \omega$, as proved before. Thus $\Delta[\omega]$ is represented by $d(\chi \wedge \omega) = d\chi \wedge \omega \in \Omega_B^{r+1}(\mathcal{F})$, i.e.

$$(6.16) \quad \Delta[\omega] = [d\chi] \cdot [\omega], \text{ where } [d\chi] \in H_B^2(\mathcal{F}).$$

In the presence of an involutive complement to the flow, it follows by part (vii) of Proposition 6.6 that $d\chi = 0$, and the coboundary map (6.16) is trivial.

As pointed out in [Sara], the vanishing of $[d\chi] \in H_B^2(\mathcal{F})$ conversely implies that the isometric flow $\mathcal{F}$ is transverse to a fibration $M \to S^1$. To see this, let $\alpha \in \Omega_B^1(\mathcal{F})$ such that $d\chi = d_B\alpha$. Then $\omega = \chi - \alpha \in \Omega^1(M)$ satisfies $d\omega = 0$ and

$$\omega(T) = \chi(T) - \alpha(T) = 1.$$

Thus ω is nonsingular, and by Tischler's Theorem 3.32 there is a fibration $f : M \to S^1$ with $\omega' = f^*d\theta$ arbitrarily close to ω. Thus $\omega'(T) > 0$ and $\mathcal{F}$ is transverse to $f : M \to S^1$.

From (6.15) we know that for an isometric (equivalently, geodesible) flow $\mathcal{F}$ on a closed oriented manifold M^{n+1} we have $H_B^n(\mathcal{F}) \cong \mathbb{R}$. Thus the following result of Molino and Sergiescu is very appealing.

6.17 THEOREM [Mo-Se]. *Let $\mathcal{F}$ be a Riemannian flow on a closed oriented manifold M^{n+1}. Then the following conditions are equivalent:*

(i) *$\mathcal{F}$ is isometric;*
(ii) *$H_B^n(\mathcal{F}) \neq 0$.*

In view of the remark above, these conditions are further equivalent to:
(ii') *$H_B^n(\mathcal{F}) \cong \mathbb{R}$.*

The interest of this characterization of isometric flows resides in the fact that it involves no metric data whatsoever.

A generalization of this result to Riemannian foliations of any dimension holds, and a proof will be discussed in Chapter 7 (Theorem 7.56).

For simply connected (and closed oriented) M, these conditions are always satisfied: see [Gh 2, Theorem B], and [K-To 10, Corollary 4.24] for the case of a bundle-like metric with basic mean curvature form.

Carrière [Ca 1,2] has given examples of Riemannian flows on oriented closed 3-manifolds with $H_B^2(\mathcal{F}) = 0$. This led him to conjecture Theorem 6.17. Carrière's examples are the torus fibrations T_A over the circle discussed in Chapter 3. A typical example is given by $A = \begin{pmatrix} 1 & 1 \\ 1 & 2 \end{pmatrix} \in SL(2, \mathbb{Z})$. The eigenspace corresponding to the eigenvalue $(3-\sqrt{5})/2$ gives rise to a Riemannian flow on $T_A = \mathbb{R} \times_A T^2$. The basic cohomology of this flow can be calculated, because the transversal structure is modeled on the affine group GA associated to A, and acting on $\mathbb{R}^2$. The relevant fact is then that $H_B(\mathcal{F})$ is isomorphic to the cohomology of the (left-)invariant forms on GA invariant under a compact subgroup K, the closure of the holonomy group in GA (see also Blumenthal [Bl 2]). Then Carrière shows explicitly, that a K-invariant 2-form on GA is the boundary of a K-invariant 1-form, which proves $H_B^2(\mathcal{F}) = 0$.

In the attempt to classify Riemannian flows (which succeeds on manifolds of low dimension), the following result of Carrière [Ca 1,2] is a key result.

6.18 THEOREM. *Let $\mathcal{F}$ be a Riemannian flow with dense orbits on a closed manifold M^{n+1}. Then M^{n+1} is diffeomorphic to the torus T^{n+1}, and the flow $\mathcal{F}$ is (smoothly) conjugate to a linear flow on T^{n+1}.*

The last statement means that there exists a smooth diffeomorphism from M^{n+1} to T^{n+1} taking the leaves of $\mathcal{F}$ to the orbits of a linear flow on T^{n+1}.

A classical example of a Riemannian flow on S^3 is given by the fibers of the Hopf fibration $S^3 \to S^2$. This example can be slightly generalized as follows. Let

$S^3 \subset \mathbb{C}^2$ with coordinates (z_1, z_2) and define $\varphi_t(z_1, z_2) = (e^{i\alpha t}z_1, e^{it}z_2)$, where $\alpha \neq 0$. For $\alpha = 1$ the orbits are precisely the fibres of the Hopf fibration. For $\alpha = 0$, the leaf through (z_1, z_2), $z_2 \neq 0$ is a circle, but the leaf through $(z_1, 0)$ is a point, so for that choice of α there are singularities.

Let $z_1 = x_1 + iy_1$, $z_2 = x_2 + iy_2$. Each leaf is confined to

$$T_{\rho_1,\rho_2} = \{(z_1, z_2) | |z_1|^2 = \rho_1^2, |z_2|^2 = \rho_2^2\},$$

where ρ_1, ρ_2 are constants and $\rho_1^2 + \rho_2^2 = 1$. For $\rho_1 \neq 0$, $\rho_2 \neq 0$ the set T_{ρ_1,ρ_2} is a 2-torus, and the orbits wind around this torus.

The examples are all transversely symplectic foliations as discussed in Chapter 4. Many problems in classical mechanics have a circular symmetry which can be used to reduce the number of degrees of freedom of the system. The appropriate concept is a circle action $S^1 \times M^{n+1} \to M^{n+1}$ with an invariant symplectic form defined on the normal bundle to the orbits of S^1.

The following nonexistence result is a typical application of the Gray-O'Neill technique described in detail in Chapter 5.

6.19 THEOREM [Ra]. *Let (M^{n+1}, g) be closed oriented with $\mathrm{Ric}^M < 0$. Then M admits no Riemannian flow $\mathcal{F}$ such that g is bundle-like for $\mathcal{F}$.*

Proof. Let $\mathcal{F}$ be such a flow generated by a unit vector field V (the letter T in the arguments below refers to O'Neill's tensor (5.25)). By (5.38b) we have then for a unit vector field $X \in \Gamma L^\perp$

(6.20) $\quad K^M(X, V) = g((\nabla_X^M T)_V V, X) - |T_V X|^2 + |A_X V|^2.$

Note that the mean curvature vector field is given by $\tau = \nabla_V^M V$. Then

$$g((\nabla_X^M T)_V V, X) = g(\nabla_X^M T_V V, X) - g(T_{\nabla_X^M V} V, X) - g(T_V \nabla_X^M V, X).$$

Since $\nabla_X^M V \in \Gamma L^\perp$, it follows from (5.26a) that $T_{\nabla_X^M V} V = 0$, and from (5.26b) that $T_V \nabla_X^M V \in \Gamma L$. Thus

$$g((\nabla_X^M T)_V V, X) = g(\nabla_X^M \tau, X).$$

We choose a local orthonormal frame with $E_1 = V$ and $E_2, \ldots, E_{n+1} \in \Gamma L^\perp$. Then by the identity (6.20) we get

(6.21)
$$\mathrm{Ric}^M(V, V) = \sum_{\alpha=2}^{n+1} K^M(E_\alpha, V)$$
$$= \sum_\alpha g(\nabla_{E_\alpha}^M \tau, E_\alpha) - \sum_\alpha |T_V E_\alpha|^2 + \sum_\alpha |A_{E_\alpha} V|^2.$$

Note that

$$\sum_\alpha g(\nabla^M_{E_\alpha}\tau, E_\alpha) = \operatorname{div}\tau - g(\nabla^M_V\tau, V) = \operatorname{div}\tau + g(\tau, \nabla^M_V V) = \operatorname{div}\tau + |\tau|^2.$$

Further

$$\sum_\alpha |T_V E_\alpha|^2 = \sum_\alpha g(V, T_V E_\alpha)^2 = \sum_\alpha g(V, \nabla^M_V E_\alpha)^2 = \sum_\alpha g(\nabla^M_V V, E_\alpha)^2$$
$$= \sum_\alpha g(\tau, E_\alpha)^2 = |\tau|^2.$$

Thus (6.21) yields

(6.22) $$\operatorname{Ric}^M(V, V) = \operatorname{div}\tau + \sum_\alpha |A_{E_\alpha} V|^2.$$

This expression at a point of M is clearly independent of the choice of the orthonormal frame. In fact let $A^V : L^\perp \to L^\perp$ be the map defined by $A^V X = A_X V$. This map is skew-symmetric. Namely

$$g(A^V X, Y) + g(X, A^V Y) = g(\pi\nabla^M_X V, Y) + g(X, \pi\nabla^M_Y V)$$
$$= -g(V, \nabla^M_X Y) - g(\nabla^M_Y X, V) = -g(V, \nabla^M_X Y + \nabla^M_Y X) = 0,$$

as $\mathcal{F}$ is Riemannian, and hence $L^\perp$ a totally geodesic subbundle (see Theorem 5.17, part (iv)). It follows that

$$\operatorname{Tr}(A^V)^2 = \sum_\alpha g((A^V)^2 E_\alpha, E_\alpha) = -\sum_\alpha g(A^V E_\alpha, A^V E_\alpha) = -\sum_\alpha |A_{E_\alpha} V|^2.$$

Thus our formula reads

(6.23) $$\operatorname{Ric}^M(V, V) = \operatorname{div}\tau - \operatorname{Tr}(A^V)^2.$$

Integration with respect to the volume form μ yields

$$\int_M \operatorname{Ric}(V, V)\mu = -\int_M \operatorname{Tr}(A^V)^2 \mu = \int_M \sum_\alpha |A_{E_\alpha} V|^2 \mu \geq 0.$$

This is incompatible with $\operatorname{Ric}^M < 0$. □

In fact the argument is clearly valid if $\operatorname{Ric}^M \leq 0$, and at least at one point $x \in M$ one has $\operatorname{Ric}^M_x < 0$.

This result applies in particular to negatively curved (M, g). In [Ca 2] it was shown moreover, that no Riemannian flow can exist on a manifold M admitting any metric of negative sectional curvature.

Note that the proof shows that for a Riemannian flow generated by a unit vector field V the condition $\operatorname{Ric}(V,V) \leq 0$ implies $A = 0$, and thus V is then orthogonal to a totally geodesic foliation.

For extensive work on special Riemannian flows we refer to the papers of J. and M. González-Dávila and L. Vanhecke [Gon 1,2], [Gon-Gon 1,2], [Gon-Gon-V 1-5], [Gon-V 1-3].

In the next chapter it is pointed out why the situation where the mean curvature one-form κ is a basic form merits special attention.

The following result characterizes such Riemannian flows on closed manifolds.

6.24 THEOREM. *Let $\mathcal{F}$ be a Riemannian flow defined by a unit vector field V on a closed manifold (M^{n+1}, g). Then the following conditions are equivalent:*

(i) $\kappa \in \Omega^1_B(\mathcal{F})$;
(ii) *the tangent bundle of $\mathcal{F}$ is locally spanned by Killing vector fields.*

Proof. (ii)$\Rightarrow$(i) is obvious. To establish (i)$\Rightarrow$(ii) we use the fact $d\kappa = 0$, proved in the next chapter for Riemannian foliations on closed manifolds satisfying the condition that κ is a basic form (see (7.5)). It follows that locally $\kappa = df$, where f is necessarily a basic function. We wish to show that $e^{-f}V$ is then a Killing vector field for g.

Obviously $(\theta(e^{-f}V)g)(V,V) = 0$. It is similarly immediate that for projectable X, Y (local sections of $L^\perp$) we have

$$(\theta(e^{-f}V)g)(X,Y) = 0.$$

It remains to prove the vanishing on mixed arguments V, $X \in \Gamma L^\perp$, X projectable. Now

$$(\theta(e^{-f}V)g)(V,X) = -g([e^{-f}V,V],X) - g(V,[e^{-f}V,X]).$$

The first term on the RHS vanishes, while the second term equals

$$-g(V, e^{-f}[V,X] - Xe^{-f}V) = -e^{-f}g(V,[V,X]) - e^{-f}Xf.$$

Note that $g(V, \nabla^M_X V) = \frac{1}{2}Xg(V,V) = 0$ so that

$$g(V,[V,X]) = g(V,\nabla^M_V X) = -g(\nabla^M_V V, X) = -g(\tau, X) = -\kappa(X).$$

It follows that

$$(\theta(e^{-f}V)g)(V,X) = e^{-f}(\kappa(X) - df(X)) = 0. \qquad \square$$

Since by a recent theorem of Domínguez [Do 1,2] the condition $\kappa \in \Omega^1_B(\mathcal{F})$ can always be achieved by modifying the metric only along the leaves, the result above actually characterizes Riemannian flows on closed manifolds.

6.25 THEOREM ([To 6]). *Let $\mathcal{F}$ be a flow on a closed manifold M. Then $\mathcal{F}$ is Riemannian if and only if the tangent bundle of $\mathcal{F}$ is locally spanned by Killing vector fields for a Riemannian metric on M.*

Carrière's example of a Riemannian flow on a closed M^3 with $H_B^2(\mathcal{F}) = 0$, together with Theorem 6.17, shows that in general this cannot be achieved by a global Killing vector field. On the other hand, if M is simply connected, the equation $\kappa = df$ in the proof of Theorem 6.24 can be solved globally. The vector field $e^{-f}V$ is then a global Killing field. But then by Theorem 6.10 the flow is also geodesible, namely for a metric renormalized by (6.9).

Chapter 7
Hodge Theory for the Transversal Laplacian

Throughout this chapter $\mathcal{F}$ denotes a transversally oriented Riemannian foliation on a closed oriented manifold M. We discuss Hodge theory and a duality theorem for the cohomology of basic forms [K-To 10,12].

Let g be a bundle-like metric inducing g_Q on $L^\perp \cong Q$. Besides the $*$-operator on $\Omega^{\cdot}(M)$ associated to g, there is also a star operator

(7.1) $$\bar{*} : \Omega_B^r(\mathcal{F}) \to \Omega_B^{q-r}(\mathcal{F}).$$

Since $\mathcal{F}$ is locally given by Riemannian submersions $f : U \to N$ into an oriented manifold N, with isometries of N as local transition functions, it follows that the star operator on the q-dimensional manifold N transports to the $\mathcal{F}$-basic forms. The relationships between $*$ in $\Omega^{\cdot}(M)$ and $\bar{*}$ in $\Omega_B^{\cdot}(\mathcal{F})$ are described by the following formulas:

(7.2) $$\bar{*}\alpha = (-1)^{p(q-r)} * (\alpha \wedge \chi_{\mathcal{F}}),$$

(7.3) $$*\alpha = \bar{*}\alpha \wedge \chi_{\mathcal{F}}$$

for $\alpha \in \Omega_B^r(\mathcal{F})$, and $\chi_{\mathcal{F}}$ the characteristic form of $\mathcal{F}$. For the transversal volume form $\nu \in \Omega_B^q(\mathcal{F})$, the last formula yields $*\nu = \chi_{\mathcal{F}}$, in accord with the orientation conventions made earlier.

In $\Omega_B^r(\mathcal{F})$ we have the natural scalar product

(7.4) $$\langle \alpha, \beta \rangle_B = \int_M \alpha \wedge \bar{*}\beta \wedge \chi_{\mathcal{F}}.$$

In view of (7.3), this is the restriction of the usual scalar product on $\Omega^r(M)$ to the subspace $\Omega_B^r(\mathcal{F})$.

A fundamental result on Riemannian foliations proved recently is as follows.

THEOREM (Domínguez [Do 1,2]). *Let $\mathcal{F}$ be a Riemannian foliation on a closed manifold M. Then there is a bundle-like metric g for $\mathcal{F}$ such that the mean curvature form κ is a basic one-form.*

We will not attempt to prove this result, which was announced during the writing of this book. But we will base the subsequent calculations in this chapter on the presence of a bundle-like metric g with mean curvature form $\kappa \in \Omega^1_B(\mathcal{F})$. Under this assumption we have the following crucial property [K-To 10]:

$$(7.5) \qquad d\kappa = 0.$$

Proof of (7.5). Since $d\kappa \in \Omega^2_B$, we have $d\kappa = \bar{*}\alpha$ for some $\alpha \in \Omega^{q-2}_B$. It suffices to show $\alpha = 0$. But for the square of the global norm we have

$$\|\alpha\|^2 = \int_M \alpha \wedge \bar{*}\alpha \wedge \chi_\mathcal{F} = \int_M \alpha \wedge d\kappa \wedge \chi_\mathcal{F}.$$

By (4.26) we have $d\chi_\mathcal{F} + \kappa \wedge \chi_\mathcal{F} = \varphi_0 \in F^2 \Omega^{p+1}$, and thus by differentiation

$$d\kappa \wedge \chi_\mathcal{F} - \kappa \wedge d\chi_\mathcal{F} = d\varphi_0.$$

Using (4.26) again this implies

$$d\kappa \wedge \chi_\mathcal{F} = \kappa \wedge (\varphi_0 - \kappa \wedge \chi_\mathcal{F}) + d\varphi_0 = \kappa \wedge \varphi_0 + d\varphi_0.$$

On the RHS the first term is a form $\psi = \kappa \wedge \varphi_0 \in F^3$, while $d\varphi_0 \in F^2$. It follows that

$$\|\alpha\|^2 = \int_M \alpha \wedge \psi + \int_M \alpha \wedge d\varphi_0.$$

On the RHS the first integrand is of filtration degree $(q-2) + 3 = q + 1$, hence vanishes. The second integrand is up to sign of the form

$$d(\alpha \wedge \varphi_0) - d\alpha \wedge \varphi_0.$$

The second term is of filtration degree $(q-1) + 2 = q + 1$, and hence trivial, while the other term vanishes by integration. Thus $\|\alpha\|^2 = 0$ and hence $\alpha = 0$. □

The cohomology class $[\kappa] \in H^1_B(\mathcal{F})$ so obtained is of great interest [K-To 10].

7.6 PROPOSITION. *Let $[\kappa] = 0$. Then $\mathcal{F}$ is taut, i.e. the bundle-like metric g can be modified to a bundle-like metric g' with minimal leaves.*

Proof. Let $\kappa = df$ with $f \in \Omega^0_B(\mathcal{F})$. This can be written $\kappa = d \log \lambda$, for $\lambda = e^f \in \Omega^0_B(\mathcal{F})$. Define

$$(7.7) \qquad g' = (\lambda^{2/p} g_L) \oplus g_Q.$$

This is a conformal modification of the metric along the leaves, but the transversal metric is left unchanged. Then the mean curvature form κ' associated to g' is calculated to be

$$(7.8) \qquad \kappa' = \kappa - d \log \lambda = 0.$$

□

By Proposition 4.1 we have an injective map $H^1_B(\mathcal{F}) \to H^1_{DR}(M)$. This yields the following result ([Gh 2], [K-To 10] for the case of a bundle-like metric with $\kappa \in \Omega^1_B(\mathcal{F})$).

7.9 COROLLARY. *Let M be closed and simply connected. Then every Riemannian foliation on M is taut.*

Next we calculate the formal adjoint
$$\delta_B : \Omega_B^r(\mathcal{F}) \to \Omega_B^{r-1}(\mathcal{F}) \text{ of } \quad d_B = d : \Omega_B^{r-1}(\mathcal{F}) \to \Omega_B^r(\mathcal{F})$$
with respect to the scalar product (7.4). It is globally defined by
$$\langle d_B \alpha, \beta \rangle_B = \langle \alpha, \delta_B \beta \rangle_B.$$

The mean curvature form appears in an essential role in the local expression for δ_B.

7.10 THEOREM [K-To 8,10]. *The formal adjoint of d_B in $\Omega_B^r(\mathcal{F})$ with respect to $\langle , \rangle_B$ is the operator*

(7.11) $$\delta_B = (d_B - \kappa \wedge)^{\bar{*}} : \Omega_B^r(\mathcal{F}) \to \Omega_B^{r-1}(\mathcal{F}),$$

where for $\beta \in \Omega_B^r(\mathcal{F})$

(7.12) $$(d_B - \kappa\wedge)^{\bar{*}}\beta \equiv (-1)^{q(r+1)+1}\bar{*}(d_B - \kappa\wedge)\bar{*}\beta.$$

Proof. Let $\alpha \in \Omega_B^{r-1}$, $\beta \in \Omega_B^r$. Then
$$\langle d_B\alpha, \beta\rangle_B = \int_M d\alpha \wedge (\bar{*}\beta \wedge \chi_\mathcal{F}) = (-1)^r \int_M \alpha \wedge d(\bar{*}\beta \wedge \chi_\mathcal{F})$$
$$= (-1)^r \int_M \alpha \wedge (d\bar{*}\beta \wedge \chi_\mathcal{F} + (-1)^{q-r}\bar{*}\beta \wedge d\chi_\mathcal{F}).$$

By (4.26) we can replace $d\chi_\mathcal{F}$ by $-\kappa \wedge \chi_\mathcal{F}$, up to a form $\varphi_0 \in F^2\Omega^{p+1}$. Since $\alpha \in F^{r-1}$ and $\bar{*}\beta \in F^{q-r}$, it follows that
$$\alpha \wedge \bar{*}\beta \wedge d\chi_\mathcal{F} = \alpha \wedge \bar{*}\beta \wedge (-\kappa \wedge \chi_\mathcal{F}),$$
because the difference is a form of filtration degree $(r-1) + (q-r) + 2 = q+1$, and hence vanishes. Thus
$$\langle d_B\alpha, \beta\rangle_B = (-1)^r \int_M \alpha \wedge [d_B\bar{*}\beta - (-1)^{q-r}\bar{*}\beta \wedge \kappa] \wedge \chi_\mathcal{F}$$
$$= (-1)^r(-1)^{(q-r+1)(r-1)} \int_M \alpha \wedge \bar{*}[\bar{*}d_B\bar{*}\beta - \bar{*}(\kappa \wedge \bar{*}\beta)] \wedge \chi_\mathcal{F}$$
$$= (-1)^{q(r+1)+1} \int_M \alpha \wedge \bar{*}[\bar{*}(d_B - \kappa\wedge)\bar{*}\beta] \wedge \chi_\mathcal{F}$$
$$= \int_M \alpha \wedge \bar{*}(d_B - \kappa\wedge)^{\bar{*}}\beta \wedge \chi_\mathcal{F},$$

which establishes the desired result. $\square$

7.13 COROLLARY. *For the transversal invariant volume ν, we obtain in particular*

$$(7.14) \qquad \delta_B \nu = \bar{*}\kappa.$$

For another illustrative example we consider an infinitesimal automorphism $Y \in V(\mathcal{F})$, which satisfies $Y \in \Gamma L^\perp$. The dual 1-form given by $\omega(E) = g(Y, E)$ for $E \in \Gamma TM$ is then basic, since for $V, W \in \Gamma L$ and $X \in \Gamma L^\perp$

$$(\theta(V)\omega)(W) = V\omega(W) - \omega[V, W] = 0,$$
$$(\theta(V)\omega)(X) = Vg(Y, X) - g(Y, [V, X]) = g([V, Y], X) = 0.$$

We further observe that in this case

$$\bar{*}\omega = i_Y \nu.$$

To verify this identity, it is no restriction to assume that Y is a multiple of E_1, in a local orthonormal frame $E_1, \ldots, E_q$ of $L^\perp$. If $Y = g(Y, E_1)E_1$, then $\omega = g(Y, E_1)\omega^1$ for the dual coframe $\omega^1, \ldots, \omega^q$. It follows that

$$\bar{*}\omega = g(Y, E_1)\bar{*}\omega^1 = g(Y, E_1)\omega^2 \wedge \cdots \wedge \omega^q = g(Y, E_1)i_{E_1}(\omega^1 \wedge \cdots \wedge \omega^q) = i_Y\nu.$$

We calculate now by (7.12)

$$\delta_B \omega = -\bar{*}(d_B - \kappa \wedge)\bar{*}\omega = -\bar{*}(d_B i_Y \nu - \kappa \wedge i_Y \nu) = -\bar{*}(\theta(Y)\nu - (i_Y \kappa)\nu)$$
$$= -\bar{*}(\operatorname{div}_B Y - \kappa(Y))\nu = -\operatorname{div}_B Y + \kappa(Y),$$

since $\bar{*}\nu = 1$. It is immediate to verify that for a vector field $Y \in \Gamma L^\perp$, $Y \in V(\mathcal{F})$ we have

$$\operatorname{div} Y = \operatorname{div}_B Y - \kappa(Y),$$

so that in this case the calculation above proves that $\delta_B \omega = -\operatorname{div} Y$. By the way, this yields another verification of the Transversal Divergence Theorem 5.24.

These calculations motivate the introduction of the twisted differential

$$(7.15) \qquad d_\kappa = d_B - \kappa \wedge.$$

Formula (7.11) reads then $\delta_B = (d_\kappa)^{\bar{*}}$. Since $d\kappa = 0$, $(d_\kappa)^2 = 0$ and hence $\delta_B^2 = 0$. Note that δ_B is obtained by modifying the codifferential associated to the transversal Riemannian metric by the operator $(-\kappa \wedge)^*$ of order zero (and degree -1). In terms of the operator d_κ we can write

$$(7.16) \qquad \langle d_B \alpha, \beta \rangle_B = \langle \alpha, (d_\kappa)^{\bar{*}} \beta \rangle_B \quad \text{for} \quad \alpha \in \Omega_B^{r-1}(\mathcal{F}),\ \beta \in \Omega_B^r(\mathcal{F}).$$

We define further

$$(7.17) \qquad d_B^{\bar{*}} \beta = (-1)^{q(r+1)+1} \bar{*} d_B \bar{*} \beta \quad \text{for} \quad \beta \in \Omega_B^r(\mathcal{F}).$$

Then by a calculation like the one establishing (7.11) we find

(7.18) $\quad \langle d_\kappa \alpha, \beta \rangle_B = \langle \alpha, d_B^{\bar{*}} \beta \rangle_B \quad \text{for} \quad \alpha \in \Omega_B^{r-1}(\mathcal{F}),\ \beta \in \Omega_B^r(\mathcal{F}).$

Formulas (7.16) and (7.18) express the fact that d_B, $\delta_B = (d_\kappa)^{\bar{*}}$ and d_κ, $\delta_\kappa = d_B^{\bar{*}}$ are two pairs of mutually adjoint operators with respect to $\langle,\rangle_B$. We have therefore two Laplacians

(7.19) $\quad\quad\quad\quad\quad\quad\quad\quad \Delta_B = d_B \delta_B + \delta_B d_B\,,$

(7.20) $\quad\quad\quad\quad\quad\quad\quad\quad \Delta_\kappa = d_\kappa \delta_\kappa + \delta_\kappa d_\kappa.$

They are related by

(7.21) $\quad\quad\quad\quad\quad\quad\quad\quad \bar{*} \Delta_B = \Delta_\kappa \bar{*},$

and thus it suffices to consider Δ_B. The harmonic basic r-forms $\mathcal{H}_B^r$ are those satisfying $\Delta_B \omega = 0$. The following generalization of the usual De Rham-Hodge decomposition holds.

7.22 THEOREM. *Let $\mathcal{F}$ be a transversally oriented Riemannian foliation on a closed oriented manifold (M, g). Assume g to be bundle-like with $\kappa \in \Omega_B^1(\mathcal{F})$. Then there is a decomposition into mutually orthogonal subspaces*

$$\Omega_B^r \cong \operatorname{im}\, d_B \oplus \operatorname{im}\, \delta_B \oplus \mathcal{H}_B^r$$

with finite-dimensional $\mathcal{H}_B^r$.

Proofs have appeared in [EK-Hc 2] and [K-To 15]. The finite-dimensionality of $\mathcal{H}_B^r$ was first established in [EK-Hc-Se]. It is also a consequence of the heat equation proof summarized in Theorem 7.46 below. The point is that the heat operator acts trivially on the cohomology, and is a compact operator. The proof in [EK-Hc 2] is based on Molino's structure theorem for Riemannian foliations [Mo 8,10] (see the statement in Chapter 10), and does not make direct use of the assumption $\kappa \in \Omega_B^1(\mathcal{F})$. That proof tracks the De Rham-Hodge decomposition valid for the base of the associated adherence foliation on the normal frame bundle through the corresponding spectral sequences.

The idea of the proof in [K-To 15] is to construct a strongly elliptic operator on all forms, which on basic forms restricts to the basic Laplacian Δ_B. The ordinary Laplacian Δ does not have this property (except for particularly simple foliations). The proper extension is an operator $\Delta - \tilde{\eta}$, where Δ is the ordinary Laplacian, and $\tilde{\eta}$ an explicitly defined operator of order (not exceeding) one, and preserving the degree of forms. This extension is not necessarily self-adjoint.

The next point is that the known coercivity of the bilinear form associated to the strongly elliptic $\Delta - \tilde{\eta}$ implies the corresponding property for the operator

Δ_B. This leads to the existence of weak solutions for the usual Poisson equations. A technical difficulty encountered at this stage is the verification of the Rellich and Sobolev property for the Sobolev chain $H_s(\Omega_B)$, $s \geq 0$, of the basic complex. The remaining part of the proof in [K-To 15] consists in establishing a regularity theorem, which leads to the actual solvability of the relevant Poisson equations. Note that one cannot simply apply the usual arguments directly, because the basic forms do not constitute all sections of a vector bundle, but rather the intersection of the kernels of Lie derivative operators within all sections of such a bundle.

Next we describe the extension of Δ_B in detail. For this purpose it is useful to introduce a bounded linear operator $\tilde{\gamma} : \Omega^r \to \Omega^{p+q-r+1}$ of order 0 defined by

$$(7.23) \qquad \tilde{\gamma}(\omega) = (-1)^{(q+1)(p+r)+1} * (\omega \wedge \chi_{\mathcal{F}}) \wedge \varphi_0$$

where φ_0 is given as in (4.26). This formula restricts by (7.2) on $\alpha \in \Omega^r_B$ to the expression

$$(7.24) \qquad \gamma(\alpha) = (-1)^{(p+1)(r+1)+qr} \bar{*}\alpha \wedge \varphi_0.$$

Since $\bar{*}\alpha \in F^{q-r}$ and $\varphi_0 \in F^2$, we have $\gamma(\alpha) \in F^{q-r+2}\Omega^{p+q-r+1}$, i.e. $\gamma(\alpha)$ is $\mathcal{F}$-trivial.

We can now compare δ_B with the usual $\delta : \Omega^r \to \Omega^{r-1}$ given by

$$\delta\alpha = (-1)^{n(r+1)+1} * d * \alpha$$

as follows. For $\alpha \in \Omega^r_B$

$$(7.25) \qquad \delta\alpha = \delta_B\alpha + *\gamma(\alpha), \text{ where } *\gamma(\alpha) \text{ is orthogonal to } \Omega^r_B.$$

As a consequence, for $\alpha \in \Omega^r_B$, $\beta \in \Omega^{r-1}_B$

$$(7.26) \qquad \langle \delta\alpha, \beta \rangle = \langle \delta_B\alpha, \beta \rangle_B.$$

This identity proves that δ_B is the adjoint of d on basic forms (note that by (7.25) the form $\delta\alpha$ need not be basic). Further for $\alpha \in \Omega^r_B$ it follows by (7.25)

$$(7.27) \qquad \gamma(\delta_B\alpha) = (-1)^{(n-r)r+1} * \delta * \gamma(\alpha) = (-1)^{n-r+1} d\gamma(\alpha).$$

These are the formulas which allow us to compare the basic Laplacian $\Delta_B = \delta_B d_B + d_B \delta_B$ and the ordinary Laplacian $\Delta = \delta d + d\delta$ restricted to basic forms. The result is as follows. For $\alpha \in \Omega^r_B$

$$(7.28) \qquad \begin{array}{c} \Delta\alpha = \Delta_B\alpha + \eta(\alpha), \\ \text{where} \quad \eta(\alpha) = *\gamma(d_B\alpha) + d*\gamma(\alpha). \end{array}$$

Note that $\eta(\alpha)$ is the restriction to $\alpha \in \Omega^r_B$ of the differential operator $\tilde{\eta} : \Omega^r \to \Omega^r$, of order one or less (and preserving degrees), given by

$$(7.29) \qquad \tilde{\eta}(\omega) = *\tilde{\gamma}(d\omega) + d*\tilde{\gamma}(\omega).$$

The content of (7.28) is that the differential operator $\Delta - \tilde{\eta} : \Omega^{\cdot} \to \Omega^{\cdot}$ is an extension of $\Delta_B : \Omega_B^{\cdot} \to \Omega_B^{\cdot}$. Since Δ is elliptic of order 2, and $\tilde{\eta}$ of lower order, $\Delta - \tilde{\eta}$ is still elliptic. The classical results applied to the elliptic operator $\Delta - \tilde{\eta}$ furnish the ingredients to conclude the desired results for the restriction Δ_B to Ω_B, for which the classical results do not directly apply. This is carried out in detail in [K-To 15].

We use the existence of the extension $\tilde{\Delta} = \Delta - \tilde{\eta} : \Omega^r \to \Omega^r$ of $\Delta_B : \Omega_B^r \to \Omega_B^r$ to prove the following result.

7.30 Proposition. *There exists a complete orthonormal system (COS) for $L^2(\Omega_B^r)$, consisting of smooth eigenforms of Δ_B in Ω_B^r.*

Proof. First we note the obvious identities

$$\langle \Delta_B \alpha, \beta \rangle = \langle \alpha, \Delta_B \beta \rangle ,$$
$$\langle \Delta_B \alpha, \alpha \rangle = \|d_B \alpha\|_{L^2}^2 + \|\delta_B \alpha\|_{L^2}^2$$

for $\alpha, \beta \in \Omega_B^r$, which imply that Δ_B defines a symmetric, positive operator

$$\Delta_B : D(\Delta_B) = \Omega_B \subset L^2(\Omega_B) \to L^2(\Omega_B).$$

We consider the symmetric operator $\Delta_1 = \Delta_B + I$

$$\Delta_1 : D(\Delta_1) = \Omega_B \subset L^2(\Omega_B) \to L^2(\Omega_B),$$

which is semi-bounded from below with lower bound 1, i.e.

$$\langle \Delta_1 \alpha, \alpha \rangle \geqq \|\alpha\|_{L^2} \quad \text{for} \quad \alpha \in \Omega_B.$$

Let Δ_1^F denote the Friedrichs extension of Δ_1

$$\Delta_1^F : D(\Delta_1^F) \subset L^2(\Omega_B) \to L^2(\Omega_B),$$

which by definition is the adjoint Δ_1^* of Δ_1 restricted to the domain $D(\Delta_1^F) = D(\Delta_1^*) \cap H_1$. Here H_1 denotes the Hilbert space completion of Ω_B with respect to the scalar product

$$\langle \alpha, \beta \rangle_1 = \langle \Delta_1 \alpha, \beta \rangle \quad \text{for} \quad \alpha, \beta \in \Omega_B.$$

Note that since

$$\|\alpha\|_1^2 = \langle \Delta_1 \alpha, \alpha \rangle = \|d_B \alpha\|_{L^2}^2 + \|\delta_B \alpha\|_{L^2}^2 + \|\alpha\|_{L^2}^2,$$

the new norm $\| \ \|_1$ is equivalent to the Sobolev H_1-norm on Ω_B, hence $H_1 \cong H_1(\Omega_B)$.

It follows from its construction that the Friedrichs extension Δ_1^F is still symmetric and surjective, and hence self-adjoint. Furthermore, since Δ_1^F has the same lower bound 1 as Δ_1, Δ_1^F is also injective. Hence we have the self-adjoint inverse

$$G_1^F = (\Delta_1^F)^{-1} : L^2(\Omega_B) \to D(\Delta_1^F) \subset L^2(\Omega_B),$$

which is a bounded map into $D(\Delta_1^F)$. Since by Rellich's lemma the inclusion $H_1 \cong H_1(\Omega_B) \hookrightarrow L^2(\Omega_B)$ is compact, the composition

$$G_1^F : L^2(\Omega_B) \to L^2(\Omega_B)$$

is a compact self-adjoint operator. Thus G_1^F has eigenvalues $|\mu_1| \geq |\mu_2| \geq \cdots \downarrow 0$ with corresponding eigenforms $\{\psi_i\}$ constituting a COS for $L^2(\Omega_B)$.

We prove now that the ψ_i are smooth eigenforms of Δ_B. First note that $G_1^F \psi_i = \mu_i \psi_i$ implies that $\psi_i \in D(\Delta_1^F) \subset H_1(\Omega_B)$ and $\Delta_1^F \psi_i = \mu_i^{-1} \psi_i$. Now consider the elliptic operator $\tilde{\Delta}_1 = \tilde{\Delta} + I$, and let $\tilde{\Delta}_1^*$ denote its formal adjoint. Then we observe by the definition of distribution derivatives that for any $\alpha \in \Omega_B$

$$\left\langle \tilde{\Delta}_1 \psi_i, \alpha \right\rangle = \left\langle \psi_i, \tilde{\Delta}_1^* \alpha \right\rangle = \left\langle \psi_i, \Delta_1 \alpha \right\rangle = \left\langle \Delta_1^F \psi_i, \alpha \right\rangle = \left\langle \mu_i^{-1} \psi_i, \alpha \right\rangle,$$

that is, $\left\langle (\tilde{\Delta}_1 - \mu_i^{-1} I)\psi_i, \alpha \right\rangle = 0$ for any $\alpha \in \Omega_B$. Note that $(\tilde{\Delta}_1 - \mu_i^{-1} I)\psi_i \in H_{-1}(\Omega_B)$, the completion of Ω_B with respect to the Sobolev H_{-1}-norm $\|\ \|_{-1}$ on Ω_B defined by $\|\alpha\|_{-1} = \sup\{|\langle \alpha, \beta\rangle | \|\beta\|_1 \leq 1, \beta \in \Omega_B\}$. Since the pairing $\langle,\rangle : H_{-1}(\Omega_B) \times H_1(\Omega_B) \to \mathbb{R}$ is continuous and nondegenerate, it then follows that ψ_i satisfies the elliptic equation

$$(\tilde{\Delta}_1 - \mu_i^{-1} I)\psi_i = 0.$$

This implies the regularity of ψ_i, that is, $\psi_i \in \Omega_M$ and hence $\psi_i \in H_1(\Omega_B) \cap \Omega_M$. Since by [K-To 15, Corollary 4.14] we have $H_1(\Omega_B) \cap \Omega_M = \Omega_B$, this implies $\psi_i \in \Omega_B$. Now $\tilde{\Delta}_1 \psi_i = \mu_i^{-1} \psi_i$ implies that $\Delta_B \psi_i = \lambda_i \psi_i$ with $\lambda_i = \mu_i^{-1} - 1$. □

Note that since Δ_B is a positive operator, we have $\lambda_i \geq 0$. From $|\mu_1| \geq |\mu_2| \geq \cdots \downarrow 0$ it follows that $0 \leq \lambda_1 \leq \lambda_2 \leq \cdots \uparrow \infty$ (except for the case of a finite-dimensional Ω_B^r, an exception we will not point out repeatedly). Furthermore, the multiplicities of the eigenvalues λ_i cannot increase too rapidly due to the following fact.

7.31 LEMMA. *Let $0 \leq \lambda_1 \leq \lambda_2 \leq \cdots$ be the eigenvalues of Δ_B in Ω_B^r. Then there exists a constant $C > 0$ and an exponent $\delta > 0$ such that $\lambda_j \geq Cj^\delta$, provided $j \geq j_0$ is large.*

Heat equation method

This proof method for Theorem 7.22 is based on the idea to find a best representative within the basic cohomology class of a closed $\alpha_0 \in \Omega_B^r$. This proof was presented in [N-Ra-To 2]. The form α_0 is modified to a form $\alpha_t = \alpha_0 + d_B \gamma_t$, where $\gamma_t \in \Omega_B^{r-1}$. The representative α_t is expected to be optimal, if $\|\alpha_t\|$ is minimal. Since

$$\frac{d}{dt}\left(\frac{1}{2}\|\alpha_t\|^2\right) = \langle \dot{\alpha}_t, \alpha_t \rangle = \langle d_B \dot{\gamma}_t, \alpha_t \rangle = \langle \dot{\gamma}_t, \delta_B \alpha_t \rangle,$$

an extremal of $\|\alpha_t\|^2$ is characterized by $\delta_B \alpha_t = 0$, or since $d_B \alpha_t = 0$, equivalently $\Delta_B \alpha_t = 0$. Moreover, the minimum of $\|\alpha_t\|^2$ is approached by steepest descent for the choice of $\dot{\gamma}_t = -\delta_B \alpha_t$, for which $\frac{d}{dt}\left(\frac{1}{2}\|\alpha_t\|^2\right) = -\|\delta_B \alpha_t\|^2$ is minimal. But then

$$\dot{\alpha}_t = d_B \dot{\gamma}_t = -d_B \delta_B \alpha_t = -\Delta_B \alpha_t,$$

since $d_B \alpha_t = 0$.

Thus for a given initial r-form $\alpha_0 \in \Omega_B^r$, we are led to consider the following initial value problem on Ω_B^r:

(7.32)
$$\frac{\partial}{\partial t}\alpha(x,t) = -\Delta_B \alpha(x,t),$$
$$\lim_{t \downarrow 0} \alpha(x,t) = \alpha_0(x).$$

We are going to prove that the heat equation problem (7.32) has a unique solution $\alpha(x,t) \in \Omega_B^r$, for $t > 0$, which is given in terms of the fundamental solution $e_B^r(x,y,t)$ of the basic heat operator $L_B = \frac{\partial}{\partial t} + \Delta_B$, to be constructed below, by

(7.33)
$$\alpha(x,t) = \int_M e_B^r(x,y,t) \wedge *\alpha_0(y).$$

For $(x,y,t) \in M \times M \times (0,\infty)$ we define

(7.34)
$$e_B^r(x,y,t) = \sum_{i=1}^{\infty} e^{-\lambda_i t} \psi_i(x) \otimes \psi_i(y) \in \Lambda^r T_x^* M \otimes \Lambda^r T_y^* M,$$

where $\{\psi_i\}_{i=1}^{\infty}$ is a COS for $L^2(\Omega_B^r)$ consisting of smooth eigenforms ψ_i of Δ_B corresponding to the eigenvalues λ_i (see Proposition 7.30).

7.35 PROPOSITION. *$e_B^r(x,y,t)$ is a well-defined smooth double form of bidegree (r,r) on $M \times M$, depending smoothly on the parameter $t > 0$.*

Proof. Let k and s be positive integers such that $2s > k + \frac{n}{2}$. Then, by the a priori coercive estimate for $\tilde{\Delta}$ combined with the Sobolev embedding lemma, we get the fundamental estimate

$$\|\psi_i\|_{C^k} \leqq C(\|\tilde{\Delta}^s \psi_i\|_{L^2} + \|\psi_i\|_{L^2}) \leqq C(1 + \lambda_i^s).$$

Note also that for $t > 0$ and $\lambda > 0$ we have the estimate

$$e^{-t\lambda} \lambda^s \leqq t^{-s} C(s) e^{-t\lambda/2}.$$

By applying these two inequalities to the RHS in (7.34), we get the following estimate

(7.36) $$\|e_B^r(x,y,t)\|_{C^k} \leqq t^{-s(k)} C(k) \sum_{i=1}^{\infty} e^{-t\lambda_i/2}.$$

But by Lemma 7.31, for large $j > j_0$, we have also the estimate

$$\lambda_j > C j^\delta$$

for some $C > 0$ and $\delta > 0$. This implies that the series on the RHS of (7.36) can be bounded by some constant multiple of the series

$$\sum_{j=1}^{\infty} e^{-tj^\delta/2},$$

which converges absolutely for each $t > 0$. Since k is arbitrary, this shows that $e_B^r(x,y,t)$ is a well-defined smooth (r,r)-form on $M \times M \times (0, \infty)$. $\square$

Note that the integrand in (7.33) is a double form of bidegree (r,n), which integrates for fixed x and t to an element in $\Lambda^r T_x^* M$. It is now verified without difficulty that $\alpha(x,t)$ in (7.33) gives rise to a solution of the initial value problem (7.32).

For $\alpha \in \Omega_B^r$ and $t > 0$ we define next

(7.37) $$[P_B(t)\alpha](x) = \int_M e_B^r(x,y,t) \wedge *\alpha(y) = \sum_{i=1}^{\infty} e^{-\lambda_i t} \langle \alpha, \psi_i \rangle_B \psi_i(x).$$

Note that $P_B(t) : \Omega_B^r \to \Omega_B^r$ for all $t > 0$. Moreover $P_B(t)$ is a symmetric operator for all $t > 0$:

(7.38) $$\langle P_B(t)\alpha, \beta \rangle_B = \langle \alpha, P_B(t)\beta \rangle_B \quad \text{for} \quad \alpha, \beta \in \Omega_B^r.$$

In fact, let $f(s,t) = \langle P_B(t)\alpha, P_B(s)\beta \rangle_B$. Then

$$\frac{\partial}{\partial t} f(t,s) = \left\langle \frac{\partial}{\partial t} P_B(t)\alpha, P_B(s)\beta \right\rangle_B = \langle -\Delta_B P_B(t)\alpha, P_B(s)\beta \rangle_B$$

$$= \langle P_B(t)\alpha, -\Delta_B P_B(s)\beta \rangle_B = \left\langle P_B(t)\alpha, \frac{\partial}{\partial s} P_B(s)\beta \right\rangle_B = \frac{\partial}{\partial s} f(t,s).$$

It follows that $f(t,s)$ is a function of $t+s$. For $s=0$ this yields the desired result in the form $f(t,0) = f(0,t)$.

For a given $\omega_0 \in \Omega_M^r$, we can also consider the initial value problem for the heat equation on Ω_M^r:

$$\text{(7.39)} \qquad \frac{\partial}{\partial t}\omega(x,t) = -\tilde{\Delta}\omega(x,t),$$
$$\lim_{t \downarrow 0} \omega(x,t) = \omega_0(x).$$

It is well-known, that the fundamental solution $e_M^r(x,y,t)$ for the heat operator $\frac{\partial}{\partial t} + \tilde{\Delta}$ exists, and that the unique solution of (7.39) is given by

$$\text{(7.40)} \qquad \omega(x,t) = \int_M e_M^r(x,y,t) \wedge *\omega_0(y).$$

For $\omega \in \Omega_M^r$ and $t > 0$ let

$$\text{(7.41)} \qquad [P(t)\omega](x) = \int_M e_M^r(x,y,t) \wedge *\omega(y).$$

This defines $P(t) : \Omega_M^r \to \Omega_M^r$. We prove the following result [N-Ra-To 2].

7.42 INVARIANCE THEOREM.
(i) If $\alpha \in \Omega_B^r$, then $P(t)\alpha \in \Omega_B^r$ for all $t > 0$.
(ii) $P(t)|\Omega_B = P_B(t)$ for all $t > 0$.

Proof. For a given $\alpha \in \Omega_B$, we have a solution $P_B(t)\alpha$ of (7.32), and a solution $P(t)\alpha$ of (7.39). Since $P_B(t)\alpha \in \Omega_B$ and $\tilde{\Delta}|\Omega_B = \Delta_B$, it follows that $P_B(t)\alpha$ is also a solution of (7.39). By the uniqueness of the solution to (7.39), it follows that $P(t)\alpha = P_B(t)\alpha$. This proves (i) and (ii). □

7.43 COROLLARY. *The solution to the initial value problem (7.32) is unique. In particular, $P_B(t)$ has the semi-group property*

$$P_B(t_1) \circ P_B(t_2) = P_B(t_1 + t_2) \quad \text{for} \quad t_1, t_2 > 0.$$

It is now easy to prove the following supplementary fact:

$$\text{(7.44)} \qquad \alpha \in \Omega_B^r, \quad d_B\alpha = 0 \Rightarrow d_B(P_B(t)\alpha) = 0 \quad \text{for all} \quad t \geq 0.$$

Proof. By assumption $d_B P_B(0)\alpha = d_B\alpha = 0$. We show that $d_B P_B(t)\alpha$ is a solution of the heat equation (7.32). Namely

$$\frac{d}{dt} d_B P_B(t)\alpha = d_B(\frac{d}{dt}P_B(t)\alpha) = d_B(-\Delta_B P_B(t)\alpha) = -\Delta_B d_B P_B(t)\alpha.$$

By the uniqueness of the solution of the initial value problem (7.32), it follows that $d_B P_B(t)\alpha = 0$ for all $t \geq 0$. □

Long time behavior of solutions

We examine next $P_B(t)\alpha$ for $\alpha \in \Omega_B$ in its dependence on t.

7.45 LEMMA. *Let $\alpha \in \Omega_B^r$. Then $\|P_B(t)\alpha\|$ is a nonincreasing function of t for $t > 0$.*

Proof. We have

$$\frac{d}{dt}\frac{1}{2}\|P_B(t)\alpha\|^2 = \left\langle \frac{d}{dt}P_B(t)\alpha, P_B(t)\alpha \right\rangle = -\langle \Delta_B P_B(t)\alpha, P_B(t)\alpha \rangle$$
$$= -\|d_B P_B(t)\alpha\|^2 - \|\delta_B P_B(t)\alpha\|^2 \leqq 0. \qquad \square$$

Next we turn to the discussion of the behavior of $P_B(t)\alpha$ for $t \to \infty$.

7.46 Theorem. *Let $\alpha \in \Omega_B^r$.*

(i) *$P_B(t)\alpha$ converges uniformly for $t \uparrow \infty$.*
(ii) *$H_B\alpha = \lim_{t\uparrow\infty} P_B(t)\alpha \in \Omega_B^r$ is Δ_B-harmonic.*

Proof. (i) By Lemma 7.45, there exists $\lim_{t\uparrow\infty} \|P(t)\alpha\|^2 = a \geqq 0$. But by Corollary 7.43 and (7.38)

$$\|P_B(t+2h)\alpha - P_B(t)\alpha\|^2$$
$$= \|P_B(t+2h)\alpha\|^2 + \|P_B(t)\alpha\|^2 - 2\langle P_B(t+2h)\alpha, P_B(t)\alpha \rangle$$
$$= \|P_B(t+2h)\alpha\|^2 + \|P_B(t)\alpha\|^2 - 2\|P_B(t+h)\alpha\|^2,$$

which approaches 0 for $t \uparrow \infty$. Thus $\{P_B(t)\alpha\}_{t\geqq 0}$ has the Cauchy property, hence converges in $L^2(\Omega_B^r)$ for $t \uparrow \infty$.

To prove uniform convergence, we fix $h > 0$. Then we have for all $t \in (0, \infty)$ and $x \in M$

$$|P_B(t+2h)\alpha - P_B(t+h)\alpha|(x) = |P_B(h)(P_B(t+h)\alpha - P_B(t)\alpha)|(x)$$
$$= \left| \int_M e_B^r(x,y,h) \wedge *(P_B(t+h)\alpha - P_B(t)\alpha)(y) \right|$$
$$\leqq \sup_{x\in M} \|e_B^r(x,\cdot,h)\| \, \|P_B(t+h)\alpha - P_B(t)\alpha\|,$$

which goes to 0 for $t \uparrow \infty$ by the preceding observation. It follows that $P_B(t)\alpha$ converges uniformly on M for $t \uparrow \infty$. We define $H_B\alpha$ as this uniform limit.

(ii) To prove that $H_B\alpha$ is Δ_B-harmonic, we observe first that for fixed $t > 0$

$$|P_B(t+h)\alpha - P_B(t)H_B\alpha|(x) = |P_B(t)(P_B(h)\alpha - H_B\alpha)|(x)$$
$$\leqq \sup_{x\in M} \|e_B^r(x,\cdot,t)\| \, \|P_B(h)\alpha - H_B\alpha\|,$$

which implies for $h \uparrow \infty$

$$0 = \lim_{h\uparrow\infty}(P_B(t+h)\alpha - P_B(t)H_B\alpha) = H_B\alpha - P_B(t)H_B\alpha.$$

Hence we have the invariance property

(7.47) $$P_B(t)H_B\alpha = H_B\alpha.$$

It now follows from (7.47) that

$$\Delta_B H_B \alpha = \Delta_B P_B(t) H_B \alpha = -\frac{\partial}{\partial t} P_B(t) H_B \alpha = -\frac{\partial}{\partial t} H_B \alpha = 0,$$

and $H_B\alpha$ is indeed Δ_B-harmonic. $\square$

Note that since $\tilde{\Delta} \mid \Omega_B = \Delta_B$, the Δ_B-harmonic forms $\mathcal{H}_B$ form a subspace of the finite-dimensional space $\mathcal{H}$ of $\tilde{\Delta}$-harmonic forms. Thus $\mathcal{H}_B$ is itself finite-dimensional.

For $\alpha \in \Omega_B^r$ define now

(7.48) $$G_B \alpha(x) = \int_0^\infty (P_B(t)\alpha - H_B\alpha)(x) dt.$$

7.49 Theorem [N-Ra-To 2]. *Let $\alpha \in \Omega_B^r$.*
(i) *$G_B\alpha$ is well-defined, and $G_B\alpha \in \Omega_B^r$.*
(ii) *The operator $G_B : \Omega_B^r \to \Omega_B^r$ satisfies $\alpha = \Delta_B G_B \alpha + H_B \alpha$.*

Proof. Let $k = \dim \mathcal{H}_B^r$. Then in (7.34) we have $\lambda_1 = \cdots = \lambda_k = 0$ and $\psi_1, \ldots, \psi_k \in \mathcal{H}_B^r$. It follows from (7.37) that

$$H_B \alpha = \lim_{t \uparrow \infty} P_B(t)\alpha = \sum_{i=1}^k \langle \psi_i, \alpha \rangle_B \psi_i \quad \text{in} \quad \Omega_B^r,$$

since the contribution of the positive eigenvalues $\lambda_{k+1} \leq \lambda_{k+2} \leq \ldots$ disappears for $t \uparrow \infty$. Hence we have

$$P_B(t)\alpha - H_B \alpha = \sum_{i=k+1}^\infty e^{-\lambda_i t} \langle \psi_i, \alpha \rangle_B \psi_i \quad \text{in} \quad \Omega_B^r.$$

For fixed $h > 0$, and each $x \in M$, $t \in (0, \infty)$ we have then

$$|P_B(t+h)\alpha - H_B\alpha|(x) = |P_B(h)(P_B(t)\alpha - H_B\alpha)|(x)$$

$$= |\int_M e_B^r(x,y,h) \wedge *(P_B(t)\alpha - H_B\alpha)(y)|$$

$$\leq (\sup_x \|e_B^r(x,\cdot,h)\|) \, \|(P_B(t)\alpha - H_B\alpha\| = c\| \sum_{i=k+1}^\infty e^{-\lambda_i t} \langle \psi_i, \alpha \rangle_B \psi_i \|$$

$$\leq c e^{-\lambda_{k+1} t} \| \sum_i \langle \psi_i, \alpha \rangle_B \psi_i \| \leq c \|\alpha\| e^{-\lambda_{k+1} t}.$$

It follows that
$$\left|\int_0^{t+h} (P_B(t+h)\alpha - H_B\alpha)(x)dt\right| \leq \int_0^{t+h} |P_B(t+h)\alpha - H_B\alpha|(x)dt$$
$$\leq c\|\alpha\| \int_0^{t+h} e^{-\lambda_{k+1}t} dt.$$

Since the last term converges as $t \uparrow \infty$, it follows that the RHS in (7.48) exists as a limit for $t \uparrow \infty$ of $\int_0^t$, uniformly in x. Since the integrand is a basic r-form for each t, it follows that $G_B\alpha \in \Omega_B^r$.

Finally we calculate, using the harmonicity of $H_B\alpha$,
$$\Delta_B G_B \alpha = \Delta_B \int_0^\infty (P_B(t)\alpha - H_B\alpha) dt = \int_0^\infty \Delta_B(P_B(t)\alpha - H_B\alpha) dt$$
$$= \int_0^\infty (\Delta_B P_B(t)\alpha) dt = \int_0^\infty \left(-\frac{d}{dt} P_B(t)\alpha\right) dt$$
$$= \lim_{t \downarrow 0} P_B(t)\alpha - \lim_{t \uparrow \infty} P_B(t)\alpha$$
$$= \alpha - H_B\alpha. \qquad \square$$

The identity
$$(7.50) \qquad \alpha = \Delta_B G_B \alpha + H_B \alpha$$
completes now the proof of Theorem 7.22.

The first application of Theorem 7.22 is the unique representability of basic cohomology classes by basic harmonic forms. It is proved in the same way as the corresponding usual result in De Rham-Hodge Theory.

7.51 THEOREM. *Let the situation be as in Theorem 7.22. Then $H_B^r(\mathcal{F}) \cong \mathcal{H}_B^r$.*

Proof. Let α be a closed basic r-form, and consider the De Rham-Hodge decomposition $\alpha = d_B\beta + \delta_B\gamma + \pi_B\alpha$, with $\pi_B : \Omega_B^r(\mathcal{F}) \to \mathcal{H}_B^r$ the orthogonal projection to harmonic forms. Then $0 = d_B\alpha = d_B\delta_B\gamma$ implies
$$\langle d_B\delta_B\gamma, \gamma\rangle_B = \langle \delta_B\gamma, \delta_B\gamma\rangle_B = \|\delta_B\gamma\|^2 = 0,$$
hence $\delta_B\gamma = 0$. It follows that $\alpha = d_B\beta + \pi_B\alpha$, and α is cohomologous to its harmonic representative $\pi_B\alpha$. If $\alpha = d_B\beta$, its harmonic representative vanishes. Thus we have a well-defined homomorphism $H_B^r(\mathcal{F}) \to \mathcal{H}_B^r$, which is clearly surjective. If $\pi_B\alpha = 0$, then $\alpha = d_B\beta$, which proves that this map is also injective. $\square$

Similarly we consider the De Rham-Hodge decomposition
$$(7.52) \qquad \Omega_B^r \cong \operatorname{im} d_\kappa \oplus \operatorname{im} \delta_\kappa \oplus \mathcal{H}_\kappa^r$$
with finite-dimensional $\mathcal{H}_\kappa^r = \ker \Delta_\kappa \subset \Omega_B^r$. The type of argument leading to Theorem 7.51 proves similarly that
$$(7.53) \qquad H_B^r(\mathcal{F}, d_\kappa) \equiv H^r(\Omega_B^r(\mathcal{F}), d_\kappa) \cong \mathcal{H}_\kappa^r.$$
We use these results to prove the following fact [K-To 10,12].

7.54 THEOREM (Twisted Duality). *Let $\mathcal{F}$ be a transversally oriented Riemannian foliation, on a closed oriented manifold M. Assume g to be bundle-like with $\kappa \in \Omega^1_B(\mathcal{F})$. Then the pairing $\alpha \otimes \beta \to \int_M \alpha \wedge \beta \wedge \chi_\mathcal{F}$ induces a nondegenerate pairing*

$$H^r_B(\mathcal{F}, d_B) \otimes H^{q-r}_B(\mathcal{F}, d_\kappa) \to \mathbb{R}$$

of finite-dimensional vector spaces.

Proof. Let $\alpha \in \Omega^r_B(\mathcal{F})$ with $d_B\alpha = 0$, and $\beta \in \Omega^{q-r}_B(\mathcal{F})$ with $d_\kappa\beta = 0$. Consider moreover $\alpha' = \alpha + d_B v$, and $\beta' = \beta + d_\kappa w$, with $v \in \Omega^{r-1}_B(\mathcal{F})$ and $w \in \Omega^{q-r-1}_B(\mathcal{F})$. Then

$$\alpha' \wedge \beta' \wedge \chi_\mathcal{F} = (\alpha + d_B v) \wedge (\beta + d_\kappa w) \wedge \chi_\mathcal{F}$$
$$= \alpha \wedge \beta \wedge \chi_\mathcal{F} + \alpha \wedge d_\kappa w \wedge \chi_\mathcal{F} + d_B v \wedge \beta' \wedge \chi_\mathcal{F}.$$

Now $d\alpha = d_B\alpha = 0$ implies

$$d(\alpha \wedge w \wedge \chi_\mathcal{F}) = (-1)^r \alpha \wedge dw \wedge \chi_\mathcal{F} + (-1)^{q-1} \alpha \wedge w \wedge d\chi_\mathcal{F}.$$

The last term differs according to (4.26) from $(-1)^{q-1} \alpha \wedge w \wedge (-\kappa \wedge \chi_\mathcal{F})$ by $(-1)^{q-1}\alpha \wedge w \wedge \varphi_0$, which is of filtration degree $r + (q-r-1) + 2 = q+1$, and hence vanishes. Thus

$$d(\alpha \wedge w \wedge \chi_\mathcal{F}) = (-1)^r \alpha \wedge dw \wedge \chi_\mathcal{F} + (-1)^q(-1)^{q-r-1} \alpha \wedge (\kappa \wedge w) \wedge \chi_\mathcal{F}$$
$$= (-1)^r \alpha \wedge (dw - \kappa \wedge w) \wedge \chi_\mathcal{F} = (-1)^r \alpha \wedge d_\kappa w \wedge \chi_\mathcal{F}.$$

Similarly $d_\kappa\beta = 0$ implies

$$d(v \wedge \beta' \wedge \chi_\mathcal{F}) = d_B v \wedge \beta' \wedge \chi_\mathcal{F} + (-1)^{r-1} v \wedge d\beta' \wedge \chi_\mathcal{F} + (-1)^{q-1} v \wedge \beta' \wedge d\chi_\mathcal{F},$$

where the last term can be replaced by $(-1)^{q-1} v \wedge \beta' \wedge (-\kappa \wedge \chi_\mathcal{F})$, since the difference $(-1)^{q-1} v \wedge \beta' \wedge \varphi_0$ is of filtration degree $(r-1) + (q-r) + 2 = q+1$, and hence vanishes. It follows that

$$d(v \wedge \beta' \wedge \chi_\mathcal{F}) = d_B v \wedge \beta' \wedge \chi_\mathcal{F} + (-1)^{r-1} v \wedge (d\beta' - \kappa \wedge \beta') \wedge \chi_\mathcal{F}$$
$$= d_B v \wedge \beta' \wedge \chi_\mathcal{F},$$

since $d_\kappa\beta' = d_\kappa\beta + d^2_\kappa w = 0$. These calculations show that

(7.55) $\qquad \alpha' \wedge \beta' \wedge \chi_\mathcal{F} - \alpha \wedge \beta \wedge \chi_\mathcal{F} = (-1)^r d(\alpha \wedge w \wedge \chi_\mathcal{F}) + d(v \wedge \beta' \wedge \chi_\mathcal{F}).$

Therefore there is indeed a cohomology pairing as stated in the theorem.

By Theorem 7.51 the cohomology spaces $H_B^r(\mathcal{F}, d_B)$ are finite-dimensional. (7.53) implies the same fact for the spaces $H_B^r(\mathcal{F}, d_\kappa)$. To complete the proof, it suffices therefore to establish the injectivity of the maps

$$H_B^r(\mathcal{F}, d_B) \to H_B^{q-r}(\mathcal{F}, d_\kappa)^*, \ H_B^{q-r}(\mathcal{F}, d_\kappa) \to H_B^r(\mathcal{F}, d_B)^*$$

into the dual spaces defined by the pairing.

The first of these maps assigns to a d_B-closed basic r-form α the functional on $H_B^{q-r}(\mathcal{F}, d_\kappa)$ given by $[\beta] \to \int_M \alpha \wedge \beta \wedge \chi_\mathcal{F}$, where $[\beta]$ is represented by a d_κ-closed $(q-r)$-form β. We can choose α to be Δ_B-harmonic, i.e. $\Delta_B \alpha = 0$. Now (7.21) implies that $\Delta_\kappa \bar{*}\alpha = \bar{*}\Delta_B \alpha = 0$, and thus $\bar{*}\alpha$ is Δ_κ-harmonic. Assuming the functional corresponding to α to be trivial, it follows in particular that

$$\|\alpha\|^2 = \langle \alpha, \alpha \rangle_B = \int_M \alpha \wedge \bar{*}\alpha \wedge \chi_\mathcal{F} = 0,$$

and hence $\alpha = 0$. The injectivity of the other map is proved similarly. □

7.56 COROLLARY. *Let the situation be as in Theorem 7.54. Then $H_B^q(\mathcal{F}, d_\kappa) \cong \mathbb{R}$. Moreover, the following conditions are equivalent:*

(i) $H_B^q(\mathcal{F}) \cong \mathbb{R}$;
(ii) $\mathcal{F}$ *is taut*;
(iii) $[\kappa] = 0$.

For $q = 1$, it was proved in [K-To 10, Theorem 3.26] that these conditions are always satisfied for a Riemannian foliation. On the other hand Carrière's example [Ca 1,2] of a Riemannian flow on a 3-manifold with $H_B^2(\mathcal{F}) = 0$ (see also Chapter 6) shows that the alternative situation can occur in higher codimension. The equivalence of conditions (i) and (ii) in Corollary 7.56 for any transversally oriented Riemannian foliation on a closed oriented manifold was first established by Masa [Ma 4]. The result of Domínguez [Do 1,2] stated at the beginning of this chapter, reduces the general case to the special case $\kappa \in \Omega_B^1(\mathcal{F})$, which can be handled by the proof described below. Note that this argument also yields a proof of Theorem 6.17. Observe further that these conditions are always satisfied for simply connected M.

Proof of 7.56. Since we have $H_B^0(\mathcal{F}, d_B) \cong \mathbb{R}$, Theorem 7.54 proves that $H_B^q(\mathcal{F}, d_\kappa) \cong \mathbb{R}$. To prove (ii) ⇒ (i), we observe that in the taut case there exists a bundle-like metric for which $\kappa = 0$. Thus $d_\kappa = d_B$ and $H_B^q(\mathcal{F}) \equiv H_B^q(\mathcal{F}, d_B) \cong \mathbb{R}$. To prove (i) ⇒ (ii), we can assume g to be a bundle-like metric such that $\kappa \in \Omega_B^1(\mathcal{F})$. Then the twisted duality implies that $H_B^0(\mathcal{F}, d_\kappa) \cong H_B^q(\mathcal{F}, d_B) \cong \mathbb{R}$. Thus there is a global nontrivial basic function $\lambda : M \to \mathbb{R}$ satisfying

$$0 = d_\kappa \lambda \equiv d\lambda - \lambda \kappa.$$

It follows that $\kappa = d \log \lambda$ (λ cannot vanish anywhere, if it is a nontrivial solution of the linear first order DE above). Since $\lambda \in \Omega_B^0(\mathcal{F})$, $[\kappa] = 0 \in H_B^1(\mathcal{F})$. We can now modify the metric g as in (7.7), so as to make the corresponding mean curvature form vanish. The argument just used shows the equivalence of (ii) and (iii). □

7.57 COROLLARY. *Let the situation be as in Theorem 7.54. Then $H_B^q(\mathcal{F}) \cong \mathbb{R}$ or $H_B^q(\mathcal{F}) = 0$. The first case occurs if and only if $\mathcal{F}$ is taut.*

Proof. The nontaut case occurs when $[\kappa] \neq 0 \in H_B^1(\mathcal{F})$. By the argument above this condition implies that the equation $d_\kappa \lambda = 0$ has only the trivial solution $\lambda = 0$, hence $H_B^0(\mathcal{F}, d_\kappa) = 0$. By Theorem 7.54 this implies $H_B^q(\mathcal{F}, d_B) = 0$. □

The redundancy of the assumption $\kappa \in \Omega_B^1$ is the statement of Domínguez' theorem. Another approach was made by Alvarez, who proved in [A 6] that all the arguments in this chapter go through, if instead of the assumption $\kappa \in \Omega_B^1$ one uses the basic component κ_B of the mean curvature form arising from the orthogonal direct sum decomposition

$$\Omega = \Omega_B(\mathcal{F}) \oplus \Omega_B(\mathcal{F})^\perp,$$
$$\kappa = \kappa_B + \kappa_0.$$

This decomposition itself is a consequence of the result discussed in [A-To 2]. It follows as in (7.5) that $d\kappa_B = 0$. It is further shown that the cohomology class $\zeta(\mathcal{F}) = [\kappa_B] \in H_B^1(\mathcal{F})$ is an invariant of the Riemannian foliation, i.e. unchanged under modifications of the bundle-like metric defining the same transversal Riemannian structure.

Another interesting contribution to this subject was made by Park and Richardson in [Par-R]. Their idea is to write the basic Laplacian of a Riemannian foliation in terms of the orthogonal projection P from $L^2(\Omega(M))$ to $L^2(\Omega_B(\mathcal{F}))$ corresponding to the inclusion $\Omega_B(\mathcal{F}) \hookrightarrow \Omega(M)$, which is isometric with respect to the L^2-metric. Using a geometric interpretation of P, this leads to a formula for Δ_B in terms of Δ^M and the mean curvature form $\kappa \in \Omega^1(M)$, as well as its projection $P(\kappa) \in \Omega_B^1(\mathcal{F})$. This formula is expected to be generalization of (7.28) to the case of arbitrary mean curvature, not necessarily basic.

We finally formulate the particular case of Theorem 7.54 for a taut foliation [K-To 8].

7.58 COROLLARY. *Let $\mathcal{F}$ be a taut and transversally oriented Riemannian foliation on a closed oriented manifold M. Then the pairing $\alpha \otimes \beta \to \int_M \alpha \wedge \beta \wedge \chi_\mathcal{F}$ induces a nondegenerate pairing*

$$H_B^r(\mathcal{F}) \otimes H_B^{q-r}(\mathcal{F}) \to \mathbb{R}$$

on finite-dimensional vector spaces.

Thus in the taut case (and in that case only) Poincaré duality in $H_B^{\cdot}(\mathcal{F})$ holds in the expected form.

Chapter 8
Cohomology Vanishing and Tautness

The Bochner technique is based on the Weitzenböck formula for the Laplacian applied to forms. In particular the cohomology vanishing theorems are based on the positivity of certain curvature expressions. The Weitzenböck formula for the transversal Laplacian Δ_B has, aside from the usual terms, correction terms involving the mean curvature, which interfere with the usual arguments leading to vanishing theorems.

To circumvent this difficulty, and also to get around the hypothesis of a basic mean curvature form, we introduce in this chapter an elliptic operator on the De Rham complex of the total space of a Riemannian foliation, such that its restriction to basic forms coincides with the naive Laplacian $d\delta + \delta d$ arising from the geometry of the model space. This construction is distinct from the construction used in Chapter 7. Its salient feature is that the local computations on basic forms in distinguished coordinate neighborhoods are identical with the computations in the usual Riemannian case. As a consequence, vanishing theorems based on the positivity of certain curvature expressions can be proved essentially as in the special case of Riemannian manifolds.

The construction of the elliptic operator mentioned above is based on a new connection $\tilde{\nabla}$ on M which is metric, but has nonvanishing torsion $\tilde{T}$. The connection $\tilde{\nabla}$ turns the leaves of $\mathcal{F}$ into totally geodesic submanifolds. The resulting differential operator $\tilde{\Delta}$ is not self-adjoint. As a consequence the space of harmonic forms may be larger than the corresponding cohomology. However, using the heat equation, we prove that if the curvature term in the Weitzenböck formula is strictly positive, then any closed basic form is exact. As illustrations, we present the case of positive transversal Ricci curvature due to Hebda [Hb 1], as well as the case of a positive transversal curvature operator which generalizes a result of Gallot-Meyer.

Let ∇^M be the Levi Civita connection on (M, g). We define a connection on M by

(8.1) $$\tilde{\nabla}_B C = \pi^\perp \nabla^M_B \pi^\perp C + \pi \nabla^M_B \pi C \quad \text{for} \quad B, C \in \Gamma TM.$$

For $V, W \in \Gamma L$ we have $\tilde{\nabla}_V W = \pi^\perp \nabla_V W = \nabla^M_V W - \alpha(V, W)$, where α is the second fundamental form of the leaves. Thus the leaves of $\mathcal{F}$ are totally geodesic with respect to the connection $\tilde{\nabla}$. For $X, Y \in \Gamma L^\perp$, we have $\tilde{\nabla}_X Y = \nabla^M_X Y - A(X, Y)$, where $A(X, Y) = \frac{1}{2}\pi^\perp [X, Y]$ is the O'Neill tensor (5.27).

$\tilde{\nabla}$ is a metric connection, but has nonzero torsion $\tilde{T}$. In fact

(8.2)
$$\tilde{T}(V,W) = 0,$$
$$\tilde{T}(V,X) = A_X V - T_V X,$$
$$\tilde{T}(X,Y) = -2A_X Y = -\pi^{\perp}[X,Y].$$

In the second formula the letter T refers to O'Neill's tensor defined in (5.25).

The difference tensor

(8.3)
$$\gamma(B,C) = \tilde{\nabla}_B C - \nabla_B^M C$$

can be expressed as follows:

(8.4)
$$\gamma(U,V) = -\pi \nabla_U^M V = -T_U V,$$
$$\gamma(X,U) = -\pi \nabla_X^M U = -A_X U,$$
$$\gamma(U,X) = -\pi^{\perp} \nabla_U^M X = -T_U X,$$
$$\gamma(X,Y) = -\pi^{\perp} \nabla_X^M Y = -A_X Y.$$

A consequence of these formulas is that $\tilde{\nabla}$ and ∇^M coincide if and only if $\mathcal{F}$ is totally geodesic ($T = 0$) and has integrable normal bundle ($A = 0$).

We take this opportunity to point out that the formulas for the curvature

$$\tilde{R}_{BC} = [\tilde{\nabla}_B, \tilde{\nabla}_C] - \tilde{\nabla}_{[B,C]}$$

listed in [To-V 5, Proposition 9] should read correctly as stated below. In these formulas $\hat{R}$ denotes the curvature of the Levi Civita connection $\hat{\nabla}$ induced on the leaves. The curvature sign is opposite to the convention used in [To-V 5], but this is compensated by defining e.g. $\tilde{R}_{XYUV} = g(\tilde{R}(X,Y)V,U)$. The curvature formulas should then read

(8.5a) $$\tilde{R}_{XYUV} = R_{XYUV}^M + g(A_X U, A_Y V) - g(A_Y U, A_X V),$$

(8.5b) $$\tilde{R}_{XUYV} = \tilde{R}_{UXVY} = \tilde{R}_{UXYV} = \tilde{R}_{XUVY} = 0,$$

(8.5c) $$\tilde{R}_{UVWX} = \tilde{R}_{UVXW} = 0,$$

(8.5d) $$\tilde{R}_{XYZU} = \tilde{R}_{XYUZ} = 0,$$

(8.5e) $$\tilde{R}_{UVXY} = R_{UVXY}^M - g(T_V X, T_U Y) + g(T_U X, T_V Y),$$

(8.5f) $$\tilde{R}_{UVWW'} = \hat{R}_{UVWW'} = R_{UVWW'}^M + g(T_U W, T_V W') - g(T_U W', T_V W),$$

(8.5g) $$\tilde{R}_{XYZZ'} = R_{XYZZ'}^M - g(A_Y Z, A_X Z') + g(A_X Z, A_Y Z'),$$

(8.5h) $$\tilde{R}_{XYUV} = R_{XYUV}^M + g(A_X U, A_Y V) - g(A_Y U, A_X V),$$

(8.5i) $\tilde{R}_{XUVW} = R^M_{XUVW} - g(T_U V, A_X W) + g(A_X V, T_U W),$

(8.5j) $\tilde{R}_{UXYZ} = R^M_{UXYZ} - g(A_X Y, T_U Z) + g(T_U Y, A_X Z).$

For the sectional curvatures of the planes determined by orthogonal unit vectors X, Y, U, V this yields the formulas

$$\tilde{K}(U,V) = \hat{K}(U,V) = K^M(U,V) + g(T_U U, T_V V) - g(T_U V, T_U V),$$
$$\tilde{K}(X,Y) = K^M(X,Y) + |A_X Y|^2,$$
$$\tilde{K}(X,U) = 0.$$

The main purpose of introducing the connection $\tilde{\nabla}$ in our context is the following property.

8.6 PROPOSITION [Mi-R-To 1]. *Let $X, Y \in \Gamma L^\perp$ be projectable vector fields on a distinguished chart U, with $\bar{X} = f_* X$, $\bar{Y} = f_* Y$ vector fields on the local base space $f(U) = B \subset N$. Then $\tilde{\nabla}_X Y$ is projectable and*

(8.7) $$f_* \tilde{\nabla}_X Y = \nabla^N_{\bar{X}} \bar{Y}.$$

Proof. Let $Z \in \Gamma L^\perp$ be a third projectable vector field on U. Then by Koszul's classical formula

$$2g(\nabla^M_X Y, Z) = X g(Y, Z) + Y g(Z, X) - Z g(X, Y)$$
$$- g(X, [Y, Z]) + g(Y, [Z, X]) + g(Z, [X, Y]).$$

Now $X g(Y, Z) = \bar{X} g_Q(\bar{Y}, \bar{Z})$, and similarly for the other terms in the first line. Further $g(X, [Y, Z]) = g_Q(\bar{X}, [\bar{Y}, \bar{Z}])$, and similarly for the other terms in the second line. It follows that $g(\nabla^M_X Y, Z) = g_Q(\nabla^N_{\bar{X}} \bar{Y}, \bar{Z})$. From this it follows that $\nabla^M_X Y$ is projectable and $f_* \nabla^M_X Y = \nabla^N_{\bar{X}} \bar{Y}$. But $f_* \nabla^M_X Y = f_* \tilde{\nabla}_X Y$, and thus (8.7) follows. $\square$

Using the connection $\tilde{\nabla}$, we define a codifferential $\tilde{\delta}$ acting on forms by the formula:

(8.8)
$$\tilde{\delta}\omega(A_2, \ldots, A_r) = -\sum_{k=1}^{n} E_k(\omega(E_k, A_2, \ldots, A_r))$$
$$+ \sum_{k=1}^{n} \omega(\tilde{\nabla}_{E_k} E_k, A_2, \ldots, A_r)$$
$$+ \sum_{k=1}^{n} \sum_{i=2}^{r} \omega(E_k, A_2, \ldots, \tilde{\nabla}_{E_k} A_i, \ldots, A_r),$$

where $\omega \in \Omega^r(M)$, $A_2, \ldots, A_r \in \Gamma TM$ and $\{E_k\}_{k=1,\ldots,n}$ is a local orthonormal frame field of TM.

Obviously, $\tilde{\delta}\omega$ is independent of the choice of the orthonormal frame $\{E_k\}_{k=1,\ldots,n}$, because it is equal to the trace of $-\tilde{\nabla}\omega$ over the appropriate entries. The operator $\tilde{\delta}$ is not the formal adjoint δ of the exterior derivative d, but differs from it by a zero-th order term determined by O'Neill's tensors T and A.

In spite of this, we can define the second order differential operator

$$\tilde{\Delta} = d\tilde{\delta} + \tilde{\delta}d, \tag{8.9}$$

which, although not self-adjoint, is still an elliptic operator with a self-adjoint principal symbol.

Next we discuss the effect of these operators on basic forms. On a distinguished chart U such a form ω can be canonically identified with a form $\bar{\omega}$ on the quotient $B - f(U)$ via the pull-back f^*. Let $\Omega_B(\mathcal{F}) \subset \Omega(M)$ be the subspace of all basic forms.

8.10 PROPOSITION [Mi-R-To 1]. (i) $\tilde{\delta}$ restricted to $\Omega_B(\mathcal{F})$ maps $\Omega_B(\mathcal{F})$ into itself, and coincides on a distinguished chart U with the formal usual codifferential δ defined on the Riemannian quotient $B = f(U)$.

(ii) The operator $\tilde{\Delta}$ restricted to $\Omega_B(\mathcal{F})$ maps $\Omega_B(\mathcal{F})$ to itself, and coincides on a distinguished chart U with the operator $\Delta = d\delta + \delta d$ defined on the Riemannian quotient $B = f(U)$.

Proof. We use a local orthonormal frame field $\{E_k\}_{k=1,\ldots n}$ of TM such that E_1, ..., $E_p \in \Gamma L$, and such that $E_{p+1}, \ldots, E_n \in \Gamma L^\perp$ are projectable vector fields. Applying formula (8.8) to $\omega \in \Omega_B^r(\mathcal{F})$ yields then

$$\tilde{\delta}\omega(A_2, \ldots, A_r) = -\sum_{k=p+1}^{n} E_k(\omega(E_k, A_2, \ldots, A_r))$$
$$+ \sum_{k=p+1}^{n} \omega(\tilde{\nabla}_{E_k} E_k, A_2, \ldots, A_r)$$
$$+ \sum_{k=p+1}^{n} \sum_{i=2}^{r} \omega(E_k, \ldots, \tilde{\nabla}_{E_k} A_i, \ldots, A_r)$$

since $\tilde{\nabla}_V W \in \Gamma L$ for $V, W \in \Gamma L$. The formula above shows that $\tilde{\delta}\omega$ is also basic for $\omega \in \Omega_B(\mathcal{F})$, because $\tilde{\nabla}_{E_k} V \in \Gamma L$ if $V \in \Gamma L$. Moreover, by Proposition 8.6, $\tilde{\nabla}_X Y$ coincides with the Levi Civita connection ∇^N of the local Riemannian quotient. This completes the proof. □

To study the operator $\tilde{\Delta}$ on M we first observe the following. Let δ denote the usual codifferential (formal adjoint of d) on M. Then (8.8) together with the usual similar formula for $\delta\omega$ yields

$$(\tilde{\delta}\omega - \delta\omega)(A_2, \ldots, A_r)$$
(8.11)
$$= \sum_{k=1}^{n}\{\omega(\gamma(E_k, E_k), A_2, \ldots, A_r) + \sum_{i=2}^{r} \omega(E_k, A_2, \ldots, \gamma(E_k, A_i), \ldots, A_r)\},$$

where γ is the difference tensor of $\tilde{\nabla}$ and the Levi Civita connection ∇^M on (M, g) defined in (8.3).

Comparing $\tilde{\Delta}$ with the usual De Rham-Hodge Laplacian $\Delta^M = d\delta + \delta d$, we have

(8.12) $$\tilde{\Delta} - \Delta^M = d(\tilde{\delta} - \delta) + (\tilde{\delta} - \delta)d$$

which is a differential operator of order one. It follows that $\tilde{\Delta}$ is an elliptic operator with self-adjoint symbol. Thus the heat equation

$$\dot{\omega} = -\tilde{\Delta}\omega \quad \text{with} \quad \lim_{t\downarrow 0} \omega_t = \omega_0 \in \Omega(M)$$

has a smooth solution $\omega_t = e^{-t\tilde{\Delta}}\omega_0$ for all $t > 0$ and for any initial value ω_0. The basic ingredient is the following standard result (Gårding inequality):

(8.13) $$\langle \tilde{\Delta}\omega, \omega \rangle \geq \frac{1}{2}\|\nabla^M \omega\|^2 - c\|\omega\|^2 \quad \text{for some constant } c.$$

The results of this chapter can now be stated as follows [Mi-R-To 1]. Let $\mathcal{F}$ be a Riemannian foliation on a closed oriented Riemannian manifold with a bundle-like metric.

8.14 THEOREM. *The heat equation $\dot{\omega} = -\tilde{\Delta}\omega$ with the initial condition $\lim_{t\downarrow 0} \omega_t = \omega_0 \in \Omega(M)$ has a solution for $0 \leq t < \infty$. If $\omega_0 \in \Omega_B(\mathcal{F})$, then $\omega_t \in \Omega_B(\mathcal{F})$ for $0 \leq t < \infty$.*

8.15 THEOREM. *Let $\omega_0 \in \Omega_B(\mathcal{F})$ satisfy $d\omega_0 = 0$. Then $d\omega_t = 0$ for $0 \leq t < \infty$ and the basic cohomology class of ω_t is the same for all $0 \leq t < \infty$.*

Transversal curvature properties below refer to curvature properties of the local Riemannian quotients. Applying the Bochner technique yields the following vanishing theorems.

8.16 THEOREM. *Assume the transversal Ricci curvature of $\mathcal{F}$ to be strictly positive. Then $H_B^1(\mathcal{F}) = 0$.*

This result is due to Hebda [Hb 1], proved by a different method. The next application involves the transversal curvature operator $\mathcal{R} : \Lambda^2 Q \to \Lambda^2 Q$.

8.17 THEOREM [Mi-R-To 1]. *Assume the transversal curvature operator to be positive. Then $H_B^r(\mathcal{F}) = 0$ for $0 < r < q = $ codim $\mathcal{F}$.*

Proof of Theorem 8.14. By Proposition 8.10, the modified Laplacian $\tilde{\Delta}$ leaves $\Omega_B(\mathcal{F})$ invariant. Since the Gårding inequality holds a fortiori for basic forms, the operator $e^{-t\tilde{\Delta}}$ is a well defined smoothing operator on $L_2(\Omega(M))$ and $L_2(\Omega_B(\mathcal{F}))$ for all $t > 0$. This holds because for $\text{Re}(\lambda)$ large enough, for instance $\text{Re}(\lambda) \geq c+2$, the operator $-(\tilde{\Delta} + \lambda I)$ is dissipative, and hence generates a semigroup by a theorem of Lumer and Phillips. The same is then true for the bounded perturbation $-\tilde{\Delta} \circ f - (\tilde{\Delta} + \lambda I)$. Uniqueness implies that the semigroup generated by $-\tilde{\Delta}$ on $L_2(\Omega(M))$ restricts on $L_2(\Omega_B(\mathcal{F}))$ to the semigroup generated by $-\tilde{\Delta} \mid \Omega_B(\mathcal{F})$. This completes the proof of Theorem 8.14. $\square$

Proof of Theorem 8.15. We first observe that $d\tilde{\Delta} = \tilde{\Delta}d$. This implies for a solution ω_t of the heat equation $\dot{\omega}_t = -\tilde{\Delta}\omega_t$ that

$$(d\omega_t)^\cdot = d\dot{\omega}_t = -d\tilde{\Delta}\omega_t = -\tilde{\Delta}d\omega_t,$$

and $d\omega_t$ is also a solution of the heat equation. But the initial condition $d\omega_0 = 0$ implies then necessarily $d\omega_t = 0$. To prove that the basic cohomology class $[\omega_t]$ is constant, it suffices to note that $d\omega_t = 0$ implies

$$\omega_t - \omega_0 = \int_0^t \dot{\omega}_\tau d\tau = -\int_0^t \tilde{\Delta}\omega_\tau d\tau$$
$$= -\int_0^t d\delta\omega_\tau d\tau = d\left(-\int_0^t \delta\omega_\tau d\tau\right),$$

and the desired conclusion follows. $\square$

Remark. The above technique provides again a proof of the finite-dimensionality of the basic cohomology. Indeed, the heat operator (say for time t) is a compact operator and acts as the identity on the cohomology. The result follows.

Proof of Theorems 8.16 and 8.17. For a distinguished chart U, $f(U) = B \subset N$ the Weitzenböck formula for $\Delta\omega$, $\omega \in \Omega(B)$ (representing a basic form) reads

$$\Delta\omega = -\text{Tr}(\nabla^N)^2\omega + S_{R^N}(\omega).$$

For $r = 1$ the operator S_{R^N} is well-known to be given as follows.

8.18 LEMMA (Bochner). *Let $\omega \in \Omega^1_B(\mathcal{F})$. Then*

$$S_{R^N}(\omega) = \omega \circ \text{Ric},$$

where $\text{Ric}: TB \to TB$ *is the Ricci curvature operator.*

For arbitrary $0 < r < q$ the self-adjoint operator S_{R^N} can be described as follows. Let $\bar{x} = f(x)$, $x \in U$. The symmetric curvature operator $R^N : \Lambda^2 T_{\bar{x}}B \to \Lambda^2 T_{\bar{x}}B$ has eigenvalues $\rho_1, \ldots, \rho_m$ ($m = \frac{1}{2}q(q-1)$) and corresponding eigenforms $A_1, \ldots, A_m$. These can be thought of as skew-symmetric endomorphisms of $T_{\bar{x}}B$, and hence they operate naturally (and skew-symmetrically) on r-forms $\Lambda^r T^*_{\bar{x}}B$. The action of A on any tensor τ at $\bar{x}$ is denoted $A \cdot \tau$.

8.19 LEMMA (Gallot-Meyer). Let $\omega \in \Omega^r_B(\mathcal{F})$, $0 < r < q$. Then
$$S_{R^N}(\omega) = -\sum_{k=1}^m \rho_k A_k \cdot A_k \cdot \omega.$$

Proof. See S. Gallot and D. Meyer, Opérateurs de courbure et Laplacien des formes différentielles d'une variété riemannienne, J. Math. Pures Appl. 54 (1975), 259–284. □

8.20 PROPOSITION. *If the curvature operator R^N is positive, then S_{R^N} on $\Lambda^r T^* B$ is positive definite for $0 < r < q = \dim B$.*

Proof. If $\rho_1 > 0$ is the smallest eigenvalue, then
$$g_Q(S_{R^N}(\omega), \omega) = -g_Q(\sum_{k=1}^m \rho_k A_k \cdot A_k \cdot \omega, \omega)$$
$$= \sum_{k=1}^m \rho_k g_Q(A_k \cdot \omega, A_k \cdot \omega) \geq \rho_1 \sum_{k=1}^m |A_k \cdot \omega|^2.$$

The sum $\sum_{k=1}^m |A_k \cdot \omega|^2$ is nonzero for nonzero ω, because the representation of the Lie algebra $\mathfrak{o}(m)$ on r-forms is irreducible for $r \neq 0, q$. In fact, $\sum_{k=1}^m |A_k \cdot \omega|^2 = r(q-r)|\omega|^2$ (see Gallot-Meyer, loc. cit., Cor. 2.6). Hence

(8.21) $$F(\omega) \equiv g_Q(S_{R^N}(\omega), \omega) \geq \rho|\omega|^2$$

with $\rho = r(q-r)\rho_1$. □

In the case of one-forms, we have by Lemma 8.18 the same inequality (8.21), except that now ρ is the minimum eigenvalue of the Ricci curvature.

We consider now the solution $\omega_t = e^{-t\tilde{\Delta}}\omega_0$ of the heat equation $\dot{\omega} = -\tilde{\Delta}\omega$ with initial condition $\lim_{t \downarrow 0} \omega_t = \omega_0$. By Theorems 8.14 and 8.15, for a closed basic form ω_0 the forms ω_t are also closed basic forms for $t \geq 0$. We compute the pointwise derivative
$$\frac{1}{2}\frac{\partial}{\partial t}|\omega_t|^2 = -g_Q(\tilde{\Delta}\omega_t, \omega_t).$$

By Proposition 8.10 and the pointwise Weitzenböck formula in the local model we have then
$$g_Q(\tilde{\Delta}\omega_t, \omega_t) = g_Q(\Delta\omega_t, \omega_t) = \frac{1}{2}\Delta|\omega_t|^2 + |\nabla\omega_t|^2 + F(\omega_t).$$

Note that $|\omega_t|^2$ for basic ω_t is a basic function. It follows that
$$\frac{1}{2}\left(\frac{\partial}{\partial t} + \Delta\right)|\omega_t|^2 = -|\nabla\omega_t|^2 - F(\omega_t)$$
$$\leq -\rho|\omega_t|^2 \quad \text{with} \quad \rho > 0.$$

At a point $x_0 \in M$ where $|\omega_t|^2$ attains its maximum value, the standard De Rham-Hodge Laplacian Δ^M of M acting on the basic function $f = |\omega_t|^2$ coincides with the Laplacian Δ of the local model acting on f. In fact Δf can by Proposition 8.10 be evaluated as $\tilde{\Delta} f = \tilde{\delta} df$. It differs from $\Delta^M f = \delta df$ by the expression given by the RHS of (8.11) applied to $\omega = df$, which vanishes at the critical point x_0 of $f = |\omega_t|^2$.

The maximum principle now implies that $e^{\rho t} \max |\omega_t|^2$ is a nonincreasing function of t, hence

$$\max |\omega_t|^2 \leq e^{-\rho t} \max |\omega_0|^2 \quad \text{for all} \quad t \geq 0.$$

This proves that $\lim_{t \to \infty} \omega_t = 0$ in the C^0-norm. By the smoothing properties of the heat semigroup $e^{-t\tilde{\Delta}}$, it follows that $\lim_{t \to \infty} \omega_t = 0$ in the C^∞-norm. Since by Theorem 8.15 the basic cohomology class of ω_t is the same for all $t \geq 0$, the proof of Theorems 8.16 and 8.17 is now complete. □

We apply now these vanishing theorems to prove the following result (see [A 6] [Mi-R-To 3]).

8.22 THEOREM. *Let $\mathcal{F}$ be a transversely oriented Riemannian foliation of codimension $q \geq 2$ on a closed oriented Riemannian manifold (M, g). Assume either the transversal Ricci operator $\rho : Q \to Q$ or the transversal curvature operator $\mathcal{R} : \Lambda^2 Q \to \Lambda^2 Q$ to be positive. Then $\mathcal{F}$ is taut.*

Proof. If the mean curvature one-form κ is basic, it defines a cohomology class in $H_B^1(\mathcal{F})$. It was shown in [A 6], that more generally for every Riemannian foliation $\mathcal{F}$ there is a well-defined cohomology class $\zeta(\mathcal{F}) \in H_B^1(\mathcal{F})$, whose vanishing characterizes the tautness of $\mathcal{F}$ (see also the end of Chapter 7). Under the curvature assumptions of Theorem 8.22, the first basic cohomology group of $\mathcal{F}$ vanishes by Theorems 8.16 and 8.17. It follows that $\zeta(\mathcal{F})$ vanishes and thus $\mathcal{F}$ is taut. □

As an illustration, consider $\mathcal{F}$ on (M, g) with positive curvature. If g is bundle-like, the transversal sectional curvature of $\mathcal{F}$ is positive as a consequence of O'Neill's formula (5.38c). It follows that the transversal Ricci curvature is positive. Theorem 8.22 implies that $\mathcal{F}$ is a taut Riemannian foliation.

Chapter 9
Lie Foliations

A Lie foliation is a foliation whose transversal structure is modeled on a Lie group. These were initially studied by Fedida [Fe 1] and Molino [Mo 5,8].

The precise definition of a Lie foliation is as follows. Let $\mathfrak{g}$ be a real Lie algebra, and $\omega \in \Omega^1(M,\mathfrak{g})$ a $\mathfrak{g}$-valued 1-form. The form ω is a Maurer-Cartan form (MC-form), if it satisfies the MC-equation $d\omega + \frac{1}{2}[\omega,\omega] = 0$. In other words, the (formal) curvature vanishes. If $\omega_x : T_xM \to \mathfrak{g}$ is surjective for all $x \in M$, then $L_x = \ker \omega_x$ defines a foliation of codimension $q = \dim \mathfrak{g}$ on M. Such a foliation is called a $\mathfrak{g}$-Lie foliation. For $\mathfrak{g} = \mathbb{R}$ (or more generally for abelian $\mathfrak{g}$), the MC-equation reduces to $d\omega = 0$, and we are back to a class of Riemannian foliations.

The involutivity of L follows immediately, namely $i(V)\omega = 0$, $i(W)\omega = 0$ implies
$$d\omega(V,W) = -\omega[V,W],$$
which vanishes by the MC-equation.

Consider a MC-form ω on M with values in $\mathfrak{g}$, and G a Lie group with Lie algebra $\mathfrak{g}$. Then a flat connection $\tilde{\omega}$ on the principal G-bundle $M \times G$ is defined by the formula
$$\tilde{\omega}_{(x,g)}(X_x, Y_g) = \mathrm{Ad}(g^{-1})\omega(X_x) + Y,$$
where $X \in \Gamma TM$, $Y \in \mathfrak{g}$, and Y_g denotes the value at g of the corresponding left-invariant vector field on G. The trivial section s of $M \to G$ pulls this form $\tilde{\omega}$ back to ω. This establishes a bijective correspondence between MC-forms on M and flat connections on $M \times G$. The Lie foliations on M are distinguished among these by the nonsingularity of $L = \ker \omega \subset TM$. This means that $\dim L = $ constant, and the codimension q equals $\dim \mathfrak{g}$.

To show that a $\mathfrak{g}$-Lie foliation is necessarily Riemannian, consider a basis $e_1, \ldots, e_q$ of $\mathfrak{g}$. Let $s_1, \ldots, s_q \in \Gamma Q$ such that for each $x \in M$ we have $\omega(s_\alpha)_x = e_\alpha$. Since $\omega_x : T_xQ \xrightarrow{\cong} \mathfrak{g}$, it follows that $s_1, \ldots, s_q$ define a framing of Q. Let Y_α be a lift of s_α to TM. Thus $\bar{Y}_\alpha = s_\alpha$ with an earlier notation. We will see in Lemma 9.1 that $Y_\alpha \in V(\mathcal{F})$, i.e. $[V, Y_\alpha] \in \Gamma L$ for $V \in \Gamma L$. The Euclidean metric on $\mathfrak{g}$, with $e_1, \ldots, e_q$ as orthonormal basis, defines a Riemannian metric g_Q on Q with

$s_1, \ldots, s_q$ as orthonormal frame. We have then for $V \in \Gamma L$

$$(\theta(V)g_Q)(s_\alpha, s_\beta) = V g_Q(s_\alpha, s_\beta) - g_Q(\pi[V, Y_\alpha], s_\beta) - g_Q(s_\alpha, \pi[V, Y_\beta])$$
$$= V g_Q(s_\alpha, s_\beta) \ .$$

Since $g_Q(s_\alpha, s_\beta) = \delta_{\alpha\beta}$, this expression vanishes, and $\mathcal{F}$ is a Riemannian foliation. We have used the following fact.

9.1 LEMMA. *With the notations above we have*
  (i) $Y_\alpha \in V(\mathcal{F})$, $\alpha = 1, \ldots, q$;
  (ii) $\omega[\bar{Y}_\alpha, \bar{Y}_\beta] = [e_\alpha, e_\beta]$.

Proof. (i) Let $V \in \Gamma L$. We have to show $\omega[Y_\alpha, V] = 0$. But

$$d\omega(Y_\alpha, V) = Y_\alpha \omega(V) - V\omega(Y_\alpha) - \omega[Y_\alpha, V] = -\omega[Y_\alpha, V],$$

since clearly $\omega(V) = 0$, and $\omega(Y_\alpha) = \omega(\bar{Y}_\alpha) =$ constant, and thus also $V\omega(Y_\alpha) = 0$. By the MC-equation

$$d\omega(Y_\alpha, V) = -\frac{1}{2}[\omega, \omega](Y_\alpha, V) = -[\omega(Y_\alpha), \omega(V)] = 0,$$

and thus $\omega[Y_\alpha, V] = 0$.

(ii) By a similar reasoning

$$\omega[\bar{Y}_\alpha, \bar{Y}_\beta] = \omega[Y_\alpha, Y_\beta] = -d\omega(Y_\alpha, Y_\beta)$$
$$= \frac{1}{2}[\omega, \omega](Y_\alpha, Y_\beta) = [\omega(Y_\alpha), \omega(Y_\beta)] = [e_\alpha, e_\beta].$$

This concludes the proof of the lemma. □
Thus we can identify $\mathfrak{g}$ with ΓQ^L and s_α with e_α.

The developing map of a $\mathfrak{g}$-Lie foliation on M is described as follows. Let $\tilde{M} \xrightarrow{p} M$ be the universal covering of M, and G the simply connected Lie group with Lie algebra $\mathfrak{g}$. Let $\tilde{x}_0$ be a basepoint in $\tilde{M}$, and $x_0 = p(\tilde{x}_0)$. For any path γ from $\tilde{x}_0$ to a point $\tilde{x}$ in $\tilde{M}$, the parallel transport τ_γ in the flat principal G-bundle $\tilde{M} \times G \to \tilde{M}$ is well-defined. It maps $(\tilde{x}_0, e)$ to a point $(\tilde{x}, g)$, and by definition

$$\mathrm{f}_\omega : \tilde{M} \to G, \ \mathrm{f}_\omega(\tilde{x}) = g.$$

This value depends only on $\tilde{x}$ (and not on the choice of γ), because another path γ' from $\tilde{x}_0$ to $\tilde{x}$ is homotopic to γ, thus $\tau_{\gamma'} = \tau_\gamma$ as the curvature is zero. The surjectivity of ω turns f_ω into a submersion. If M is closed, this map can be shown to be a locally trivial fibration. The crucial point is the existence of a Riemannian metric, turning f_ω into a Riemannian submersion. Such a metric is constructed by defining first a holonomy invariant Riemannian metric g_Q on Q as before, and setting $g = g_L \oplus g_Q$ with g_L an arbitrary bundle metric on L.

The holonomy homomorphism
$$h_\omega : \pi_1(M, x_0) \to G$$
is similarly defined in the flat principal bundle $M \times G \to M$, and for loops in M based at x_0. The developing map f_ω is equivariant with respect to h_ω, i.e.
$$f_\omega(\tilde{x}\gamma) = f_\omega(\tilde{x}) h_\omega(\gamma).$$
It follows that f_ω induces a map $M \to G/\operatorname{im} h_\omega$.

For a general $\mathfrak{g}$-Lie foliation $\mathcal{F}$ on M with developing map $\tilde{M} \to G$, it follows that $\mathcal{F}$ can be defined by an atlas $\mathcal{U} = \{U_\alpha\}$ of distinguished charts, submersions $f_\alpha : U_\alpha \to G$, and transition functions which are left translations in G. Thus we have a G-foliation, where G is identified with its group of left translations, which in turn is a subgroup of the affine automorphism group of the left parallelism of G. To see that $G \subset GL(q)$, $q = \dim G$, observe that an affine automorphism is completely determined by its value and the derivative at one point.

The Lie group structure is conveniently characterized by the Maurer-Cartan equation, or equivalently, by the vanishing of the curvature of a certain Cartan connection.

If the Cartan curvature is small, we call the space an almost Lie group. An almost Lie foliation therefore is a foliation whose transversal structure is defined by a Cartan connection with small curvature. In the remainder of this chapter the model group is a compact semi-simple Lie group. The next purpose is to set up a heat flow which transforms the transversal almost Lie structure into a Lie foliation. The prime examples for almost Lie foliations are certain principal bundles over foliated Riemannian manifolds. The general technique of the proof was first explained in [R-To]. Here we base the heat flow on the elliptic operator introduced in [Mi-R-To 1], and discussed in Chapter 8. This has several advantages. One advantage is that the restriction of this operator to basic forms coincides with the Laplace operator introduced by Reinhart, and therefore the well-known vanishing theorems of Riemannian geometry apply as long as they are based on a maximum principle. Another advantage is that, in contrast to [R-To], the mean curvature form no longer plays any role. The removal of this restriction turns the assumptions of the theorem into an open condition, which is desirable for comparison theorems.

9.2 THEOREM [Mi-R-To 2]. *Let $\mathcal{F}$ be a transversally oriented Riemannian foliation of codimension $q \geq 3$ on a closed oriented Riemannian manifold (P, g_P). Let G be a compact semi-simple Lie group of dimension q. There exists a constant $A > 0$ depending only on the Lie algebra $\mathfrak{g}$ of G and curvature bounds on P with the following property. If $\omega : TP \to \mathfrak{g}$ is a basic Cartan connection form on P with Cartan curvature Ω, then $\|\Omega\|_{1,\infty} < A$ implies that $\mathcal{F}$ is a Lie foliation.*

The norm is defined below in (9.3). In the proof we construct a Cartan connection $\bar{\omega}$ with vanishing curvature near the original Cartan connection. The flat Cartan connection $\bar\omega$ yields then a developing map $\Phi : \tilde P \to G$ on the universal covering $\tilde P$ of P, which is equivariant with respect to a homomorphism $\varphi : \pi_1(P) \to G$ of the fundamental group of P.

Let $\mathfrak{g}$ be a Lie algebra of dimension $q = \dim Q$, and $\omega : TP \to \mathfrak{g}$ a Cartan connection. We assume that the metrics on Q defined by $\mathfrak{g}$ and by the Riemannian metric on P agree. The Cartan curvature Ω of ω is defined by

$$\Omega = d\omega + \frac{1}{2}[\omega,\omega]$$

where $[\,,\,]$ is the bracket induced by $\mathfrak{g}$ on $\mathfrak{g}$-valued forms. A connection is said to be adapted to the foliation $\mathcal{F}$ if ω restricted to L vanishes. It is a Cartan connection if ω is nondegenerate on Q. An adapted Cartan connection is basic if $i(V)\Omega = 0$ for all $V \in \Gamma L$.

To prepare for the proof of Theorem 9.2, let $\nabla = \nabla^P$ be the Levi Civita connection of (P, g_P). We define as in (8.1) a connection $\tilde\nabla_A B$ on P. Using the connection $\tilde\nabla$ we define a codifferential $\tilde\delta$ acting on forms by formula (8.8), where now $\omega \in \Omega^r(P)$, $A_2, \ldots, A_r \in \Gamma TP$ and $\{E_k\}_{k=1,\ldots,n}$ is a local orthonormal frame field of TP.

The Sobolev norms are defined as follows. Let

$$\exp_u : T_u P \to P$$

denote the exponential map. We pull back the differential forms via $\exp_u$ restricted to a ball $B_r(0)$ of suitable radius r and center $0 \in T_u P$, and define

$$(9.3) \qquad \|\beta\|_{s,m}^m = \sup_{u \in P} \sum_{|\mu| \le s} \int_{B_r(0)} |\nabla^\mu \beta|^m,$$

where $m \in \mathbb{N}$ is the exponent, and $s \in \mathbb{N}$ denotes the number of derivatives taken into account in the definition of the Sobolev space $W_{s,m}$. As usual, $|\mu|$ is the degree of the multi-index μ, and $m = \infty$ indicates the essential supremum.

Proof of Theorem 9.2. The data of the theorem specifies an adapted Cartan connection $\omega : TP \to \mathfrak{g}$ with small basic curvature $\Omega = d\omega + \frac{1}{2}[\omega,\omega]$. We need to establish the existence of an adapted Cartan connection $\bar\omega$ on P with vanishing Cartan curvature $\bar\Omega$.

We will obtain $\bar\omega$ as the limit for $t \to \infty$ of the solution of a certain heat equation, whose initial connection $\omega = \omega_0$ is the Cartan connection provided by the data of the theorem. To define this heat equation, let $E = P \times \mathfrak{g}$ denote the trivial vector bundle over P whose fiber is the Lie algebra $\mathfrak{g}$. We define the linear connection

$$(9.4) \qquad D_X s = Xs + [\omega(X), s],$$

where s is a section of E, Xs is the derivative of s in direction X, and ω is a Cartan connection. The curvature R^D of D is

(9.5) $$R^D(X,Y)s = [\Omega(X,Y), s].$$

For reasons of formal symmetry we define an E-valued function Θ which we call co-curvature. Then

(9.6) $$\Omega = d\omega + \frac{1}{2}[\omega, \omega], \quad \Theta = \tilde{\delta}\omega.$$

where $\tilde{\delta}$ is the operator defined in (8.8). Under the assumption of Theorem 9.2 both Ω and Θ are in the basic complex

$$\Omega_B(\mathcal{F}, E) = \{\beta \in \Omega(P, E) | i(V)\beta = 0, \quad \theta(V)\beta = 0 \text{ for } V \in \Gamma L\}.$$

As usual we define the exterior derivative relative to the covariant derivative $D \otimes \nabla$, where D is as above and ∇ is the Levi Civita connection on P. To define $\tilde{\delta}$ we use the connection $D \otimes \tilde{\nabla}$. As in Chapter 8, the operators d and $\tilde{\delta}$ restrict to the basic E-valued complex $\Omega_B(\mathcal{F}, E)$. We denote the restrictions by d_B and $\tilde{\delta}_B$.

The evolution equation is defined as follows

(9.7) $$\frac{\partial}{\partial t}\omega(x,t) = -(\tilde{\delta}_B\Omega + d_B\Theta)(x,t) \quad \omega(x,0) = \omega(x),$$

where ω is the Cartan connection provided by the data of the theorem.

To prove short time existence of a solution we observe that (9.7) is the restriction of a parabolic equation on the Riemannian manifold P to the basic complex of differential forms. To prove existence of the solution for $t \in (0, \infty)$ as well as the convergence of $\omega(x,t)$ to a flat Cartan connection as $t \to \infty$, we derive a maximum principle.

We observe first that (9.4) and (9.5) imply

$$\dot{\Omega} = d\dot{\omega} + \frac{1}{2}[\omega, \omega]\dot{} = d_B\dot{\omega}.$$

We obtain thus from (9.7)

$$\dot{\Omega} = d_B\dot{\omega} = -d_B\tilde{\delta}_B\Omega - d_Bd_B\Theta = -\tilde{\Delta}_B\Omega + \tilde{\delta}_Bd_B\Omega - d_Bd_B\Theta$$

and

(9.8) $$g_Q(\dot{\Omega}, \Omega) = -g_Q(\tilde{\Delta}_B\Omega, \Omega),$$

where the remaining two terms vanish. The first term vanishes because of the Bianchi equation $d_B\Omega = 0$. The second term vanishes because $d_Bd_B = [\Omega, \cdot]$, which is skew-symmetric.

To control Θ it is not necessary to compute an evolution equation. The fact is that Θ measures essentially the divergence of the D-parallel vector fields on P which is computed in terms of Ω. To prove that $\max |\Omega|^2$ decreases exponentially in time we compute a Bochner-Weitzenböck formula for the operator $\tilde{\Delta}$. We observe that the calculations can be done in an adapted coordinate system for a point at which $|\Omega|^2$ takes it maximum. The operator $\tilde{\Delta}_B$ is constructed such that, in an adapted coordinate system, the corresponding formula from Riemannian geometry is valid. Therefore we can use the formula in [M. Min-Oo and E. Ruh, Ann. Sci. Ecole Norm. Sup. 12 (1979), 335–353; (4.17)] to obtain at a point where $|\Omega|^2$ takes its maximum

$$(9.9) \qquad g_Q(\tilde{\Delta}_B \Omega, \Omega) > \frac{1}{5}|\Omega|^2,$$

provided the constant Λ in the theorem is sufficiently small. We remark that in the model Lie group the factor is $\frac{1}{4}$. The smaller factor $\frac{1}{5}$ is chosen to absorb various nonlinear contributions of Ω and its derivatives. Here, there is one additional summand to be absorbed because the operators d_B and $\tilde{\delta}_B$ vary slightly from those of [M. Min-Oo and E. Ruh, loc. cit.]. However, the assumptions of the theorem are such that this contribution is smaller than the constant A of the theorem. To finish the proof we observe that (9.8) and (9.9) imply

$$\frac{d}{dt}\|\Omega\|^2_{0,\infty} \leq -\frac{2}{5}\|\Omega\|^2_{0,\infty},$$

and

$$\|\Omega\|^2_{0,\infty}(t) \leq \|\Omega\|^2_{0,\infty}(0) e^{-\frac{2}{5}t}.$$

Now, the usual parabolic estimates show that $\omega(t)$ converges to a flat Cartan connection $\bar{\omega}$ near $\omega(0)$ as $t \to \infty$. □

Chapter 10
Structure of Riemannian Foliations

For Riemannian foliations on closed manifolds, Molino has found a remarkable structure theorem [Mo 8,10]. This theorem is based on several fundamental observations. The first is that the canonical lift $\hat{\mathcal{F}}$ of a Riemannian foliation $\mathcal{F}$ to the bundle $\hat{M}$ of orthonormal frames of Q is a transversally parallelizable Riemannian foliation. The canonical lift $\hat{\mathcal{F}}$ on $\hat{M}$ is a foliation of the same dimension as $\mathcal{F}$ on M, and invariant under the action of the orthogonal structural group of $\hat{M}$. Now let M be closed and oriented, and consider on $\hat{M}$ the closures of the leaves of $\hat{\mathcal{F}}$. The second fundamental observation is that these closures form the fibers of a fibration $X_0 \to \hat{M} \xrightarrow{\hat{\pi}} \hat{W}$ over the space $\hat{W}$ of orbit closures, with typical fiber X_0. The foliation $\hat{\mathcal{F}}$ induces on X_0, and on each fiber of $\hat{\pi}$, a Lie foliation with dense leaves. The Lie algebra of the model group G of this Lie foliation is another structural invariant of the foliation. With the help of this structure theorem, many questions on Riemannian foliations can be reduced to questions on Lie foliations, by passing to the bundle of transversal orthonormal frames. We state all this in more detail, but without proof, which is the main topic of Molino's book [Mo 10].

Transverse parallelization

A foliation $\mathcal{F}$ has a transversal parallelization, if there exists a global frame $s_1, \ldots, s_q \in \Gamma Q$ by L-invariant sections, i.e. $\theta(V)s_\alpha = 0$ for $V \in \Gamma L$ and $\alpha = 1, \ldots, q$.

Examples are given by Lie foliations. Let G be a Lie group with MC-form $\omega \in \Omega^1(G, \mathfrak{g})$. Let $f : M \to G$ be a submersion, and define $\hat{\omega} = f^*\omega$. Then $d\hat{\omega} + \frac{1}{2}[\hat{\omega}, \hat{\omega}] = 0$. The foliation with tangent bundle $L = \ker \hat{\omega}$ is a foliation. A basis of $\mathfrak{g}$ gives rise to a transverse parallelization of $\mathcal{F}$ by Lemma 9.1.

An important example for our purpose is the canonical lift $\hat{\mathcal{F}}$ of $\mathcal{F}$ on M to the bundle $\hat{M} \xrightarrow{p} M$ of orthonormal frames of Q. For a submersion $f_\alpha : U_\alpha \to B_\alpha$ defining $\mathcal{F}$ locally, consider the bundle $\hat{B}_\alpha \to B_\alpha$ of orthonormal frames of the local model space. Then the diagram

$$\begin{array}{ccc} \hat{M} & \xrightarrow{p} & M \\ \cup & & \cup \\ p^{-1}(U_\alpha) & \longrightarrow & U_\alpha \\ \downarrow \hat{f}_\alpha & & \downarrow f_\alpha \\ \hat{B}_\alpha & \longrightarrow & B_\alpha \end{array}$$

is commutative. The local submersions $\hat{f}_\alpha$ define a foliation $\hat{\mathcal{F}}$ on $\hat{M}$ of the same dimension as the foliation $\mathcal{F}$ on M. The structure of complete Riemannian foliations (i.e. with infinitely extendable orthogonal geodesics) can then be described as follows.

10.1 THEOREM (Molino [Mo 8,10]). *Let $\mathcal{F}$ be a complete Riemannian foliation on (M, g). Let $\hat{\mathcal{F}}$ be the canonical lift of $\mathcal{F}$ to the bundle $\hat{M} \to M$ of orthonormal frames of the normal bundle Q of $\mathcal{F}$.*

(i) *$\hat{\mathcal{F}}$ is transverse parallelizable.*
(ii) *Consider the closures of the fibers of $\hat{\mathcal{F}}$ on $\hat{M}$. They are the fibers of a smooth fibration $\hat{\pi} : \hat{M} \to \hat{W}$, where $\hat{W}$ is a smooth manifold with a smooth $O(q)$-action for which $\hat{\pi}$ is equivariant.*
(iii) *Consider the closures of the leaves of $\mathcal{F}$ on M, and let $M \xrightarrow{\pi} W$ be the quotient map to the space of closures. Then $W = \hat{W}/O(q)$, and the diagram*

$$\begin{array}{ccc} \hat{M} & \xrightarrow{p} & M \\ \downarrow \hat{\pi} & & \downarrow \pi \\ \hat{W} & \longrightarrow & W \end{array}$$

is commutative.
(iv) *The restriction $\hat{\mathcal{F}}_0$ of $\hat{\mathcal{F}}$ to the closure $\hat{\pi}^{-1}(w)$ of a leaf is a $\mathfrak{g}$-Lie foliation with dense holonomy group $\Gamma \subset G$, a Lie group with Lie algebra $\mathfrak{g}$. If U is a contractible neighborhood of w in $\hat{W}$, there exists a diffeomorphism $\hat{\pi}^{-1}(U) \cong U \times \hat{\pi}^{-1}(w)$, which transports the restriction of $\hat{\mathcal{F}}$ to $\hat{\pi}^{-1}(U)$ onto the product of $\hat{\mathcal{F}}_0$ by U.*

The proof of this result is the main topic of [Mo 10]. This structure theorem associates to a Riemannian foliation a space $\hat{W}$ and a Lie algebra $\mathfrak{g}$. They represent invariants attached to the foliation, which are generally difficult to determine. This theorem has many important consequences.

An example of such a consequence is the following topological invariance result for the basic cohomology $H_B(\mathcal{F})$ of a Riemannian foliation. Let $h : (M, \mathcal{F}) \to (M', \mathcal{F}')$ be a homeomorphism $M \to M'$ which sends the leaves of $\mathcal{F}$ onto the leaves of $\mathcal{F}'$. Such a homeomorphism is called a foliated homeomorphism.

10.2 Theorem ([EK-N 2]). *Let $(M, \mathcal{F})$ and $(M', \mathcal{F}')$ be complete Riemannian foliations of codimension q, and $h : (M, \mathcal{F}) \to (M', \mathcal{F}')$ a foliated homeomorphism. Then h induces an algebra isomorphism $h_* : H_B(\mathcal{F}) \to H_B(\mathcal{F}')$.*

This property does not hold for general foliations even on closed manifolds.

Molino has generalized his structure theorem to a class of singular Riemannian foliations. As a basic model, he takes the orbits of a connected Lie group acting isometrically on a Riemannian manifold M. Thus there is a given partition $\mathcal{F}$ of M into connected submanifolds, the leaves, not necessarily of the same dimension, together with a Riemannian metric g adapted to $\mathcal{F}$, in the sense that geodesics perpendicular to one leaf are perpendicular to all the leaves they meet. Moreover, the module $L(\mathcal{F})$ of vector fields tangent to $\mathcal{F}$ is required to be transitive on each leaf. These conditions on $\mathcal{F}$ are paraphrased by saying that $\mathcal{F}$ is a transnormal system on M. What Molino proved in [Mo 9] for such a triple $(M, \mathcal{F}, g)$ is the following:

(i) M is stratified by leaf dimension into embedded submanifolds;
(ii) the closures of the leaves form a new transnormal system $\bar{\mathcal{F}}$;
(iii) there is a locally constant sheaf $C(M, \mathcal{F})$ of Lie algebras, independent of the choice of the adapted metric, which in each stratum projects to a sheaf of germs of transverse Killing vector fields whose orbits are the closures of the leaves.

Molino conjectured that $\bar{\mathcal{F}}$ itself is a singular Riemannian foliation. This theme is taken up again later in [Mo 13] under the name of orbit-like foliations. See in particular Theorem 2 on p. 102-103 of [Mo 13]. An ideal concept of singular Riemannian foliation should at least embrace this type of examples, and perhaps more. In any case, it might be advisable to follow what Strichartz has called in another context the doctrine of (micro-)local myopia: pay attention to the singularities, and other issues will take care of themselves.

Chapter 11
Spectral Geometry of Riemannian Foliations

Throughout this chapter M is oriented, closed and Δ_g is the Laplacian associated to a metric g on M. All this will not be explicitly stated as a hypothesis in our theorems below.

The study of the spectrum of Δ_g acting on functions or forms has attracted a lot of attention and a lot of work has been done about the following question: What kind of data of (M,g) can be "heard" and what data cannot be "heard"? In particular, the problem whether two isospectral manifolds are isometric or not has been intensively studied.

In this chapter we treat a similar problem for Riemannian foliations $\mathcal{F}$ on (M,g). In this case there exists aside from Δ_g another natural differential operator, the Jacobi operator J_∇ (see [K-To 6] and below for more details), which is a second order elliptic operator acting on sections of the normal bundle. Its spectrum is discrete as a consequence of the compactness of M. Hence one has two spectra, $\mathrm{spec}(M, g) =$ spectrum of Δ_g (acting on functions), and $\mathrm{spec}(\mathcal{F}, J_\nabla) =$ spectrum of J_∇. We will deal with the following problem: *Which geometric properties of a foliation $\mathcal{F}$ on a Riemannian manifold (M,g) are determined by the two types of spectral invariants?* Moreover, we will focus on the following question: *Let $(M^n, g, \mathcal{F})$ and $(M_0^{n_0}, g_0, \mathcal{F}_0)$ be two Riemannian foliations and assume both spectra to be the same for these data. What can we conclude about corresponding geometric properties of these data?* This was discussed in [N-To-V].

We start with some preliminaries and general results. Then we consider first properties about the integrability of the normal bundle of the foliation. Moreover we derive a result about the characterization of some particular Hopf fibrations. Next we focus on the relation between the two spectra and the transversal geometry of a Riemannian foliation.

For a Riemannian foliation $\mathcal{F}$ with metric g_Q and canonical connection ∇ on Q the usual calculus for Q-valued forms on M applies. Let $\Delta = d_\nabla^* d_\nabla$ be the Laplacian acting on sections of ΓQ. Then the Jacobi operator of a Riemannian foliation $\mathcal{F}$ is given by [K-To 6]

$$J_\nabla u = (\Delta - \rho_\nabla) u \quad \text{for } u \in \Gamma Q.$$

With respect to the natural scalar product on ΓQ it is strongly elliptic of the second order with leading symbol g. It occurs naturally as the second variation operator for a Riemannian and harmonic foliation (critical foliation) for the 2-parameter variations discussed in [K-To 6]. It has a discrete spectrum with finite multiplicities.

Consider the case of a transversally oriented codimension one foliation $\mathcal{F}$. Then the transversal Ricci operator ρ_∇ vanishes. Sections of Q can be identified with functions on M, and it is easy to see that then an eigenfunction of the Jacobi operator on ΓQ corresponds to an eigenfunction of the ordinary Laplacian on M associated to the same eigenvalue. Thus no new information is encoded in $\mathrm{spec}(\mathcal{F}, J_\nabla)$. Throughout the rest of the chapter we assume therefore that the codimension of the foliation is greater than or equal to 2.

Consider the semigroup $e^{-t\Delta_g}$, and the semigroup e^{-tJ_∇} given by

$$e^{-tJ_\nabla}u(x) = \int_M K(t, x, y, J_\nabla)u(y)\mu(y),$$

where $K(t, x, y, J_\nabla) \in \mathrm{Hom}(Q_y, Q_x)$ is the kernel function. We have asymptotic expansions for the corresponding L^2-trace of $e^{-t\Delta_g}$ and the L^2-trace

$$\mathrm{Tr}\ e^{-tJ_\nabla} = \int_M \mathrm{tr}_{Q_x} K(t, x, x, J_\nabla)\mu(x)$$

for $t \downarrow 0$

(11.1)
$$\mathrm{Tr}\ e^{-t\Delta_g} = \sum_{i=1}^{\infty} e^{-t\lambda_i} \underset{t\downarrow 0}{\sim} (4\pi t)^{-\frac{n}{2}} \sum_{k=0}^{\infty} t^k a_k(\Delta_g),$$

$$\mathrm{Tr}\ e^{-tJ_\nabla} = \sum_{i=1}^{\infty} e^{-t\mu_i} \underset{t\downarrow 0}{\sim} (4\pi t)^{-\frac{n}{2}} \sum_{k=0}^{\infty} t^k b_k(J_\nabla)$$

where

$$a_k(\Delta_g) = \int_M a_k(x, \Delta_g)\mu(x),$$
$$b_k(J_\nabla) = \int_M b_k(x, J_\nabla)\mu(x)$$

are invariants of Δ_g and J_∇ depending only on the discrete spectra

$$\mathrm{spec}(M, g) = \{0 \leq \lambda_1 \leq \lambda_2 \leq \cdots \leq \lambda_i \leq \ldots \uparrow \infty\},$$
$$\mathrm{spec}(\mathcal{F}, J_\nabla) = \{\mu_1 \leq \mu_2 \leq \cdots \leq \mu_i \leq \ldots \uparrow \infty\}.$$

We state the classical formulas for $a_k(\Delta_g)$ given by Patodi. Using the local formulas for $b_k(x, J_\nabla)$ given by Gilkey, we also obtain the $b_k(J_\nabla)$. The curvature data associated to (M, g) are denoted by R^M, ρ^M, and τ^M. We have then by Gilkey the following result.

11.2 THEOREM. *Let $\mathcal{F}$ be a smooth Riemannian foliation of codimension $q \geq 2$ on (M, g) Then*

(11.3)
$$a_0(\Delta_g) = \text{vol } M,$$
$$a_1(\Delta_g) = \frac{1}{6} \int_M \tau^M \mu,$$
$$a_2(\Delta_g) = \frac{1}{360} \int_M \left[2|R^M|^2 - 2|\rho^M|^2 + 5(\tau^M)^2 \right] \mu$$

and

(11.4)
$$b_0(J_\nabla) = q \text{ vol } M,$$
$$b_1(J_\nabla) = q a_1(\Delta_g) + \int_M \tau^\nabla \mu,$$
$$b_2(J_\nabla) = q a_2(\Delta_g) + \frac{1}{12} \int_M \left[2\tau^M \tau^\nabla + 6|\rho^\nabla|^2 - |R^\nabla|^2 \right] \mu.$$

Note that $\int_M \tau^M \mu$ is the total scalar curvature of (M, g), and we call $\int_M \tau^\nabla \mu$ the *total (transversal) scalar curvature* of the foliation.

DEFINITION. *The Riemannian foliations $(M, g, \mathcal{F})$ and $(M_0, g_0, \mathcal{F}_0)$ are said to be isospectral if*
$$\text{spec}(M, g) = \text{spec}(M_0, g_0),$$
$$\text{spec}(\mathcal{F}, J_\nabla) = \text{spec}(\mathcal{F}_0, J_{\nabla_0}).$$

From (11.1) and Theorem 11.2 we get then the following results ([N-To-V, Theorem 2]).

11.5 THEOREM. *Let $(M, g, \mathcal{F})$ and $(M_0, g_0, \mathcal{F}_0)$ be isospectral Riemannian foliations. Then we have*

(i) $\dim M = \dim M_0$;
(ii) $\text{vol } M = \text{vol } M_0$;
(iii) (M, g) *and* (M_0, g_0) *have equal total scalar curvature;*
(iv) $\text{codim } \mathcal{F} = \text{codim } \mathcal{F}_0$, *and hence $\mathcal{F}$ and $\mathcal{F}_0$ have the same energy;*
(v) $\mathcal{F}$ *and* $\mathcal{F}_0$ *have equal total transversal scalar curvature;*
(vi) $\int_M \left[2|R^M|^2 - 2|\rho^M|^2 + 5(\tau^M)^2 \right] \mu = \int_{M_0} \left[2|R^{M_0}|^2 - 2|\rho^{M_0}|^2 + 5(\tau^{M_0})^2 \right] \mu_0$;
(vii) $\int_M \left[2\tau^M \tau^\nabla + 6|\rho^\nabla|^2 - |R^\nabla|^2 \right] \mu = \int_{M_0} \left[2\tau^{M_0} \tau^{\nabla_0} + 6|\rho^{\nabla_0}|^2 - |R^{\nabla_0}|^2 \right] \mu_0$.

For (iv) note that the energy $E(\mathcal{F}) = \frac{1}{2} q \text{ vol } M$ [K-To 5, p. 116].

Clearly isometric (congruent) data are isospectral. To see that the converse does not hold we consider for a moment again the case of $q = 1$. Then the two 16-dimensional nonisometric flat Milnor tori carry trivial Riemannian hypersurface foliations obtained by projecting the lattices onto, say, the line generated by the first of the sixteen independent basic vectors of the lattice. By the remark made above, for isospectral tori we will obtain isospectral Riemannian foliations. But the tori are not isometric, hence the Riemannian foliations a fortiori not congruent.

On the other hand, Theorem 11.19 below yields an affirmative answer to the isometry problem in the case of Riemannian fibrations on S^3 and S^5 with fiber dimension one.

We describe a particular case.

EXAMPLE. Let $(M, g) = (S^3(1), g)$ be the standard sphere and $\mathcal{F}$ the foliation defined by the fibers of the Hopf fibration $S^3(1) \to S^2(1/2)$. This is a Riemannian foliation on M and g is bundle-like for $\mathcal{F}$. For this foliation we have then:

(i) $\rho_\nabla = 4$ id, $J_\nabla = \Delta - 4$ id;

(ii)
$$\mathrm{spec}(\mathcal{F}, J_\nabla) = \begin{cases} \frac{1}{2}l(l+1) + i & \text{with multiplicity } 2l+1 \\ \frac{1}{2}l(l+1) - i & \text{with multiplicity } 2l+1 \end{cases}$$

where $l \geq 0$, $l \in \frac{1}{2}\mathbb{Z}$, and $i \in \{-l, -l+1, \ldots, l-1, l\}$. This result is due to Urakawa [Trans. Amer. Math. Soc. 301 (1987), 557–558; see Corollary 8.12]. In particular, index $\mathcal{F} = 2$ and nullity $\mathcal{F} = 8$.

Integrability of $L^\perp$

To state the next results we consider O'Neill's integrability tensor A given by (5.27), and the generalized Gauss equation for $L^\perp$ (5.37f)

$$\begin{aligned}(11.6) \quad g_M(R^M(X,Y)Z, Z') =& g_M(R^\nabla(X,Y)Z, Z') + g_M(A_X Z, A_Y Z') \\ &- g_M(A_Y Z, A_X Z') + 2 g_M(A_X Y, A_Z Z')\end{aligned}$$

for $X, Y, Z, Z' \in \Gamma L^\perp$.

First we prove ([N-To-V, Theorem 8])

11.7 THEOREM. Let $(M^n, g, \mathcal{F})$ and $(M_0^n, g_0, \mathcal{F}_0)$ be Riemannian flows on Einstein spaces with the same total scalar curvature, and assume that the flows have the same total (transversal) scalar curvature. Then $L^\perp$ is integrable if and only if $L_0^\perp$ is integrable.

Proof. Let $x \in M$ and let P be a 2-plane of $L_x^\perp$. For an orthonormal basis $\{e_\alpha, e_\beta\}$ of P we get from (11.6)

$$K^\nabla(e_\alpha, e_\beta) = K^M(e_\alpha, e_\beta) + 3|A_{e_\alpha}e_\beta|^2.$$

Hence

$$\tau^\nabla = \sum_{\alpha \neq \beta} K^\nabla(e_\alpha, e_\beta) = \sum_{\alpha \neq \beta} K^M(e_\alpha, e_\beta) + 3|A|^2, \tag{11.8}$$

where $\{e_\alpha, \alpha = 1, \ldots, q\}$ is an orthonormal basis of $L_x^\perp$. But

$$\tau^M = 2\rho_{00}^M + \sum_{\alpha \neq \beta} K^M(e_\alpha, e_\beta), \tag{11.9}$$

where e_0 is a unit vector of L_x. Then (11.8) and (11.9) yield

$$\tau^\nabla = \tau^M - 2\rho_{00}^M + 3|A|^2,$$

and since (M, g) is an Einstein manifold, we obtain

$$\tau^\nabla = \frac{n-2}{n}\tau^M + 3|A|^2.$$

This, and the hypotheses lead to

$$\int_M |A|^2 \mu = \int_{M_0} |A_0|^2 \mu_0.$$

Thus $A = 0$ if and only if $A_0 = 0$. $\square$

From Theorem 11.5 and Theorem 11.7 we have

11.10 COROLLARY. *Let $(M, g, \mathcal{F})$ and $(M_0, g_0, \mathcal{F}_0)$ be isospectral oriented Riemannian foliations on Einstein spaces such that $\mathcal{F}_0$ is a Riemannian flow. Then $\dim M = \dim M_0$, $\mathcal{F}$ is also a Riemannian flow, and $L^\perp$ is integrable if and only if $L_0^\perp$ is integrable.*

Now, we extend this result to Riemannian foliations of arbitrary codimension for the case of totally geodesic foliations. We have the following result.

11.11 THEOREM [N-To-V, Theorem 5]. *Let $(M^m, g, \mathcal{F})$ and $(M_0^m, g_0, \mathcal{F}_0)$ be totally geodesic Riemannian foliations of the same codimension on Einstein spaces and assume that they have the same total scalar curvature, and the same total transversal scalar curvature. Then $L^\perp$ is integrable if and only if $L_0^\perp$ is integrable.*

Proof. We start from a relation between the different types of scalar curvatures:

(11.12) $$\tau^M = \tau^\nabla + \hat{\tau} - |A|^2$$

(see [Be, p. 244]), where $\hat{\tau}$ denotes the scalar curvature of the leaf through $x \in M$. Now, let $\{e_i,\ i = 1, \ldots, p\}$ be an orthonormal basis of L_x and $\{e_\alpha,\ \alpha = p+1, \ldots, m = p+q\}$ an orthonormal basis of $L_x^\perp$. Then we have

$$\hat{\tau} = \sum_{i \neq j} \hat{R}_{ijij} = \sum_{i \neq j} R^M_{ijij},$$

since the leaf through x is a totally geodesic submanifold. Hence

(11.13) $$\hat{\tau} = \sum_i \rho^M_{ii} - \sum_{i,\alpha} K^M(e_i, e_\alpha).$$

But

(11.14) $$\sum_{i,\alpha} K^M(e_i, e_\alpha) = |A|^2$$

[Be, p. 241] and so, for an Einstein space, we get from (11.13) and (11.14)

(11.15) $$\hat{\tau} = \frac{n-q}{n}\tau^M - |A|^2.$$

Finally, from (11.12) and (11.15) we obtain

$$\tau^\nabla = \frac{q}{m}\tau^M + 2|A|^2$$

and the desired result follows as in Theorem 11.7. □

11.16 COROLLARY. *Let $(M, g, \mathcal{F})$ and $(M_0, g_0, \mathcal{F}_0)$ be isospectral, totally geodesic Riemannian foliations on Einstein spaces. Then $L^\perp$ is integrable if and only if $L_0^\perp$ is integrable.*

It is possible to delete the "totally geodesic" condition if we restrict the Riemannian manifolds further. We have ([N-To-V, Theorem 7])

11.17 THEOREM. *Let $\mathcal{F}$ and $\mathcal{F}_0$ be isospectral Riemannian foliations on a Riemannian manifold (M, g) with bundle-like metric of constant curvature. Then $L^\perp$ is integrable if and only if $L_0^\perp$ is integrable.*

Proof. For $K^M = c$ and $\operatorname{codim} \mathcal{F} = q$, (11.8) leads to
$$\tau^\nabla = cq(q-1) + 3|A|^2.$$
Similarly,
$$\tau^{\nabla_0} = cq_0(q_0 - 1) + 3|A_0|^2.$$
The result follows then at once using Theorem 11.5. □

In fact, we have proved in [N-To-V, Theorem 8]

11.18 THEOREM. *Let $\mathcal{F}$ and $\mathcal{F}_0$ be Riemannian foliations on a Riemannian manifold (M, g) with bundle-like metric of constant curvature. Assume $\mathcal{F}$ and $\mathcal{F}_0$ have the same codimension and the same total (transversal) scalar curvature. Then $L^\perp$ is integrable if and only if $L_0^\perp$ is integrable.*

We shall apply the results we have so far to some of the Hopf fibrations. Let $\mathcal{H}_1 = (S^3, g_0, \mathcal{F}_0)$ and $\mathcal{H}_2 = (S^5, g_0', \mathcal{F}_0')$ be the Hopf fibrations $\mathcal{F}_0 \colon S^3 \to \mathbb{C}P^1$, $\mathcal{F}_0' \colon S^5 \to \mathbb{C}P^2$ on the Euclidean spheres (S^3, g_0), (S^5, g_0'). Then we have ([N-To-V, Theorem 9])

11.19 THEOREM. *Let $(M, g, \mathcal{F})$ and $\mathcal{H}_1$, respectively $(M', g', \mathcal{F}')$ and $\mathcal{H}_2$, be isospectral Riemannian fibrations. Then there exist isometries $\varphi \colon (M, g) \to (S^3, g_0)$ and $\psi \colon (M', g') \to (S^5, g_0')$ such that $\varphi(\mathcal{F}) = \mathcal{F}_0$ and $\psi(\mathcal{F}') = \mathcal{F}_0'$.*

Proof. Following a result of Tanno (M, g) and (S^3, g_0) as well as (M', g') and (S^5, g_0') are isometric. Further, the isospectrality implies that the fibrations have fiber dimension one. The result then follows from [G-G 1, Corollary 2.1]. □

Remark. It might be interesting to determine if all the classical and generalized Hopf fibrations [Be] are characterized by their spectra.

Transversal curvature invariants

Invariants of curvature tensors play a fundamental role in many aspects of Riemannian geometry. In particular, there exist several inequalities between some of the quadratic curvature invariants, and several particular Riemannian spaces may be characterized by special relations between these invariants. All this is related to the decomposition of the space of curvature tensors under the action of some particular groups and hence relates to representation theory. As can be seen from the formulas in Theorem 11.2, the curvature invariants of different order play a role in spectral theory since the $a_n(\Delta_g)$ and $b_n(J_\nabla)$ are expressed as integrals of linear combinations of these invariants.

We will now use the integrals of quadratic invariants in Theorem 11.5 to derive some results about the transversal geometry of a Riemannian foliation. Before starting on this we note that

(11.20) $$|\rho^\nabla|^2 \geq \frac{1}{q}(\tau^\nabla)^2,$$

with equality sign valid if and only if $(M, g, \mathcal{F})$ is an Einstein foliation and $q \geq 3$. For $q = 2$ we always have equality and then the foliation is Einstein if and only if τ^∇ is constant. Next,

$$(11.21) \qquad |R^\nabla|^2 \geq \frac{2}{q-1}|\rho^\nabla|^2 \text{ and } |R^\nabla|^2 \geq \frac{2}{q(q-1)}(\tau^\nabla)^2,$$

with equality sign valid if and only if we have a foliation of constant curvature and $q \geq 3$. For $q = 2$ we have again always the equalities. Finally, let C^∇ be the Weyl tensor associated to R^∇. Then

$$(11.22) \qquad |C^\nabla|^2 = |R^\nabla|^2 - \frac{4}{q-2}|\rho^\nabla|^2 + \frac{2}{(q-1)(q-2)}(\tau^\nabla)^2$$

for $q > 2$. For $q = 3$, $C^\nabla = 0$ and for $q > 3$, $C^\nabla = 0$ if and only if the foliation is (transversally) conformally flat. Note that for $q \geq 4$, $(M, g, \mathcal{F})$ has constant transversal sectional curvature if and only if it is conformally flat and Einstein.

With all these remarks out of the way we are ready to state our results. We start with ([N-To-V, Theorem 10])

11.23 THEOREM. *Let $(M, g, \mathcal{F})$ and $(M_0, g_0, \mathcal{F}_0)$ be isospectral Einstein foliations. Then $(M, g, \mathcal{F})$ has constant sectional curvature c if and only if $(M_0, g_0, \mathcal{F}_0)$ has constant sectional curvature c.*

Proof. First we have

$$|\rho^\nabla|^2 = \frac{1}{q}(\tau^\nabla)^2, \quad |\rho^{\nabla_0}|^2 = \frac{1}{q_0}(\tau^{\nabla_0})^2,$$

where τ^∇ and τ^{∇_0} are constant. Using Theorem 11.5 we get

$$q = q_0, \quad \tau^\nabla = \tau^{\nabla_0}$$

and hence

$$|\rho^\nabla|^2 = |\rho^{\nabla_0}|^2.$$

Since the total scalar curvatures are also equal, Theorem 11.5, (vii) implies

$$\int_M |R^\nabla|^2 \mu = \int_{M_0} |R^{\nabla_0}|^2 \mu_0,$$

and hence

$$\int_M \left[|R^\nabla|^2 - \frac{2}{q(q-1)}(\tau^\nabla)^2\right]\mu = \int_{M_0}\left[|R^{\nabla_0}|^2 - \frac{2}{q(q-1)}(\tau^{\nabla_0})^2\right]\mu_0.$$

Then the result follows using (11.21) and the fact that for constant curvature c we have

$$\tau^{\nabla_0} = cq(q-1). \qquad \square$$

11.24 THEOREM ([N-To-V, Theorem 11]). *Let $(M, g, \mathcal{F})$ and $(M_0, g_0, \mathcal{F}_0)$ be isospectral Riemannian foliations of codimension 2 and suppose (M, g) and (M_0, g_0) have constant scalar curvature. Then $(M, g, \mathcal{F})$ has constant sectional curvature c if and only if $(M_0, g_0, \mathcal{F}_0)$ has constant sectional curvature c.*

Proof. For $q = 2$ we have
$$|R^\nabla|^2 = 2|\rho^\nabla|^2 = (\tau^\nabla)^2 = 4(K^\nabla)^2, \quad \tau^\nabla = 2K^\nabla.$$
Now, suppose $(M_0, g_0, \mathcal{F}_0)$ has constant sectional curvature. Then, from Theorem 10.5, we get
$$\int_M (\tau^\nabla)^2 \mu = \int_{M_0} (\tau^{\nabla_0})^2 \mu_0,$$
and hence
$$\text{vol } M \int_M (\tau^\nabla)^2 \mu = \left(\int_{M_0} \tau^{\nabla_0} \mu_0 \right)^2 = \left(\int_M \tau^\nabla \mu \right)^2.$$
Similarly we get
$$\text{vol } M \int_M \tau^\nabla \tau^{\nabla_0} \mu = \left(\int_M \tau^\nabla \mu \right)^2,$$
$$\text{vol } M \int_M (\tau^{\nabla_0})^2 \mu = \left(\int_M \tau^\nabla \mu \right)^2.$$
Hence
$$\int_M \left(\tau^\nabla - \tau^{\nabla_0} \right)^2 \mu = 0,$$
which implies the required result. $\square$

11.25 THEOREM ([N-To-V, Theorem 12]). *Theorem 11 also holds for $q = 3$.*

Proof. Note that for $q = 3$, $(M, g, \mathcal{F})$ has constant (transversal) sectional curvature if and only if it is Einstein. Further, since $C_\nabla = 0$ for $q = 3$, (11.22) yields
$$(11.26) \qquad |R^\nabla|^2 = 4|\rho^\nabla|^2 - (\tau^\nabla)^2.$$

Assume $(M_0, g_0, \mathcal{F}_0)$ has constant sectional curvature. Then, from Theorem 11.5 and (11.26) we get
$$(11.27) \qquad \int_M \left[2|\rho^\nabla|^2 + (\tau^\nabla)^2 \right] \mu = \int_{M_0} \left[2|\rho^{\nabla_0}|^2 + (\tau^{\nabla_0})^2 \right] \mu_0.$$
Further, the Cauchy-Schwarz inequality yields
$$(11.28)$$
$$\text{vol } M \int_M (\tau^\nabla)^2 \mu \geq \left(\int_M \tau^\nabla \mu \right)^2 = \left(\int_{M_0} \tau^{\nabla_0} \mu_0 \right)^2 = \text{vol } M \int_{M_0} (\tau^{\nabla_0})^2 \mu_0.$$
So, (11.27) and (11.28) give
$$\int_M \left[|\rho^\nabla|^2 - \frac{1}{3}(\tau^\nabla)^2 \right] \mu \leq \int_{M_0} \left[|\rho^{\nabla_0}|^2 - \frac{1}{3}(\tau^{\nabla_0})^2 \right] \mu_0,$$
and with (11.20) we obtain
$$|\rho^\nabla|^2 = \frac{1}{3}(\tau^\nabla)^2,$$
which means that $(M, g, \mathcal{F})$ is an Einstein foliation. The rest follows from
$$\tau^\nabla = \tau^{\nabla_0} = 6c.$$
$\square$

11.29 THEOREM ([N-To-V, Theorem 13]). *Let $(M, g, \mathcal{F})$ and $(M_0, g_0, \mathcal{F}_0)$ be isospectral and conformally flat foliations of codimension ≥ 4 on manifolds with constant scalar curvature. Then $(M, g, \mathcal{F})$ has constant sectional curvature c if and only if $(M_0, g_0, \mathcal{F}_0)$ has constant sectional curvature c.*

Proof. From Theorem 11.5 we easily get

$$(11.30) \qquad \int_M \left[|R^\nabla|^2 - 6|\rho^\nabla|^2\right] \mu = \int_{M_0} \left[|R^{\nabla_0}|^2 - 6|\rho^{\nabla_0}|^2\right] \mu_0$$

and since $C_\nabla = C_{\nabla_0} = 0$, we use (11.22) to write (11.30) in the form

$$(11.31) \quad \frac{2(3q-8)}{q-2} \int_M \left[|\rho^\nabla|^2 - \frac{1}{q}(\tau^\nabla)^2\right] \mu + \frac{2(3q-4)}{q(q-1)} \int_M (\tau^\nabla)^2 \mu =$$
$$\frac{2(3q-8)}{q-2} \int_{M_0} \left[|\rho^{\nabla_0}|^2 - \frac{1}{q}(\tau^{\nabla_0})^2\right] \mu_0 + \frac{2(3q-4)}{q(q-1)} \int_{M_0} (\tau^{\nabla_0})^2 \mu_0.$$

Now, assume again that $(M_0, g_0, \mathcal{F}_0)$ has constant sectional curvature c. Then (11.31) yields
$$(11.32)$$
$$\frac{2(3q-8)}{q-2} \int_M \left[|\rho^\nabla|^2 - \frac{1}{q}(\tau^\nabla)^2\right] \mu + \frac{2(3q-4)}{q(q-1)} \int_M (\tau^\nabla)^2 \mu = \frac{2(3q-4)}{q(q-1)} \int_{M_0} (\tau^{\nabla_0})^2 \mu_0.$$

We use the Cauchy-Schwarz inequality to get
$$(11.33)$$
$$\operatorname{vol} M \int_M (\tau^\nabla)^2 \mu \geq \left(\int_M \tau^\nabla \mu\right)^2 = \left(\int_{M_0} \tau^{\nabla_0} \mu_0\right)^2 = \operatorname{vol} M \int_{M_0} (\tau^{\nabla_0})^2 \mu_0.$$

So, from (11.32), (11.33) and $q \geq 4$, we get

$$\int_M \left[|\rho^\nabla|^2 - \frac{1}{q}(\tau^\nabla)^2\right] \mu \leq 0$$

which, with (11.20), implies
$$|\rho^\nabla|^2 = \frac{1}{q}(\tau^\nabla)^2.$$

So $(M, g, \mathcal{F})$ is an Einstein foliation and since it is also conformally flat, it has constant sectional curvature. This constant equals c since

$$\tau^\nabla = \tau^{\nabla_0} = q(q-1)c. \qquad \square$$

Remark. The classical Hopf fibrations

$$S^1 \to S^3 \to S^2, \ S^3 \to S^7 \to S^4, \ S^7 \to S^{15} \to S^8$$

provide examples of Riemannian manifolds with constant (transversal) sectional curvature.

Contributions to the spectral geometry of the basic Laplacian have been made by Craioveanu and Puta in [Cr-Pt 5][Cr-Pt 6], and more recently by Park and Richardson in [Par-R]. The starting point is to prove the existence and uniqueness of the heat kernel, as was done in Chapter 7 for the case of basic mean curvature. The next step is to establish asymptotic expansion formulas for the trace of the heat operator, and find explicit formulas for the coefficients in the asymptotic expansion. Ken Richardson showed the author an example of such an impressive explicit computation for the case of the basic Laplacian on basic functions. Again the mean curvature form distorts the coefficient formulas one is familiar with for the usual Laplacian.

Chapter 12
Foliations as Noncommutative Spaces

We begin this chapter with the description of the concept of the graph of a foliation, and then describe Connes' view of foliations in the context of noncommutative spaces.

The graph of a foliation
This is the space G consisting of triples $(x, y, [\alpha])$, where x and y lie on the same leaf $\mathcal{L}$ of $\mathcal{F}$, α is a path from x to y in $\mathcal{L}$, and $[\alpha]$ its homotopy class. The space G comes with a diagonal embedding $\Delta : M \to G$ and submersions $p_1, p_2 : G \to M$ projecting on the first and second factor. The fiber $p_1^{-1}(x)$, for $x \in M$, is the holonomy covering $\tilde{\mathcal{L}}_x$ of the leaf $\mathcal{L}_x$ through x. In this sense this construction is a simultaneous unwinding of all leaves with respect to their holonomy. For many foliations this space is itself a $(n + p)$-dimensional manifold. This construction is due to Ehresmann, Reeb, Thom and Winkelnkemper, and can be useful for many purposes. It is a central concept in the study of noncommutative spaces as proposed by Connes since 1978.

Foliations as noncommutative spaces
An ordinary space is completely described by the algebra of complex valued functions on it, continuous functions for a topological space, and smooth functions for a smooth manifold. These are commutative C^*-algebras. A theorem of Gelfand implies that the category of commutative C^*-algebras and $*$-homomorphisms is dual to the category of locally compact spaces and proper continuous maps. Connes has proposed to consider noncommutative C^*-algebras as the proper equivalent for noncommutative spaces. One reason for this generalization of the space concept are singular spaces like the leaf space of a foliation. The quotient topology on this space is a poor reflection of its complexity, while the proper generalization of its function algebra is much more interesting. The question is, functions on which space should be considered? Connes' answer is to consider the graph of $\mathcal{F}$.

The strategy of noncommutative geometry is first to reformulate as much as possible of differential geometry in terms of its structure algebra, and then secondly to generalize these results to the case of noncommutative C^*-algebras. When passing from the commutative to the nonncommutative case, the concept of a point is lost. It has been said facetiously that noncommutative geometry is pointless geometry. Important examples are the quantized phase spaces of nonrelativistic quantum mechanics. This arises from the fact that in quantum mechanics and quantum field theory as well as classical and quantum statistical physics the primary focus is on the algebra of observables, while the state vector itself is a secondary derived object.

Before elaborating on Connes' point of view in the case of foliations, let us return briefly to Haefliger's cocycle definition of a foliation $\mathcal{F}$ explained in Chapter 1. There is a natural way to associate a semi-simplicial complex to the Haefliger cocycle. It generalizes the usual construction of the nerve $N(\mathfrak{U})$ associated to an open covering $\mathfrak{U}$ of a manifold M. It is a well-known fact that the geometric realization $|N(\mathfrak{U})|$ of the nerve of $\mathfrak{U}$ is homotopy equivalent to M. In the same fashion the homotopy properties of the groupoid of the foliation are embodied in the geometric realization of the simplicial complex naturally associated to the holonomy groupoid of $\mathcal{F}$. This geometric realization can be thought of as the classifying space of the holonomy groupoid.

The C^*-algebra $C^*(\mathcal{F})$ of a foliation $\mathcal{F}$ on M considered by Connes is a completion of the C^*-algebra $C^*(G)$ of $\mathbb{C}$-valued functions on the graph G of $\mathcal{F}$ with a convolution product, which turns it into a noncommutative C^*-algebra. The composition in $C^*(G)$ reflects the composition of the holonomy transformations of $\mathcal{F}$ for leafwise paths. This C^*-algebra plays a key role in the analysis and geometry of the foliation $\mathcal{F}$ by Connes and his co-workers. As explained above, it is useful to think of it as a noncommutative replacement for the algebra of functions on the leaf space of $\mathcal{F}$. Each property of the foliation $\mathcal{F}$ should on principle be characterized by a property of the C^*-algebra $C^*(\mathcal{F})$. As pointed out by Connes, this approach fits with Grothendieck's point of view that a space should be given by the sheaf of germs of functions defined on it.

This point of view might also be profitably applied to the space of gauge equivalence classes of basic connections $\mathcal{A}_B/\mathcal{G}_\mathcal{F}$ discussed briefly in Chapter 3. It comes equipped with a natural set of structural functions, the Wilson loop functions, whose values at a given basic connection are obtained by calculating the trace of the corresponding holonomy along a loop. Considering the C^*-algebra generated by these as the basic algebra, this might lead to a study of $\mathcal{A}_B/\mathcal{G}_\mathcal{F}$ from Connes' point of view (see work of Ashtekar et al).

Index theory

The Atiyah-Singer theorem provides an explicit geometrico-topological evaluation formula for the analytically defined index of an elliptic operator. Already Atiyah generalized this to operators which are only elliptic in the directions transverse to the orbits of a compact group action. The operator itself is assumed to be invariant under the given group action. Connes and Skandalis [Con-Sk] considered operators which are elliptic along the leaves of a foliation. Both resulting index theorems are best understood when the index is viewed as a K-homology class in the context of Connes' theory. The main point is that the homotopy invariance of the analytic index is a built-in property of the K-theory of C^*-algebras, which is the natural range for the index function. From this point of view, the K-theory of $C^*(\mathcal{F})$ plays the role of the analytical K-theory of $\mathcal{F}$. To the groupoid $G = G(\mathcal{F})$ is on the other hand associated a classifying space BG (namely the geometric realization of the semi-simplicial complex associated to G), and thus a K-theory $K_*(BG)$, playing the role of a topological K-theory for $\mathcal{F}$. The conjecture of Baum-Connes affirms the isomorphism of the resulting two constructions, if the holonomy group is torsion free. The proof of this conjecture for special Riemannian foliations is the subject of Hector's paper [Hc 2]. The special assumption allowing a proof is the triviality of the holonomy, or the assumption that the foliation be almost without holonomy.

For an index theory one needs to evaluate the index in terms of cohomology classes. For this purpose Connes has defined a cohomology theory, cyclic cohomology, whose cocycles play for a noncommutative algebra the role that De Rham currents play for smooth manifolds. Cyclic cohomology is easier to compute than the K-groups, which explains its interest. For current work on these ideas in the Riemannian context we refer to [D-G-K-Y], [G-K 1] to [G-K 4], and [K].

Lefschetz theorem

The Atiyah-Bott-Lefschetz formula for elliptic complexes on a closed manifold has been generalized by Heitsch and Lazarov to Dirac complexes defined along the leaves of a transversally oriented foliation $\mathcal{F}$ of a closed oriented manifold M. A smooth diffeomorphism $f : M \to M$ fixing each leaf, and covered by an endomorphism of the given complex of Dirac operators leads to a global Lefschetz number $L(f)$ via the usual alternating trace construction. The problem is the local evaluation of this global invariant. We refer to the work of Heitsch and Lazarov in [Hi-L 1-3, cf. Ch. 6, p. 15] for several interesting applications of their general evaluation formula.

Chapter 13
Infinite-dimensional Riemannian Foliations

There are many infinite-dimensional contexts, where the geometric concepts developed for the study of Riemannian foliations can be applied meaningfully. A good example is provided by gauge theory, where the space of connections on a bundle P is foliated by the orbits of the gauge group $\mathcal{G}$ of the bundle. The L^2-metric on the space $\mathcal{A}$ of connections is invariant under the action of the gauge group $\mathcal{G}$. Thus $\mathcal{A} \to \mathcal{A}/\mathcal{G}$ has many aspects of a Riemannian foliation.

For a submanifold of a Riemannian manifold the first fundamental form, and the shape operator associated to the second fundamental form give rise to the mean curvature. As explained earlier, it represents up to first order the variation of the volume of the submanifold for variations in direction of the mean curvature vector field. It is of interest to observe the natural occurrence of this situation in several infinite-dimensional contexts. The vanishing of the mean curvature is a priori a criterion for the presence of a minimal submanifold. The difficulty arising in an infinite-dimensional context is that the mean curvature, defined as usual by the trace of the symmetric operator, may not be finite. Further, the volume of the submanifold may not be defined. The idea is then to make sense of these concepts by using zeta function regularizations in the sense of Ray-Singer [Adv. in Math. 7 (1971), 145–210]. This leads in some cases to the proof of existence of minimal submanifolds. We describe first a few possible such infinite-dimensional situations, and then sketch the explicit results obtained in the case of gauge orbits.

Some infinite-dimensional situations of interest
(a) $\mathfrak{A}$ is the space of gauge fields on a principal bundle $G \to P \to M$. The orbits under the action of the gauge groups are the submanifolds considered in [M-Ro-To 1]. In the case of the trivial G-bundle over the interval $[0,1]$, the minimal orbits are identified as isoparametric submanifolds in [C. King and S. L. Terng, The Palais Festival Volume, Publish or Perish, 1993, 253–281].

(b) $\mathfrak{M}$ is the space of Riemannian metrics on a closed smooth manifold M^n. The diffeomorphisms $\mathfrak{D} = \text{Diff}(M^n)$ act on $\mathfrak{M}$. This situation is discussed e.g. in [To 1], [M-Ro-To 3].

(c) Let (M, μ) be a closed smooth manifold with a volume form μ. Let $\mathfrak{A}$ be the space of all probability measures on (M, μ) which are completely continuous with respect to μ. The Fisher metric on $\mathfrak{A}$ is preserved by the natural action of the volume preserving diffeomorphisms of (M, μ). The geometry of this context has been described in [T. Friedrich, Math. Nachr. 153 (1991), 273–296]. As pointed out by Friedrich, this is also the framework for the discussions of Gromov's minimal volumes in [G. Besson, C. Courtois and S. Gallot, Invent. Math. 103 (1991), 417–455].

Gauge orbits and traces

We turn now to a more detailed discussion of the connection orbits under the action of the gauge group. Consider a principal G-bundle $P \to M$ over a closed Riemannian manifold M, with G a compact Lie group. We look for minimal gauge orbits within the space $\mathcal{A}$ of connections, endowed with its L^2-metric. Here we think of a minimal (infinite-dimensional) gauge orbit as being extremal with respect to variations of the regularized volume element as defined by the Faddeev-Popov ghost determinant.

The tangent space $T_A \mathcal{A}$ of $\mathcal{A}$ at a connection $A \in \mathcal{A}$ can be identified with $\Omega^1(M, \operatorname{Ad} P)$, where $\operatorname{Ad} P \cong P \times_{\operatorname{Ad}} \mathfrak{g}$. This is seen by observing that the difference of two connections is an element of $\Omega^1(M, \operatorname{Ad} P)$. Thus $\mathcal{A}$ is an affine space. Let $\mathcal{G}$ be the gauge group of P. Its Lie algebra $L(\mathcal{G})$ can be identified with $\Omega^0(M, \operatorname{Ad} P)$. The covariant derivative

$$d_A : \Omega^0(M, \operatorname{Ad} P) \to \Omega^1(M, \operatorname{Ad} P)$$

allows to identify the tangent space $T_A \mathcal{O}_A$ to the $\mathcal{G}$-orbit $\mathcal{O}_A$ of A with $\operatorname{im} d_A$.

The scalar product on $T_A \mathcal{A}$ is defined by

$$\langle \eta, \eta' \rangle_A = \int (\eta, \eta') \mu_g$$

for $\eta, \eta' \in T_A \mathcal{A}$, which involves the metric g on M and its volume form μ_g, as well as a biinvariant metric on G. The resulting L^2-metric on $\mathcal{A}$ is $\mathcal{G}$-invariant. Let

$$d_A^* : \Omega^1(M, \operatorname{Ad} P) \to \Omega^0(M, \operatorname{Ad} P)$$

denote the formal adjoint of d_A. Then

$$T_A \mathcal{A} \cong \operatorname{im} d_A \oplus \ker d_A^*$$

gives the orthogonal decomposition into tangent space $T_A \mathcal{O}_A$ and normal space $N_A \mathcal{O}_A$ to the $\mathcal{G}$-orbit of A.

For a vector field X on $\mathcal{A}$ this leads to the decomposition formulas

$$\begin{aligned} X(A) &= X^T(A) + X^N(A), \\ X^T(A) &= d_A G_A d_A^* X(A), \\ X^N(A) &= X(A) - d_A G_A d_A^* X(A) \end{aligned}$$

where G_A is Green's operator for the Laplacian $\Delta_A = d_A^* d_A$ acting on $\Omega^o(M, \mathrm{Ad}P)$. With the natural connection D on $\mathcal{A}$ one can now define the shape operator of an orbit $\mathcal{O}_A$ by

$$W_N(X) = -(D_X \tilde{N})^T$$

for $X \in T_A \mathcal{O}_A$, where $\tilde{N}$ is the normal vector field on $\mathcal{O}_A$ associated to $N \in N_A \mathcal{O}_A$. The shape operator is conjugate to the operator $G_A(\delta_N d_A^*) d_A : \Omega^0(M, \mathrm{Ad}\, P) \to \Omega^0(M, \mathrm{Ad}\, P)$, which we also denote by W_N. Here as below, δ_N denotes variation in the direction of the normal vector N. This operator is an integral operator

$$W_N(\Phi)(x) = \int_M w_N(x, y) \Phi(y)\, dy.$$

However $\mathrm{Tr}\, W_N$ cannot be defined as $\int_M w_N(x, x)\, dx$, as a computation of the asymptotic expansion of w_N near the diagonal in $M \times M$ shows that $w_N(x,x)$ is infinite in general. So we use the following regularization

$$\mathrm{Tr}\, W_N = -\int_0^\infty t^s \mathrm{Tr}\left(e^{-t\Delta_A} (\delta_N d_A^*)\, d_A\right)\, dt\Big|_{s=0}.$$

If d_A were a map with trivial kernel between finite-dimensional spaces, this formula would be an identity by directly setting $s = 0$, and using the formula

$$G_A = \int_0^\infty e^{-t\Delta_A}\, dt.$$

The problem with the proposed definition of $\mathrm{Tr}\, W_N$ is that $s = 0$ is not clearly a regular value of the meromorphic function on $\mathbb{C}$ defined by the integral formula. This could be corrected by subtracting the infinite part. But in any case the regularization procedure succeeds in the following cases.

13.1 THEOREM [M-Ro-To 1]. *Let $A \in \mathcal{A}$ be an irreducible connection. Then $\mathrm{Tr}\, W_N < \infty$ in the following cases for $n = \dim M$:*

(i) *$n = 2$;*
(ii) *n odd;*
(iii) *for all n within the space of flat connections.*

The zeta function of A

For a connection A we have as explained above a nonnegative Laplacian $\Delta_A = d_A^* d_A$ with eigenvalues $0 \leq \lambda_1 \leq \cdots \leq \lambda_i \leq \cdots$, and thus a zeta function $\zeta_A(s) = \sum_{\lambda_i > 0} \lambda_i^{-s}$. This gives rise to $\det \Delta_A$ via the identity

$$\zeta_A'(0) = -\log \det \Delta_A$$

in the usual fashion [Ray-Singer, loc. cit.].

This allows to define a formal volume form for the gauge orbit $\mathcal{O}_A$, or a ghost volume in the parlance of Faddeev and Popov, by the following procedure.

For an orthonormal set of eigenvectors $\Phi_1, \ldots, \Phi_i, \ldots$ of Δ_A corresponding to the eigenvalues $\lambda_1, \ldots, \lambda_i, \ldots$, the "volume form of the orbit $\mathcal{O}_A$" is given in the form

$$d_A\Phi_1 \wedge \cdots \wedge d_A\Phi_i \wedge \ldots$$
$$= \sqrt{\det(d_A^* d_A)}\Phi_1 \wedge \cdots \wedge \Phi_i \wedge \ldots$$
$$= (\det \Delta_A)^{\frac{1}{2}} \Phi_1 \wedge \cdots \wedge \Phi_i \wedge \ldots$$

where $\det \Delta_A$ is given via ζ_A.

Extremal gauge orbits

The main result of these considerations in [M-Ro-To 1] is as follows.

13.2 THEOREM. *Let $A \in \mathcal{A}$ be an irreducible connection. Then the following conditions are equivalent:*

(i) *the regularized volume of the gauge orbit $\mathcal{O}_A$ is extremal in the sense that the variation $\delta_N \det \Delta_A = 0$ for N a normal vector to $\mathcal{O}_A$ at A;*
(ii) $\operatorname{Tr} W_N = 0$.

A gauge orbit $\mathcal{O}_A$ is minimal if one of these conditions holds for all N normal to $\mathcal{O}_A$ at A. The proof of this equivalence is based on a functional equation, relating the regularized $\operatorname{Tr} W_N$ with normal variations of the zeta function leading to the definition of the regularized volume of the gauge orbits.

EXAMPLE. As an application of this criterion, one obtains the following result.

13.3 THEOREM [M-Ro-To 1]. *Let Σ be a Seifert homology 3-sphere, with at least 4 exceptional orbits. Consider the space $\mathcal{A}$ of flat $SU(2)$-connections on the trivial $SU(2)$-bundle over Σ. Then $\mathcal{A}$ contains at least 2 minimal gauge orbits.*

The proof is based on the work of Fintushel and Stern in [Proc. London Math. Soc. 61 (1990), 109–137]. The moduli space of irreducible flat $SU(2)$-connections over such a Σ has at least one connected component which is a closed manifold of dimension $2m - 6$, where m is the number of exceptional orbits. On this component, the function $\det \Delta_A$ has at least two critical points. The critical points correspond by the result above to minimal gauge orbits.

It is intriguing to try to relate the number of minimal orbits to the Casson invariant $\lambda(\Sigma)$. According to Taubes [Casson's invariant and gauge theory, J. Diff. Geom. 31 (1990), 547–599], this number is one half the Euler characteristic $\chi(\Sigma)$ of the Floer homology of Σ. If one thinks of $\det \Delta_A$ as a Morse function on the moduli space, then the Morse inequalities lead to the following

13.4 CONJECTURE. *Let $\mathcal{A}$ be the space of flat $SU(2)$-connections on the trivial $SU(2)$-bundle over a Seifert homology 3-sphere Σ, with at least 4 exceptional orbits. Let M be the number of minimal gauge orbits in $\mathcal{A}$, and $\lambda(\Sigma)$ the Casson invariant of Σ. Then*

$$M \geqq 2\lambda(\Sigma).$$

A proof would require a better understanding of the second normal variations of $\det \Delta_A$.

A similar study of the space $\mathcal{M}$ of Riemannian metrics on a closed manifold M is carried out in [M-Ro-To 3]. The group of diffeomorphisms of M acts isometrically on $\mathcal{M}$ with its L^2-metric. Once again one can define minimal orbits for this action by a zeta function regularization. As a sample result, it is shown in [M-Ro-To 3] that odd-dimensional isotropy irreducible homogeneous spaces give rise to minimal orbits. Another result is the exhibition of a flat 2-torus giving rise to a stable minimal orbit.

References on Riemannian Foliations

[Alc] F. Alcalde Cuesta, Groupoïde d'homotopie d'un feuilletage riemannien et réalisation symplectique de certaines variétés de Poisson, Publicaciones del Departamento de Geometria y Topologia, Universidad de Santiago de Compostela, 57 (1982), 395–410.

[Al-Mo] R. Almeida and P. Molino, Flots riemanniens sur les 4-variétés compactes, Tôhoku Math. J. 38 (1986), 313–326.

[A1] J. A. Alvarez López, Sucesion espectral asociada a foliaciones riemannianas, Publicaciones del Departamento de Geometria y Topologia, Universidad de Santiago de Compostela, 72 (1987).

[A2] J. A. Alvarez López, A finiteness theorem for the spectral sequence of a Riemannian foliation, Illinois J. Math. 33 (1989), 79–92.

[A3] J. A. Alvarez López, Duality in the spectral sequence of Riemannian foliations, Amer. J. Math. 111 (1989), 905–926.

[A4] J. A. Alvarez López, On Riemannian foliations with minimal leaves, Ann. Inst. Fourier (Grenoble) 40 (1990), 163–176.

[A5] J. A. Alvarez López, A decomposition theorem for the spectral sequence of a foliation, Trans. Amer. Math. Soc. 329 (1992), 173–184.

[A6] J. A. Alvarez López, The basic component of the mean curvature of Riemannian foliations, Ann. Global Anal. Geom. 10 (1992), 179–194.

[A7] J. A. Alvarez López, Morse inequalities for pseudogroups of local isometries, J. Differential Geom. 37 (1993), 603–638.

[A8] J. A. Alvarez López, Modified Laplacians in foliated manifolds, preprint.

[A9] J. A. Alvarez López, An index formula for transversally elliptic operators with respect to pseudogroups of local isometries, preprint.

[A-Hc] J. A. Alvarez López and G. Hector, On the dimension of the leafwise reduced cohomology, preprint.

[A-Hu] J. A. Alvarez López and S. Hurder, Pure-point spectrum for foliation geometric operators, preprint.

[A-To1]   J. A. Alvarez López and Ph. Tondeur, The heat flow along the leaves of a Riemannian foliation, Geometric and topological invariants of elliptic operators (Brunswick, 1988), Contemp. Math. 105 (1990), 271–280.

[A-To2]   J. A. Alvarez López and Ph. Tondeur, Hodge decomposition along the leaves of a Riemannian foliation, J. Funct. Anal. 99 (1991), 443–458.

[Ao-Yo]   T. Aoki and S. Yorozu, L^2-transverse conformal and Killing fields on complete foliated Riemannian manifolds, Yokohama Math. J. 36 (1988), 27–41.

[As-G]   D. Asimov and H. Gluck, Morse-Smale fields of geodesics, Lecture Notes in Math. 819, Springer, 1980, 1–17.

[Bai-Wo]   P. Baird and J. C. Wood, The geometry of a pair of Riemannian foliations by geodesics and associated harmonic morphisms, Bull. Soc. Math. Belg. Sér. B 44 (1992), 115–139.

[B-K-O]   J. Barbosa, K. Kenmotsu and G. Oshikiri, Foliations by hypersurfaces with constant mean curvature, Math. Z. 207 (1991), 97–108.

[B]   R. Barre, De quelques aspects de la théorie des Q-variétés, Ann. Inst. Fourier (Grenoble) 23 (1973), 227–312.

[Bas-Wals]   A. Basmajian and G. Walschap, Metric flows in space forms of nonpositive curvature, preprint.

[Ba]   M. Bauer, Feuilletages presque réguliers, C. R. Acad. Sci. Paris 299 (1984), 387–390.

[B-E]   M. Berger and D. Ebin, Some decompositions of the space of symmetric tensors on a Riemannian manifold, J. Differential Geom. 3 (1969), 376–392.

[B-G-M]   M. Berger, P. Gauduchon and E. Mazet, Le spectre d'une variété riemannienne, Lecture Notes in Math. 194, Springer, 1971.

[Be]   A. Besse, Einstein manifolds, Ergeb. Math. Grenzgeb. 3. Folge 10, Springer-Verlag, Berlin, Heidelberg, New York, 1987.

[Be-Bo]   G. Besson and M. Bordoni, On the spectrum of Riemannian submersions with totally geodesic fibers, Atti Accad. Naz. Lincei Cl. Sci. Fis. Mat. Natur. Rend. (9) Mat. Appl. 1 (1990), 335–340.

[Bi]   R. Bishop, Clairaut submersions, Differential geometry (in honor of K. Yano), Kinokuniya, Tokyo, 1972, 21–31.

[Bi-ON]   R. Bishop and B. O'Neill, Manifolds of negative curvature, Trans. Amer. Math. Soc. 145 (1969), 1–49.

[Bl1]   R. A. Blumenthal, Transversely homogeneous foliations, Ann. Inst. Fourier (Grenoble) 29 (1979), 143–158.

[Bl2]   R. A. Blumenthal, The base-like cohomology of a class of transversely homogeneous foliations, Bull. Sci. Math. 104 (1980), 301–303.

[Bl3] R. A. Blumenthal, Riemannian homogeneous foliations without holonomy, Nagoya Math. J. 83 (1981), 197–201.

[Bl4] R. A. Blumenthal, Riemannian foliations with parallel curvature, Nagoya Math. J. 90 (1983), 145–153.

[Bl5] R. A. Blumenthal, Transverse curvature of foliated manifolds, Astérisque 116 (1984), 25–30.

[Bl-Hb1] R. A. Blumenthal and J. Hebda, De Rham decomposition theorems for foliated manifolds, Ann. Inst. Fourier (Grenoble) 33 (1983), 183–198.

[Bl-Hb2] R. A. Blumenthal and J. Hebda, Complementary distributions which preserve the leaf geometry and applications to totally geodesic foliations, Quart. J. Math. Oxford 35 (1984), 383–392.

[Bl-Hb3] R. A. Blumenthal and J. Hebda, An analogue of the holonomy bundle for a foliated manifold, preprint.

[Bor-Y] A. Borisenko and A. Yampolski, Riemannian geometry of bundles, Russian Math. Surveys 46 (1991), 55–106.

[Bo1] R. Bott, On a topological obstruction to integrability, Proc. Sympos. Pure Math., Amer. Math. Soc., 16 (1970), 127–131.

[Bo2] R. Bott, Lectures on characteristic classes and foliations, Lecture Notes in Math. 279, Springer, 1972, 1–94.

[Bou1] H. Boualem, Feuilletages riemanniens singuliers transversalement intégrables, C. R. Acad. Sci. Paris 314 (1992), 547–550.

[Bou2] H. Boualem, Théorème de décomposition d'un feuilletage riemannien transversalement intégrable, C. R. Acad. Sci. Paris 316 (1993), 59–62.

[Bou3] H. Boualem, Feuilletages riemanniens singuliers transversalement intégrables, Compositio Math. 95(1995), 101–125.

[Bou-Mo] H. Boualem and P. Molino, Modèles locaux saturés de feuilletages riemanniens singuliers, C. R. Acad. Sci. Paris 316 (1993), 913–916.

[Br] F. Brito, Une obstruction géométrique à l'existence de feuilletages de codimension un totalement géodésiques, J. Differential Geom. 16 (1981), 675–684.

[Br-Ln-R] F. Brito, R. Langevin and H. Rosenberg, Intégrales de courbure sur les variétés feuilletées, J. Differential Geom. 16 (1981), 19–50.

[Brs] R. Brooks, Some Riemannian and dynamical invariants of foliations, Differential Geometry, Proc. Maryland, 1981-82, Birkhäuser, Progr. Math. 32 (1983), 56–72.

[Bru-Gh] M. Brunella and E. Ghys, Umbilical foliations and transversely holomorphic flows, J. Differential Geom. 41 (1995), 1–19.

[B-K] J. Brüning and F. Kamber, On the spectrum and index of transversal Dirac operators associated to Riemannian foliations, preprint.

[Bu] K. Bugajska, Structure of a leaf of some codimension one Riemannian foliation, Ann. Inst. Fourier (Grenoble) 38 (1988), 169–174.

[C1] G. Cairns, Géométrie globale des feuilletages totalement géodésiques, C. R. Acad. Sci. Paris 297 (1983), 525–527.

[C2] G. Cairns, Feuilletages totalement géodésiques de dimension 1, 2 ou 3, C. R. Acad. Sci. Paris 298 (1984), 341–344.

[C3] G. Cairns, Une remarque sur la cohomologie basique d'un feuilletage riemannien, Séminaire de géométrie différentielle 1984-1985, Univ. Sci. Tech. Languedoc, Montpellier, 1985, 1–7.

[C4] G. Cairns, A general description of totally geodesic foliations, Tôhoku Math. J. 38 (1986), 37–55.

[C5] G. Cairns, Some properties of a cohomology group associated to a totally geodesic foliation, Math. Z. 192 (1986), 391–403.

[C6] G. Cairns, Feuilletages géodésibles sur les variétés simplement connexes, Séminaire Sud-Rhodanien de Géométrie VII, Vol. II, ed. N. Desolneux-Moulis, P. Dazord, Travaux en Cours, Hermann, Paris, 1987.

[C7] G. Cairns, Feuilletages géodésibles, Ph. D. Thesis, Université des Sciences et Techniques du Languedoc, Montpellier, 1987.

[C8] G. Cairns, Feuilletages totalement géodésiques sur les variétés simplement connexes, Feuilletages riemanniens, quantification géométrique et mécanique (Lyon, 1986), Travaux en Cours 26, Hermann, Paris, 1988, 1–14.

[C9] G. Cairns, The duality between Riemannian foliations and geodesible foliations, in P. Molino, Riemannian foliations, Birkhäuser, 1988, 249–263.

[C10] G. Cairns, Totally umbilic Riemannian foliations, Michigan Math. J. 37 (1990), 145–159.

[C11] G. Cairns, Compact 4-manifolds that admit totally umbilic metric foliations, Differential geometry and its applications (Brno, 1989), World Sci. Publishing, 1990, 9–16.

[C-Es1] G. Cairns and R. H. Escobales, Further geometry of the mean curvature one-form and the normal plane field one-form on a foliated Riemannian manifold, preprint.

[C-Es2] G. Cairns and R. H. Escobales, Bundle-like flows on curved manifolds, preprint.

[C-Gh] G. Cairns and E. Ghys, Totally geodesic foliations on 4-manifolds, J. Differential Geom. 23 (1986), 241–254.

[Cl] E. Calabi, An intrinsic characterization of harmonic 1-forms, Global Analysis, Papers in honor of K. Kodaira, ed. D. C. Spencer and S. Iyanaga, Princeton Math. Series 29 (1969), 101–117.

[C-Co] J. Cantwell and L. Conlon, The dynamics of open, foliated manifolds and a vanishing theorem for the Godbillon-Vey class, Adv. in Math. 53 (1984), 1–27.

[Cn-Ca] P. Caron and Y. Carrière, Flots transversalements de Lie $\mathbb{R}^n$, flots de Lie minimaux, C. R. Acad. Sci. Paris 280 (1980), 477–478.

[Ca1] Y. Carrière, Flots riemanniens et feuilletages géodésibles de codimension un, Ph. D. Thesis, Université des Sciences et Techniques de Lille I, Lille, 1981.

[Ca2] Y. Carrière, Flots riemanniens, Astérisque 116 (1984), 31–52.

[Ca3] Y. Carrière, Les propriétés topologiques des flots riemanniens retrouvées à l'aide du théorème des variétés presque plates, Math. Z. 186 (1984), 393–400.

[Ca4] Y. Carrière, Sur la croissance des feuilletages de Lie, Publ. IRMA, Lille 6 (1984).

[Ca5] Y. Carrière, Feuilletages riemanniens à croissance polynomiale, Comment. Math. Helv. 63 (1988), 1–20.

[Ca6] Y. Carrière, Variations on Riemannian flows, Appendix A in P. Molino, Riemannian foliations, Birkhäuser, 1988, 217–234.

[Ca-Gh1] Y. Carrière and E. Ghys, Feuilletages totalement géodésiques, Anais Acad. Bras. Ciencias 53 (1981), 427–432.

[Ca-Gh2] Y. Carrière and E. Ghys, Relations d'équivalences moyennables sur les groupes de Lie, C. R. Acad. Sci. Paris 300 (1985), 677–680.

[Ce] D. Cerveau, Distributions involutives singulières, Ann. Inst. Fourier (Grenoble) 29 (1979), 261–294.

[Ch] P. R. Chernoff, Essential self-adjointness of powers of generators of hyperbolic equations, J. Funct. Anal. 12 (1983), 401–404.

[C-K-Pk] K. Cho, J. Kwon and J. Pak, Transverse conformal mappings of complete foliated Riemannian manifolds with harmonic foliation, Math. J. Toyama Univ. 15 (1992), 43–58.

[C-Pk-S] K. Cho, J. Pak and W. Sohn, A note on spectral characterizations of Sasakian foliations, Kyungpook Math. J. 34 (1994), 283–291.

[Ci-Moz1] W. Cieslak and W. Mozgawa, Euclidean plane foliations, Ann. Univ. Mariae Curie - Sklodowska Sect. A 41 (1987), 1–7.

[Ci-Moz2] W. Cieslak and W. Mozgawa, Quelques remarques sur les flots riemanniens singuliers, An. Univ. Timisoara Ser. Stiint. Mat. 25 (1987), 15–20.

[Co] L. Conlon, Transversally parallelizable foliations of codimension 2, Trans. Amer. Math. Soc. 194 (1974), 79–102.

[Con-Sk] A. Connes and G. Skandalis, The longitudinal index theorem for foliations, Publ. RIMS, Kyoto Univ. 20 (1984), 1139–1183.

[Cor-Wo] L. Cordero and R. Wolak, Properties of the basic cohomology of transversely Kähler foliations, Rend. Circ. Mat. Palermo (2) 40 (1991), 177–188.

[Cr-Pt1] M. Craioveanu and M. Puta, DeRham-type currents on Riemannian foliated manifolds, Colloq. Math. Soc. János Bolyai, Budapest (1979), 159–165.

[Cr-Pt2] M. Craioveanu and M. Puta, Cohomology on a Riemannian manifold with coefficients in the sheaf of germs of foliated currents, Math. Nachr. 99 (1980), 43–53.

[Cr-Pt3] M. Craioveanu and M. Puta, Some remarks concerning the basic Laplacian, An. Univ. Timisoara Ser. Stiint. Mat. 25 (1987), 3–13.

[Cr-Pt4] M. Craioveanu and M. Puta, Cohomology classes and foliated manifolds, Nonlinear analysis, World Sci. Publishing, Singapore, 1987, 137–159.

[Cr-Pt5] M. Craioveanu and M. Puta, On the basic Laplacian of a Riemannian foliation, The XVIIIth National Conference on Geometry and Topology (Oradea, 1987), 49–52.

[Cr-Pt6] M. Craioveanu and M. Puta, Asymptotic properties of eigenvalues of the basic Laplacian associated to certain Riemannian foliations, Bull. Math. Soc. Sci. Math. Roumanie (N.S.) 35 (1991), 61–65.

[D-Pk] S. Dal Jung and J. Pak, A transversal Dirac operator and some vanishing theorems on a complete foliated Riemannian manifold, Math. J. Toyama Univ. 16 (1993), 97–108.

[Da] P. Dazord, Feuilletages à singularités, Indag. Math. (N.S.) 47 (1985), 21–39.

[Di] C. Diop, Sur les feuilletages singuliers presque isométriques, Ph. D. Thesis, Dakar, 1993.

[Do1] D. Domínguez, A tenseness theorem for Riemannian foliations, C. R. Acad. Sci. Paris 320 (1995), 1331–1335.

[Do2] D. Domínguez, Finiteness and tenseness theorems for Riemannian foliations, preprint.

[D-G-K-Y] R. Douglas, J. Glazebrook, F. Kamber and G. Yu, Index formulas for geometric Dirac operators in Riemannian foliations, K-Theory 9 (1995), 407–441.

[Du] T. Duchamp, Characteristic invariants of G-foliations, Ph. D. Thesis, University of Illinois at Urbana-Champaign, Urbana, 1976.

[D-K1] J. Dupont and F. Kamber, On a generalization of Cheeger-Chern-Simons classes, Illinois J. Math. 34 (1990), 221–255.

[D-K2] J. Dupont and F. Kamber, Cheeger-Chern-Simons classes of transversally symmetric foliations: dependence relations and η-invariant, Math. Ann. 295 (1993), 449–468.

[E-Rb]   C. Ehresmann and G. Reeb, Sur les champs d'éléments de contact de dimension p complètement intégrables dans une variété continuement différentiable V_n, C. R. Acad. Sci. Paris 216 (1944), 628–630.

[EK1]   A. El Kacimi, Equation de la chaleur sur les espaces singuliers, C. R. Acad. Sci. Paris 303 (1986), 243–246.

[EK2]   A. El Kacimi, Dualité pour les feuilletages transversalement holomorphes, Manuscripta Math. 58 (1987), 417–433.

[EK3]   A. El Kacimi, Stabilité des V-variétés kählériennes, Holomorphic dynamics (Mexico City, 1986), Lecture Notes in Math. 1345, Springer, 1988, 111–123.

[EK4]   A. El Kacimi, Opérateurs transversalement elliptiques sur un feuilletage riemannien et applications, Compositio Math. 73 (1990), 57–106.

[EK5]   A. El Kacimi, Examples of foliations and problems in transverse complex analysis, Functional analytic methods in complex analysis and applications to partial differential equations (Trieste, 1988), World Sci. Publishing, 1990, 341–364.

[EK-Hc1]   A. El Kacimi and G. Hector, Décomposition de Hodge sur l'espace des feuilles d'un feuilletage riemannien, C. R. Acad. Sci. Paris 298 (1984), 289–292.

[EK-Hc2]   A. El Kacimi and G. Hector, Décomposition de Hodge basique pour un feuilletage riemannien, Ann. Inst. Fourier (Grenoble) 36 (1986), 207–227.

[EK-Hc-Se]   A. El Kacimi, G. Hector and V. Sergiescu, La cohomologie basique d'un feuilletage riemannien est de dimension finie, Math. Z. 188 (1985), 593–599.

[EK-N1]   A. El Kacimi and M. Nicolau, Structures géométriques invariantes et feuilletages de Lie, Indag. Math. (N.S.) 1 (1990), 323–334.

[EK-N2]   A. El Kacimi and M. Nicolau, On the topologial invariance of the basic cohomology, Math. Ann. 295 (1993), 627–634.

[EK-T]   A. El Kacimi and A. Tihami, Cohomologie bigraduée de certains feuilletages, Bull. Soc. Math. Belg. Sér. B 38 (1986), 144–156; Errata, ibid. 39 (1987), 379.

[Ep1]   D. B. A. Epstein, Periodic flows on three-dimensional manifolds, Ann. of Math. 95 (1972), 66–82.

[Ep2]   D. B. A. Epstein, Foliations with all leaves compact, Ann. Inst. Fourier (Grenoble) 26 (1976), 265–282.

[Ep3]   D. B. A. Epstein, Transversally hyperbolic 1-dimensional foliations, Astérisque 116 (1984), 53–69.

[Es1]   R. H. Escobales, Riemannian submersions from complex projective space, J. Differential Geom. 13 (1978), 93–107.

[Es2] R. H. Escobales, Sufficient conditions for a bundle-like foliation to admit a Riemannian submersion onto its leaf space, Proc. Amer. Math. Soc. 84 (1982), 280–284.

[Es3] R. H. Escobales, The integrability tensor for bundle-like foliations, Trans. Amer. Math. Soc. 270 (1982), 333–339.

[Es4] R. H. Escobales, The mean curvature cohomology class for foliations and the infinitesimal geometry of the leaves, Differential Geom. Appl. 2 (1992), 167–178.

[Es-P] R. H. Escobales and Ph. Parker, Geometric consequences of the normal curvature cohomology class in umbilic foliations, Indiana Univ. Math. J. 37 (1988), 389–408.

[Fe1] E. Fedida, Sur les feuilletages de Lie, C. R. Acad. Sci. Paris 272 (1971), 999–1002.

[Fe2] E. Fedida, Feuilletages du plan, Feuilletages de Lie, Ph. D. Thesis, Strasbourg, 1983.

[Fo1] R. Forman, Adiabatic limits, small eigenvalues and spectral sequences, XXth Int. Conference on Differential Geometric Methods in Theoretical Physics (New York, 1991), World Sci. Publishing, 1992, 306–315.

[Fo2] R. Forman, Hodge theory and spectral sequences, Topology 33 (1994), 591–611.

[Fo3] R. Forman, Spectral sequences and adiabatic limits, Comm. Math. Phys. 168 (1995), 57–116.

[G-G-Hc-Rev] E. Gallego, L. Gualandri, G. Hector and A. Reventós, Groupoïdes riemanniens, Publ. Mat. 33 (1989), 417–422.

[G-Rev] E. Gallego and A. Reventós, Lie flows of codimension 3, Trans. Amer. Math. Soc. 326 (1991), 529–541.

[G-M] S. Gallot and D. Meyer, Opérateurs de courbure et Laplacien des formes différentielles d'une variété riemannienne, J. Math. Pures Appl. 54 (1975), 259–284.

[Ga1] H. Gauchman, An integral inequality for normal contact Riemannian manifolds and its applications, Geom. Dedicata 23 (1987), 53–58.

[Ga2] H. Gauchman, Basic cohomology classes of compact Sasakian manifolds, Acta Sci. Math. (Szeged) 56 (1992), 269–285.

[Gh1] E. Ghys, Classification des feuilletages totalement géodésiques de codimension un, Comment. Math. Helv. 58 (1983), 543–572.

[Gh2] E. Ghys, Feuilletages riemanniens sur les variétés simplement connexes, Ann. Inst. Fourier (Grenoble) 34 (1984), 203–223.

[Gh3] E. Ghys, Flots d'Anosov sur les 3-variétés fibrées en cercles, Ergodic Theory Dynamical Systems 4 (1984), 67–80.

[Gh4] E. Ghys, Groupes d'holonomie des feuilletages de Lie, Indag. Math. (N.S.) 47 (1985), 173–182.

[Gh5] E. Ghys, Un feuilletage analytique dont la cohomologie basique est de dimension infinie, Publ. IRMA, Lille 7 (1985).

[Gh6] E. Ghys, Flots d'Anosov dont les feuilletages stables sont différentiables, Ann. Sci. École Norm. Sup. (4) 20 (1987), 251–270.

[Gh7] E. Ghys, Riemannian foliations: examples and problems, in P. Molino, Riemannian foliations, Birkhäuser, 1988, 297–314.

[Gh-Se] E. Ghys and V. Sergiescu, Stabilité et conjugaison différentiable pour certains feuilletages, Topology 19 (1980), 179–197.

[Gi-Pr1] P. Gilkey and J. H. Park, Eigenvalues of the Laplacian and Riemannian submersions, Yokohama Math. J. 43 (1995), 7–11.

[Gi-Pr2] P. Gilkey and J. H. Park, Riemannian submersions which preserve the eigenforms of the Laplacian, Illinois J. Math., to appear.

[G-Hu-K] J. Glazebrook, S. Hurder and F. Kamber, Higher transverse index theory for Riemannian foliations, preprint.

[G-K1] J. Glazebrook and F. Kamber, Transversal Dirac families in Riemannian foliations, Comm. Math. Phys. 140 (1991), 217–240.

[G-K2] J. Glazebrook and F. Kamber, On spectral flow of transversal Dirac operators and a theorem of Vava-Witten, Ann. Global Anal. Geom. 9 (1991), 27–35.

[G-K3] J. Glazebrook and F. Kamber, Determinant line bundles for Hermitian foliations and a generalized Quillen metric, Proc. Sympos. Pure Math., Amer. Math. Soc., 52 (1991), Part 2, 225–232.

[G-K4] J. Glazebrook and F. Kamber, Chiral anomalies and Dirac families in Riemannian foliations, preprint.

[G-K-P-S] J. Glazebrook, F. Kamber, H. Pedersen and A. Swann, Foliation reduction and self-duality, Geometric Study of Foliations (Tokyo, 1993), World Sci. Publishing, 1994, 219–249.

[G-S] J. Glazebrook and D. Sundararaman, Deformation of foliated quaternionic structures and the twistor correspondence, preprint.

[Gl1] H. Gluck, Can space be filled by geodesics, and if so, how?, open letter, 1979.

[Gl2] H. Gluck, Dynamical behavior of geodesic fields, Lecture Notes in Math. 819, Springer, 1980, 190–215.

[Go] C. Godbillon, Feuilletages: Etudes géométriques I, II, Publ. IRMA Strasbourg (1985-86), Birkhäuser, Progr. Math. 98 (1991).

[Go-V] C. Godbillon and J. Vey, Un invariant des feuilletages de codimension un, C. R. Acad. Sci. Paris 273 (1971), 92–95.

[Gon-Gon1] J. González-Dávila and M. González-Dávila, Relations of curvatures in locally transversally Killing symmetric spaces, Geometry and Topology, Proc. 15th Port.-Span. Meet. Math. (Évora, 1990), 1991, Vol. III, 155–160.

[Gon-Gon2] J. González-Dávila and M. González-Dávila, The Gelfand Theorem in Killing transversally symmetric spaces, Geometry and Topology, Proc. 15th Port.-Span. Meet. Math. (Évora, 1990), 1991, Vol. III, 185–190.

[Gon-Gon-V1] J. González-Dávila, M. González-Dávila and L. Vanhecke, The Gelfand theorem in flow geometry, C. R. Math. Rep. Acad. Sci. Canada 15 (1993), 281–285.

[Gon-Gon-V2] J. González-Dávila, M. González-Dávila and L. Vanhecke, Reflections and isometric flows, Kyungpook Math. J. 35 (1995), 113–144.

[Gon-Gon-V3] J. González-Dávila, M. González-Dávila and L. Vanhecke, Classification of Killing-transversally symmetric spaces, Tsukuba J. Math., to appear.

[Gon-Gon-V4] J. González-Dávila, M. González-Dávila and L. Vanhecke, Normal flow space forms and their classification, Publ. Math. Debrecen 47 (1995), to appear.

[Gon-Gon-V5] J. González-Dávila, M. González-Dávila and L. Vanhecke, Invariant submanifolds in flow geometry, preprint.

[Gon-V1] J. González-Dávila and L. Vanhecke, Geodesic spheres and isometric flows, Colloq. Math. 67 (1994), 223–240.

[Gon-V2] J. González-Dávila and L. Vanhecke, Geometry of tubes and isometric flows, Math. J. Toyama Univ. 18 (1995), to appear.

[Gon-V3] J. González-Dávila and L. Vanhecke, New examples of weakly symmetric spaces, preprint.

[Gon1] M. González-Dávila, Espacios transversalmente simetricos de tipo Killing, Ph. D. Thesis, La Laguna (Tenerife), 1992.

[Gon2] M. González-Dávila, KTS-spaces and natural reductivity, Nihonkai Math. J. 5 (1994), 115–129.

[Got1] T. Gotoh, Harmonic foliations on a complex projective space, Tsukuba J. Math. 14 (1990), 99–106.

[Got2] T. Gotoh, A remark on foliations on a complex projective space with complex leaves, Tsukuba J. Math. 16 (1992), 169–172.

[Gr] A. Gray, Pseudo-Riemannian almost product manifolds and submersions, J. Math. Mech. 16 (1967), 715–737.

[Gr-V1] A. Gray and L. Vanhecke, Riemannian geometry as determined by the volumes of small geodesic balls, Acta Math. 142 (1979), 157–198.

[Gr-V2] A. Gray and L. Vanhecke, The volume of tubes about curves in a Riemannian manifold, Proc. London Math. Soc. 44 (1982), 215–243.

[G-G1] D. Gromoll and K. Grove, One-dimensional metric foliations in constant curvature spaces, Differential Geometry and Complex Analysis, H. E. Rauch memorial volume, ed. I. Chavel and H. M. Farkas, 1985, 165–168.

[G-G2] D. Gromoll and K. Grove, The low-dimensional metric foliations of Euclidian spheres, J. Differential Geom. 28 (1988), 143–156.

[Ha1] A. Haefliger, Structures feuilletées et cohomologie à valeur dans un faisceau de groupoïdes, Comment. Math. Helv. 32 (1958), 249–329.

[Ha2] A. Haefliger, Variétés feuilletées, Ann. Scuola Norm. Sup. Pisa 16 (1962), 249–329.

[Ha3] A. Haefliger, Sur les classes caractéristiques des feuilletages, Sém. Bourbaki 412–01 to 412–21 (1972), Lecture Notes in Math. 317, Springer, 1973.

[Ha4] A. Haefliger, Some remarks on foliations with minimal leaves, J. Differential Geom. 15 (1980), 269–284.

[Ha5] A. Haefliger, Groupoïdes d'holonomie et classifiants, Astérisque 116 (1984), 70–97.

[Ha6] A. Haefliger, Pseudogroups of local isometries, Res. Notes in Math. 131, Pitman, Boston, 1985, 174–197.

[Ha7] A. Haefliger, Leaf closures in Riemannian foliations, A fête of topology, Papers dedicated to I. Tamura, Academic Press, Boston, 1988, 3–32.

[Ha8] A. Haefliger, Feuilletages riemanniens, Séminaire Bourbaki, Vol. 1988/89, Astérisque 177–178 (1989), 183–197.

[Ha-S1] A. Haefliger and E. Salem, Pseudogroupes d'holonomie des feuilletages riemanniens sur des variétés compactes 1-connexes, Géométrie différentielle (Paris, 1986), Travaux en Cours 33, Hermann, Paris, 1988, 141–160.

[Ha-S2] A. Haefliger and E. Salem, Riemannian foliations on 1-connected manifolds and actions of tori on orbifolds, Illinois J. Math. 34 (1990), 706–730.

[Ha-S3] A. Haefliger and E. Salem, Actions of tori on orbifolds, Ann. Global Anal. Geom. 9 (1991), 37–59.

[Hb1] J. Hebda, Curvature and focal points in Riemannian foliations, Indiana Univ. Math. J. 35 (1986), 321–331.

[Hb2] J. Hebda, An example relevant to curvature pinching theorems for Riemannian foliations, Proc. Amer. Math. Soc. 114 (1992), 195–199.

[Hc1] G. Hector, Cohomologies transversales des feuilletages riemanniens I, Feuilletages riemanniens, quantification géometrique et mécanique (Lyon, 1986), Travaux en Cours 26, Hermann, Paris, 1988.

[Hc2] G. Hector, Groupoïdes, feuilletages et C^*-algèbres (quelques aspects de la conjecture de Baum-Connes), Geometric Study of Foliations (Tokyo, 1993), World Sci. Publishing, 1994, 3–34.

[Hc-M] G. Hector and E. Macias, Sur le théorème de DeRham pour les feuilletages de Lie, C. R. Acad. Sci. Paris 311 (1990), 633–636.

[Hc-Sara] G. Hector and M. Saralegi, Intersection cohomology of S^1-actions, preprint.

[Hi] J. L. Heitsch, A cohomology for foliated manifolds, Comment. Math. Helv. 50 (1975), 197–218.

[Hi-L1] J. L. Heitsch and C. Lazarov, Homotopy invariance of foliation Betti numbers, Invent. Math. 104 (1991), 321–347.

[Hi-L2] J. L. Heitsch and C. Lazarov, Rigidity theorems for foliations by surfaces and spin manifolds, Michigan Math. J. 38 (1991), 285–297.

[Hi-L3] J. L. Heitsch and C. Lazarov, Spectral asymptotics of foliated manifolds, Illinois J. Math. 38 (1994), 653–678.

[He1] R. Hermann, A sufficient condition that a map of Riemannian manifolds be a fiber bundle, Proc. Amer. Math. Soc. 11 (1960), 236–242.

[He2] R. Hermann, The differential geometry of foliations; I, Ann. of Math. 72 (1960), 445–457; II, J. Math. Mech. 11 (1962), 303–316.

[H-P-Te] W. Y. Hsiang, R. S. Palais and C. L. Terng, The topology of isoparametric submanifolds, J. Differential Geom. 27 (1988), 423–460.

[Hu1] S. Hurder, On the homotopy and cohomology of the classifying space of Riemannian foliations, Proc. Amer. Math. Soc. 81 (1981), 484–489.

[Hu2] S. Hurder, Vanishing of secondary classes for compact foliations, J. London Math. Soc. 28 (1983), 175–183.

[Hu3] S. Hurder, Global invariant for measured foliations, Trans. Amer. Math. Soc. 280 (1983), 367–391.

[Hu4] S. Hurder, The classifying space of smooth foliations, Illinois J. Math. 29 (1985), 108–133.

[Hu5] S. Hurder, Foliation dynamics and leaf invariants, Comm. Math. Helv. 60 (1985), 319–335.

[Hu6] S. Hurder, The Godbillon measure for amenable foliations, J. Differential Geom. 23 (1986), 347–365.

[Hu7] S. Hurder, Spectral theory for foliation geometric operators, preprint.

[Hu-K1] S. Hurder and A. Katok, Secondary classes and transverse measure theory of a foliation, Bull. Amer. Math. Soc. 11 (1984), 347–350.

[Hu-K2] S. Hurder and A. Katok, Ergodic theory and Weil measures for foliations, Ann. of Math. 126 (1987), 221–275.

[H-To] M. Hvidsten and Ph. Tondeur, A characterization of harmonic foliations by variations of the metric, Proc. Amer. Math. Soc. 98 (1986), 359–362.

[J] D. Johnson, Kaehler submersions and holomorphic connections, J. Differential Geom. 15 (1980), 71–79.

[J-N] D. Johnson and A. Naveira, A topological obstruction to the geodesibility of a foliation of odd dimension, Geom. Dedicata 11 (1981), 347–357.

[J-W1] D. Johnson and L. Whitt, Totally geodesic foliations on 3-manifolds, Proc. Amer. Math. Soc. 76 (1979), 355–357.

[J-W2] D. Johnson and L. Whitt, Totally geodesic foliations, J. Differential Geom. 15 (1980), 225–235.

[Ka] H. Kamada, Foliations on manifolds with positive constant curvature, Tokyo J. Math. 16 (1993), 49–60.

[K] F. Kamber, Transversal Index theory for Riemannian foliations, preprint.

[K-R-To1] F. Kamber, E. Ruh and Ph. Tondeur, Almost transversally symmetric foliations, Proc. of the II. Int. Symp. on differential geometry (Peniscola, 1985), Lecture Notes in Math. 1209, Springer, 1986, 184–189.

[K-R-To2] F. Kamber, E. Ruh and Ph. Tondeur, Comparing Riemannian foliations with transversally symmetric foliations, J. Differential Geom. 26 (1987), 461–475.

[K-To1] F. Kamber and Ph. Tondeur, Characteristic invariants of foliated bundles, Manuscripta Math. 11 (1974), 51–89.

[K-To2] F. Kamber and Ph. Tondeur, Foliated bundles and characteristic classes, Lecture Notes in Math. 494, Springer, 1975.

[K-To3] F. Kamber and Ph. Tondeur, G-foliations and their characteristic classes, Bull. Amer. Math. Soc. 84 (1978), 1086–1124.

[K-To4] F. Kamber and Ph. Tondeur, Feuilletages harmoniques, C. R. Acad. Sci. Paris 291 (1980), 409–411.

[K-To5] F. Kamber and Ph. Tondeur, Harmonic foliations, Proc. NSF Conference on Harmonic Maps (Tulane, 1980), Lecture Notes in Math. 949, Springer, 1982, 87–121.

[K-To6] F. Kamber and Ph. Tondeur, Infinitesimal automorphisms and second variation of the energy for harmonic foliations, Tôhoku Math. J. 34 (1982), 525–538.

[K-To7] F. Kamber and Ph. Tondeur, Dualité de Poincaré pour les feuilletages harmoniques, C. R. Acad. Sci. Paris 294 (1982), 357–359.

[K-To8] F. Kamber and Ph. Tondeur, Duality for Riemannian foliations, Proc. Sympos. Pure Math., Amer. Math. Soc., 40 (1983), Part 1, 609–618.

[K-To9] F. Kamber and Ph. Tondeur, The index of harmonic foliations on spheres, Trans. Amer. Math. Soc. 275 (1983), 257–263.

[K-To10] F. Kamber and Ph. Tondeur, Foliations and metrics, Proc. of a Year in Differential Geometry, University of Maryland, Birkhäuser, Progr. Math. 32 (1983), 103–152.

[K-To11] F. Kamber and Ph. Tondeur, Curvature properties of harmonic foliations, Illinois J. Math. 18 (1984), 458–471.

[K-To12] F. Kamber and Ph. Tondeur, Duality theorems for foliations, Astérisque 116 (1984), 108–116.

[K-To13] F. Kamber and Ph. Tondeur, The Bernstein problem for foliations, Proc. of the Conference on Global Differential Geometry and Global Analysis (Berlin, 1984), Lecture Notes in Math. 1156, Springer, 1985, 216–218.

[K-To14] F. Kamber and Ph. Tondeur, De Rham-Hodge theory for Riemannian foliations, Math. Ann. 277 (1987), 415–431.

[K-To15] F. Kamber and Ph. Tondeur, Foliations and harmonic forms, Harmonic mappings, twistors and σ-models, Adv. Ser. Math. Phys. 4, World Sci. Publishing, Singapore, 1988, 15–25.

[K-To-T] F. Kamber, Ph. Tondeur and G. Toth, Transversal Jacobi fields for harmonic foliations, Michigan Math. J. 34 (1987), 261–266.

[Kan-Kit-Pa] T. Kang, H. Kitahara and J. Pak, A formula for the radial part of the Laplace-Beltrami operator on the Riemannian foliation, Ann. Sci. Kanazawa Univ. 28 (1991), 1–8.

[Kea-Ke] M. S. Keane and M. Kellum, On topologically Riemannian foliations, preprint.

[Ke1] M. Kellum, Uniformly quasi-isometric foliations, Ergodic Theory Dynamical Systems 13 (1993), 101–122.

[Ke2] M. Kellum, Orbit equivalence for uniformly quasi-isometric foliations, preprint.

[Ke3] M. Kellum, Transverse homogeneous structures and deformation spaces for foliated manifolds with bundle-like metrics, preprint.

[Ke4] M. Kellum, Transverse geometric structures and deformation spaces of foliated manifolds with bundle-like metrics, preprint.

[Kib] B. Kim, On Escobales-Parker's theorem, Proc. Japan Acad. Ser. A Math. Sci. 65 (1989), 151–153.

[K-K-Kwo] B. Kim, N. Kim and J. Kwon, Fibred Riemannian spaces with critical Riemannian metrics, J. Korean Math. Soc. 31 (1994), 205–211.

[Kih] H. Kim, Geometric and dynamical properties of Riemannian foliations, Ph. D. Thesis, University of Illinois at Urbana-Champaign, Urbana, 1990.

[Kih-To] H. Kim and Ph. Tondeur, Riemannian foliations on manifolds with non-negative curvature, Manuscripta Math. 74 (1992), 39–45.

[Kih-Wal] H. Kim and G. Walschap, Riemannian foliations on compact hyperbolic manifolds, Indiana Univ. Math. J. 41 (1992), 37–42.

[Kit1] H. Kitahara, The existence of complete bundle-like metrics, Ann. Sci. Kanazawa Univ. 9 (1972), 37–40.

[Kit2] H. Kitahara, The existence of complete bundle-like metrics II, Ann. Sci. Kanazawa Univ. 10 (1973), 51–54.

[Kit3] H. Kitahara, The completeness of a Clairaut foliation, Ann. Sci. Kanazawa Univ. II (1974), 37–40.

[Kit4] H. Kitahara, Remarks on square-integrable basic cohomology spaces on a foliated Riemannian manifold, Kodai Math. J. 2 (1979), 187–193.

[Kit5] H. Kitahara, On a parametrix form in a certain V-submersion, Geometry and differential geometry (Proc. Conf., Univ. Haifa, Haifa, 1979), Lecture Notes in Math. 792, Springer, 1980, 264–298.

[Kit6] H. Kitahara, Non existence of nontrivial $\Box''$-harmonic 1-forms on a complete foliated Riemannian manifold, Trans. Amer. Math. Soc. 262 (1980), 429–435.

[Kit-Pak1] H. Kitahara and H. Pak, Construction of a complete negatively curved singular Riemannian foliation, J. Korean Math. Soc. 32 (1995), 609–614.

[Kit-Pak2] H. Kitahara and H. Pak, Some remarks on basic L^2-cohomology, Analysis and Geometry in foliated manifolds (Santiago de Compostela, 1994), World Sci. Publishing, 1995, 99–111.

[Kit-Yo1] H. Kitahara and S. Yorozu, A formula for the normal part of the Laplace-Beltrami operator on the foliated manifold, Pacific J. Math. 69 (1977), 425–432.

[Kit-Yo2] H. Kitahara and S. Yorozu, On some differential geometric characterizations of a bundle-like metric, Kodai Math. J. 2 (1979), 130–138.

[Ko1] Y. Kordyukov, L^p-theory of elliptic differential operators on manifolds of bounded geometry, Acta Appl. Math. 23 (1991), 223–260.

[Ko2] Y. Kordyukov, Transversally elliptic operators on G-manifolds of bounded geometry, Russian J. Math. Phys. 2 (1994), 175–198.

[Ln] R. Langevin, Feuilletages, énergies et cristaux liquides, Astérisque 107–108 (1983), 201–213.

[La] D. Lappas, On the integrable G-invariant metrics, Yokohama Math. J. 42 (1994), 87–94.

[L1] H. Lawson, Foliations, Bull. Amer. Math. Soc. 80 (1974), 369–418.

[L2] H. Lawson, Lectures on the quantitative theory of foliations, CBMS Regional Conference Series in Mathematics 27 (1977).

[L] C. Lazarov, Transverse index and periodic orbit, preprint.

[L-P] C. Lazarov and J. Pasternack, Secondary characteristic classes for Riemannian foliations, J. Differential Geom. 11 (1976), 365–385.

[Li] Z. Li, Harmonic foliations on the sphere, Tsukuba J. Math. 15 (1991), 397–407.

[Ll-Rev1] M. Llabres and A. Reventós, Unimodular Lie foliations, Ann. Fac. Sci. Toulouse 9 (1988), 243–255.

[Ll-Rev2] M. Llabres and A. Reventós, Some remarks on Lie flows, Publicaciones del Departamento de Geometria y Topologia, Universidad de Santiago de Compostela 57 (1989), 517–531.

[L-Mi-R] M. Lovrić, M. Min-Oo and E. Ruh, Deforming transverse Riemannian metrics on foliations, preprint.

[Lu] D. Lu, Homogeneous foliations of spheres, Trans. Amer. Math. Soc. 340 (1993), 95–102.

[Mc] E. Macias, Non-closed Lie subgroups of Lie groups, Ann. Global Anal. Geom. 11 (1993), 35–40.

[Mc-San1] E. Macias and E. Sanmartin, Minimal foliations on Lie groups, Indag. Math. (N.S.) 3 (1992), 41–46.

[Mc-San2] E. Macias and E. Sanmartin, The manifold of bundle-like metrics, Foliations (Tokyo, 1993), to appear.

[M-Ro-To1] Y. Maeda, S. Rosenberg and Ph. Tondeur, The mean curvature of gauge orbits, Global Analysis in Modern Mathematics, The Palais Festival Volume, Publish or Perish, 1993, 171–217.

[M-Ro-To2] Y. Maeda, S. Rosenberg and Ph. Tondeur, Minimal submanifolds in infinite dimensions, Analysis and geometry in foliated manifolds (Santiago de Compostela, 1994), World Sci. Publishing, 1995, 177–182.

[M-Ro-To3] Y. Maeda, S. Rosenberg and Ph. Tondeur, Minimal orbits of metrics and elliptic operators, preprint.

[Mag] M. Magid, Submersions from anti-de Sitter space with totally geodesic fibers, J. Differential Geom. 16 (1981), 323–331.

[M-Zh] E. Malysheva and N. Zhukova, On minimal sets of Riemannian foliations, Izv. Vyssh. Uchebn. Zaved. Mat. 1986, 38–45, 83.

[M-Mi-R] P. March, M. Min-Oo and E. Ruh, Mean curvature of Riemannian foliations, preprint.

[Mar] P. Marty, Feuilletages Lagrangiens et métriques quasi cotangentes complètes, C. R. Acad. Sci. Paris 309 (1989), 187–190.

[Ma1] X. M. Masa, Sucesion espectral de cohomologia asociada a variedades foliadas. Aplicaciones, Publicaciones del Departamento de Geometria y Topologia, Universidad de Santiago de Compostela, 19 (1973).

[Ma2] X. M. Masa, Quelques propriétés des feuilletages de codimension 1 à connexion transverse projetable, C. R. Acad. Sci. Paris 284 (1977), 811–812.

[Ma3] X. M. Masa, Cohomology of Lie foliations, Res. Notes in Math. 131, Pitman, Boston, 1985, 211–214.

[Ma4] X. M. Masa, Duality and minimality in Riemannian foliations, Comment. Math. Helv. 67 (1992), 17–27.

[Me] J. Meyer, e-foliations of codimension two, J. Differential Geom. 12 (1977), 583–594.

[Mi-Moz] A. Miernowski and W. Mozgawa, On Molino lifting of Riemannian vector fields, Acta Math. Hungar. 55 (1990), 185–191.

[Mi-R] M. Min-Oo and E. Ruh, Comparison theorems for compact symmetric space, Ann. Sci. École Norm. Sup. (4) 12 (1979), 335–353.

[Mi-R-To1] M. Min-Oo, E. Ruh and Ph. Tondeur, Vanishing theorems for the basic cohomology of Riemannian foliations, J. Reine Angew. Math. 415 (1991), 167–174.

[Mi-R-To2] M. Min-Oo, E. Ruh and Ph. Tondeur, A comparison theorem for almost Lie foliations, Ann. Global Anal. Geom. 9 (1991), 61–66.

[Mi-R-To3] M. Min-Oo, E. Ruh and Ph. Tondeur, Transversal curvature and tautness for Riemannian foliations, Proc. of the Conference on Global Analysis and Global Differential Geometry, (Berlin, 1990), Lecture Notes in Math. 1481, Springer, 1991, 145–146.

[Mo1] P. Molino, Connexions et G-structures sur les variétés feuilletées, Bull. Sci. Math. 92 (1968), 59–63.

[Mo2] P. Molino, Classe d'Atiyah d'un feuilletage et connexions transverses projetables, C. R. Acad. Sci. Paris 272 (1971), 779–781.

[Mo3] P. Molino, Classes caractéristiques et obstructions d'Atiyah pour les fibrés principaux feuilletés, C. R. Acad. Sci. Paris 272 (1971), 1376–1378.

[Mo4] P. Molino, Propriétés cohomologiques et propriétés topologiques des feuilletages à connexions transverses projetables, Topology 12 (1973), 317–325.

[Mo5] P. Molino, Feuilletages transversement parallélisables et feuilletages de Lie, C. R. Acad. Sci. Paris 282 (1976), 99–101.

[Mo6] P. Molino, Feuilletages transversalement complets et applications, Ann. Sci. École Norm. Sup. 70 (1977), 289–307.

[Mo7] P. Molino, Feuilletages riemanniens sur les variétés compactes; champs de Killing transverses, C. R. Acad. Sci. Paris 289 (1979), 421–423.

[Mo8] P. Molino, Géométrie globale des feuilletages riemanniens, Indag. Math. (N.S.) 44 (1982), 45–76.

[Mo9] P. Molino, Feuilletages riemanniens réguliers et singuliers, Geométrie différentielle (Paris, 1986), Travaux en Cours 33, Hermann, Paris, 1988, 173–201.

[Mo10] P. Molino, Riemannian foliations, Birkhäuser, Progr. Math. 73 (1988).

[Mo11] P. Molino, Dualité symplectique, Feuilletages et géométrie du moment, Publ. Mat. 33 (1989), 533–541.

[Mo12] P. Molino, La géométrie différentielle des feuilletages dans l'œuvre de B. Reinhart, Ann. Global Anal. Geom. 9 (1991), 5–7.

[Mo13] P. Molino, Orbit-like foliations, Geometric Study of Foliations (Tokyo, 1993), World Sci. Publishing, 1994, 97–119.

[Mo14] P. Molino, Feuilletages presque-isométriques et structures de Poisson-Riemann, preprint.

[Mo-P] P. Molino and M. Pierrot, Théorèmes de slice et holonomie des feuilletages riemanniens, Ann. Inst. Fourier (Grenoble) 37 (1987), 207–223.

[Mo-Se] P. Molino and V. Sergiescu, Deux remarques sur les flots riemanniens, Manuscripta Math. 51 (1985), 145–161.

[Mor] A. Morgan, Holonomy and metric properties of foliations in higher codimension, Proc. Amer. Math. Soc. 58 (1976), 255–261.

[Moz1] W. Mozgawa, Feuilletages de Killing, Collect. Math. 36 (1985), 285–290.

[Moz2] W. Mozgawa, Riemannian vector fields and Pontrjagin numbers, Ann. Univ. Mariae Curie-Sklodowska Sect. A 39 (1985), 113–115.

[Mz1] H. Münzner, Isoparametrische Hyperflächen in Sphären, Math. Ann. 251 (1980), 57–71.

[Mz2] H. Münzner, Isoparametrische Hyperflächen in Sphären, II, Über die Zerlegung der Sphäre in Ballbündel, Math. Ann. 256 (1981), 215–232.

[Nag] S. Nagamine, Totally umbilical Riemannian foliations on Lie groups, Math. J. Toyama Univ. 18 (1995), 195–198.

[Nak-Ta] H. Nakagawa and R. Takagi, Harmonic foliations on a compact Riemannian manifold of non-negative constant curvature, Tôhoku Math. J. 40 (1988), 465–471.

[Na1] F. Narita, The integrability tensors of Riemannian submersions and submanifolds, Res. Rep. Akita Nat. College Tech. 22 (1987), 66–74.

[Na2] F. Narita, Riemannian submersions and isometric reflections with respect to submanifolds, Math. J. Toyama Univ. 15 (1992), 83–94.

[Na3] F. Narita, Riemannian submersions with isometric reflections relative to the fibers, Kodai Math. J. 16 (1993), 416–427.

[Na4] F. Narita, Riemannian submersions of locally conformal Kaehler manifolds, preprint.

[Nav] A. Naveira, Variedades foliadas con metrica casi-fibrada, Collect. Math. 21 (1970), 5–61.

[N-Ra-To1] S. Nishikawa, M. Ramachandran and Ph. Tondeur, Heat conduction for Riemannian foliations, Bull. Amer. Math. Soc. 21 (1989), 265–267.

[N-Ra-To2] S. Nishikawa, M. Ramachandran and Ph. Tondeur, The heat equation for Riemannian foliations, Trans. Amer. Math. Soc. 319 (1990), 619–630.

[N-S] S. Nishikawa and H. Sato, On characteristic classes of Riemannian, conformal and projective foliations, J. Math. Soc. Japan 28 (1976), 223–241.

[N-To1] S. Nishikawa and Ph. Tondeur, Transversal infinitesimal automorphisms for harmonic Kähler foliations, Tôhoku Math. J. 40 (1988), 599–611.

[N-To2] S. Nishikawa and Ph. Tondeur, Transversal infinitesimal automorphisms of harmonic foliations on complete manifolds, Ann. Global Anal. Geom. 7 (1989), 47–57.

[N-To-V] S. Nishikawa, Ph. Tondeur and L. Vanhecke, Spectral geometry for Riemannian foliations, Ann. Global Anal. Geom. 10 (1992), 291–304.

[N-Yo] S. Nishikawa and S. Yorozu, Transversal infinitesimal automorphisms for compact Riemannian foliations, preprint.

[Noe] S. Noelker, Isometric immersion of warped products, Differential Geom. Appl., to appear.

[Nz] K. Nomizu, Some results in E. Cartan's theory of isoparametric families of hypersurfaces, Bull. Amer. Math. Soc. 79 (1973), 1184–1188.

[No] S. Novikov, Topology of foliations, Trudy Moskov. Mat. Obsc. 14 (1965), 248–278; AMS Translation, Trans. Moscow Math. Soc. 14 (1967), 268–304.

[Oh-P1] G. Oh and J. Pak, A note on L^2-transverse conformal fields on complete foliated Riemannian manifolds, Math. J. Toyama Univ. 13 (1990), 51–62.

[Oh-P2] G. Oh and J. Pak, Transverse conformal fields on foliated Riemannian manifolds, Math. J. Toyama Univ. 13 (1990), 63–75.

[ON] B. O'Neill, The fundamental equations of a submersion, Michigan Math. J. 13 (1966), 459–469.

[Or-R] L. Ornea and G. Romani, The fundamental equations of conformal submersions, Beiträge Algebra Geom. 34 (1993), 233–243.

[Os1] G. Oshikiri, A remark on minimal foliations, Tôhoku Math. J. 33 (1981), 133–137.

[Os2] G. Oshikiri, Totally geodesic foliations and Killing fields, Tôhoku Math. J. 35 (1983), 387–392.

[Os3] G. Oshikiri, Totally geodesic foliations and Killing fields, II, Tôhoko Math. J. 38 (1986), 351–356.

[Pa1] H. Pak, On the transversal conformal curvature tensor on Hermitian foliations, Bull. Korean Math. Soc. 28 (1991), 231–241.

[Pa2] H. Pak, On one-dimensional metric foliations in Einstein spaces, Illinois J. Math. 36 (1992), 594–599.

[Pa3] H. Pak, λ-automorphisms of a Riemannian foliation, Ann. Global Anal. Geom. 13 (1995), 281–288.

[Pa-Pk1] H. Pak and J. Pak, Stability theorem for a holomorphic Riemannian foliation, Math. Rep. Toyama Univ. 12 (1989), 181–187.

[Pa-Pk2] H. Pak and J. Pak, Normal holonomy group of a Riemannian foliation, Bull. Korean Math. Soc. 30 (1993), 17–23.

[Pa-Pk3] H. Pak and J. Pak, Notes on dense leaves of Riemannian foliations, Math. J. Toyama Univ. 18 (1995), 199–208.

[Pk-Sh-Yoo] J. Pak, Y. Shin and H. Yoo, L^2-transverse fields preserving the transverse Ricci field of a foliation, J. Korean Math. Soc. 32 (1995), 51–60.

[Pk-Yoo1] J. Pak and H. Yoo, L^2-transverse harmonic fields on complete foliated Riemannian manifolds, Kyungpook Math. J. 31 (1991), 253–262.

[Pk-Yoo2] J. Pak and H. Yoo, Transverse harmonic fields on Riemannian manifolds, Bull. Korean Math. Soc. 29 (1992), 73–80.

[Pk-Yo1] J. Pak and S. Yorozu, Transverse fields on foliated Riemannian manifolds, J. Korean Math. Soc. 25 (1988), 83–92.

[Pk-Yo2] J. Pak and S. Yorozu, The Laplace-Beltrami operator on a Riemannian manifold with a Clairaut foliation, Ann. Sci. Kanazawa Univ. 26 (1989), 13–15.

[P-Te] R. S. Palais and C. L. Terng, Critical point theory and submanifold geometry, Lecture Notes in Math. 1353, Springer, 1988.

[Pan1] P. Pang, Minimal models and Riemannian foliations, Ph. D. Thesis, University of Illinois at Urbana-Champaign, Urbana, 1988.

[Pan2] P. Pang, Basic dual homotopy invariants of Riemannian foliations, Trans. Amer. Math. Soc. 322 (1990), 189–199.

[Pan3] P. Pang, On the signature of generalised Seifert fibrations, Bull. Austral. Math. Soc. 46 (1992), 55–58.

[Par-R] E. Park and K. Richardson, The basic Laplacian of a Riemannian foliation, preprint.

[Pr] J. H. Park, The Laplace-Beltrami operator and Riemannian submersion with minimal and not totally geodesic fibers, Bull. Korean Math. Soc. 27 (1990), 39–47.

[Pr-Yo1] J. H. Park and S. Yorozu, Transverse fields preserving the transverse Ricci field of a foliation, J. Korean Math. Soc. 27 (1990), 167–175.

[Pr-Yo2] J. H. Park and S. Yorozu, Transversal conformal fields of foliations, Nihonkai Math. J. 4 (1993), 73–85.

[Pas] J. Pasternack, Foliations and compact group actions, Comment. Math. Helv. 46 (1971), 467–477.

[P-S] H. Pedersen and A. Swann, Riemannian submersions, four-manifolds and Einstein-Weyl geometry, Proc. London Math. Soc. 66 (1993), 381–399.

[Per] M. do Socorro Pereira, Obstruction à la représentation d'un feuilletage comme intersection transverse de deux feuilletages riemanniens, Bull. Soc. Math. Belg. Sér. B 43 (1991), 199–210.

[Prl] G. Perelman, Proof of the soul conjecture of Cheeger and Gromoll, J. Differential Geom. 40 (1994), 209–212.

[Pic] P. Piccinni, A Weitzenböck formula for the second fundamental form of a Riemannian foliation, Atti Accad. Naz. Lincei Rend. Cl. Sci. Fis. Mat. Natur. 77 (1984), 102–110.

[Pi] M. Pierrot, Orbites des champs feuilletés pour un feuilletage riemannien sur une variété compacte, C. R. Acad. Sci. Paris 301 (1985), 443–445.

[Pl] J. Plante, Foliations with measure preserving holonomy, Ann. of Math. 102 (1975), 327–361.

[Po-R] R. Ponge and H. Reckziegel, Twisted products in pseudo-Riemannian geometry, Geom. Dedicata 48 (1993), 15–25.

[Pop] P. Popescu, Sur une classe de groupoïdes riemanniens, preprint.

[Pt1] M. Puta, Some remarks on the cohomology of a real foliated manifold, Rend. Mat. 5 (1985), 189–201.

[Pt2] M. Puta, A remark on the basic cohomology of de Rham currents, Proceedings of the National Conference on Geometry and Topology (Tirgoviste, 1986), Univ. Bucuresti, Bucharest, 1988, 235–238.

[Ra] A. Ranjan, Structural equations and an integral formula for foliated manifolds, Geom. Dedicata 20 (1986), 85–91.

[Rk1] H. Reckziegel, A fiber bundle theorem, Manuscripta Math. 76 (1992), 105–110.

[Rk2] H. Reckziegel, Twisted products in Pseudo-Riemannian geometry, Geom. Dedicata 48 (1993), 15–25.

[Rb1] G. Reeb, Sur la courbure moyenne des variétés intégrales d'une equation de Pfaff $\omega = 0$, C. R. Acad. Sci. Paris 231 (1950), 101–102.

[Rb2] G. Reeb, Sur certaines propriétés topologiques des variétés feuilletées, Actualités Sci. Indust., Hermann, Paris, 1952.

[Rb3] G. Reeb, Structures feuilletées, Lecture Notes in Math. 652, Springer, 1978, 104–113.

[Re1] B. Reinhart, Harmonic integrals on almost product manifolds, Trans. Amer. Math. Soc. 88 (1958), 243–276.

[Re2] B. Reinhart, Foliated manifolds with bundle-like metrics, Ann. of Math. 69 (1959), 119–132.

[Re3] B. Reinhart, Harmonic integrals on foliated manifolds, Amer. J. Math. 81 (1959), 529–536.

[Re4] B. Reinhart, Closed metric foliations, Michigan Math. J. 8 (1961), 7–9.

[Re5] B. Reinhart, Structures transverse to a vector field, Internat. Sympos. Nonlinear Differential Equations and Nonlinear Mechanics, Academic Press, New York, 1963, 442–444.

[Re6] B. Reinhart, A metric formula for the Godbillon-Vey invariant for foliations, Proc. Amer. Math. Soc. 38 (1973), 427–430.

[Re7] B. Reinhart, The second fundamental form of a plane field, J. Differential Geom. 12 (1977), 619–627.

[Re8] B. Reinhart, Foliations and second fundamental form, Fourth Colloquium on differential geometry, Santiago de Compostela, 1978, 246–253.

[Re9] B. Reinhart, Differential geometry of foliations, Ergeb. Math. Grenzgeb. 99, Springer-Verlag, Berlin, New York, 1983.

[Ro] J. Roe, Elliptic operators, topology and asymptotic methods, Pitman Research Notes in Mathematics Series 179, Longman Scientific and Technical, Harlow, 1988.

[Rov1] V. Rovenskii, Metric decomposition of foliations with nonnegative curvature, Dokl. Akad. Nauk 334 (1994), 699–701; Russ Acad. Sci. Dokl. Math. 49 (1994), 202–205.

[Rov2] V. Rovenskii, Classes of submersions of Riemannian manifolds with compact fibers, Siberian Math. J. 35 (1994), 1027–1035.

[Rov-Top] V. Rovenskii and V. Topogonov, Great sphere foliations and manifolds with curvature bounded above, preprint.

[R-To] E. Ruh and Ph. Tondeur, Almost Lie foliations and the heat equation method, Proceedings of the Sixth International Colloquium on Differential Geometry (Santiago de Compostela, 1988), Cursos Congr. Univ. Santiago de Compostela 61, Univ. Santiago de Compostela, 1989, 239–246.

[Ruk1] Ph. Rukimbira, The dimension of leaf closures of K-contact flows, Ann. Global Anal. Geom. 12 (1994), 103–108.

[Ruk2] Ph. Rukimbira, Vertical sectional curvature and K-contactness, J. Geom. 53 (1995), 163–166.

[Ru1] H. Rummler, Quelques notions simples en géométrie riemannienne et leurs applications aux feuilletages compacts, Comment. Math. Helv. 54 (1979), 224–239.

[Ru2] H. Rummler, Kompakte Blätterungen durch Minimalflächen, Habilitations-Schrift, Universität Freiburg i. Ue., 1979.

[Ry] T. Rybicki, On Lie algebras of vector fields related to Riemannian foliations, Ann. Polon. Math. 58 (1993), 111–122.

[Sal1] E. Salem, Feuilletages riemanniens et pseudogroupes d'isométries, Ph. D. Thesis, Geneva, 1987.

[Sal2] E. Salem, Une généralisation du théorème de Myers-Steenrod aux pseudogroupes d'isométries locales, Ann. Inst. Fourier (Grenoble) 38 (1988), 185–200.

[Sal3] E. Salem, Riemannian foliations and pseudogroups of isometries, in P. Molino, Riemannian foliations, Birkhäuser, 1988, 254–296.

[San] M. Sanmartin, La variedad de las metricas casi-fibradas para una foliacion riemanniana, Publicaciones del Departamento de Geometria y Topologia, Universidad de Santiago de Compostela, 83 (1995), to appear.

[Sara] M. Saralegi, The Euler class for flows of isometries, Res. Notes in Math. 131, Pitman, Boston, 1985, 220–227.

[Sark] K. S. Sarkaria, A finiteness theorem for foliated manifolds, J. Math. Soc. Japan 30 (1978), 687–696.

[Sau] J. Sauvageot, Semi-groupe de la chaleur transverse sur la C*-algèbre d'un feuilletage riemannien, C. R. Acad. Sci. Paris 310 (1990), 531–536.

[Schu] C. Schultes, Characteristic classes of totally geodesic and Riemannian foliations, preprint.

[Schw] G. Schwarz, On the De Rham cohomology of the leaf space of a foliation, Topology 13 (1974), 185–187.

[Sc1] P. Scofield, Symplectic and complex foliations, Ph. D. Thesis, University of Illinois at Urbana-Champaign, Urbana, 1990.

[Sc2] P. Scofield, Some deformations of the Hopf foliation are also Kähler, Proc. Amer. Math. Soc. 119 (1993), 251–253.

[Se1] V. Sergiescu, Cohomologie basique et dualité des feuilletages riemanniens, Ann. Inst. Fourier (Grenoble) 35 (1985), 137–158.

[Se2] V. Sergiescu, Sur la suite spectrale d'un feuilletage riemannien, Proceedings of the XIXth National Congress of the Mexican Mathematical Society (Guadalajara, 1986), Vol. 2, 33–39; Aportaciones Math.: Comun. 4, Soc. Mat. Mexicana, Mexico City, 1987.

[Se3] V. Sergiescu, Basic cohomology and tautness of Riemannian foliations, Appendix B in P. Molino, Riemannian foliations, Birkhäuser, 1988, 235–248.

[So] A. Solovev, Riemannian foliations with constant transversal curvature, Sibirsk. Mat. Zh. 28 (1987), 160–166; English transl. in Siberian Math. J. 28 (1987).

[St1] S. Stepanov, A clan of Riemannian almost-product structures, Izv. Vyssh. Uchebn. Zaved. Mat. 1989, 40–46.

[St2] S. Stepanov, A geometric obstruction to the existence of completely umbilical distribution on a compact manifold, Webs and quasigroups, Kalinin. Gos. Univ., Kalinin, 1990, 135–137.

[St3] S. Stepanov, The Bochner technique in the theory of Riemannian almost product structures, Math. Notes 48 (1990), 778–781.

[St4] S. Stepanov, An integral formula for a compact manifold with a Riemannian almost product structure, Russ. Math. 38 (1994), No. 7, 66–70.

[St5] S. Stepanov, An integral formula for a Riemannian almost-product manifold, Tensor 55 (1994), 209–214.

[Su1] D. Sullivan, Cycles for the dynamical study of foliated manifolds and complex manifolds, Invent. Math. 36 (1976), 225–255.

[Su2] D. Sullivan, A foliation of geodesics is characterized by having no "tangent homologies", J. Pure Appl. Algebra 13 (1978), 101–104.

[Su3] D. Sullivan, A homological characterization of foliations consisting of minimal surfaces, Comment. Math. Helv. 54 (1979), 218–223.

[Ta-Yo1] R. Takagi and S. Yorozu, Minimal foliations on Lie groups, Tôhoku Math. J. 36 (1984), 541–554.

[Ta-Yo2] R. Takagi and S. Yorozu, Notes on the Laplace-Beltrami operator on a foliated Riemannian manifold with a bundle-like metric, Nihonkai Math. J. 1 (1990), 89–106.

[Tan-Yo] T. Tanemura and S. Yorozu, Green's theorem on a foliated Riemannian manifold and its applications, Acta Math. Hungar. 56 (1990), 239–245.

[Te1] C. L. Terng, Isoparametric submanifolds and their Coxeter groups, J. Differential Geom. 21 (1985), 79–107.

[Te2] C. L. Terng, A convexity theorem for isoparametric submanifolds, Invent. Math. 85 (1986), 487–492.

[Th1] W. Thurston, The theory of foliations of codimension greater than one, Comment. Math. Helv. 49 (1974), 214–231.

[Th2] W. Thurston, A generalization of the Reeb stability theorem, Topology 13 (1974), 347–352.

[Th3] W. Thurston, Foliations and groups of diffeomorphisms, Bull. Amer. Math. Soc. 80 (1974), 304–307.

[Th4] W. Thurston, Existence of codimension one foliations, Ann. of Math. 104 (1976), 249–268.

[Ti] D. Tischler, On fibering certain foliated manifolds over S^1, Topology 9 (1970), 153–154.

[To1] Ph. Tondeur, The mean curvature of Riemannian foliations, Feuilletages riemanniens, quantification géométrique et mécanique (Lyon, 1986), Travaux en Cours 26, Hermann, Paris, 1988, 41–52.

[To2] Ph. Tondeur, Foliations on Riemannian manifolds, Universitext, Springer Verlag, New York, 1988.

[To3] Ph. Tondeur, Riemannian foliations and tautness, Proc. Sympos. Pure Math., Amer. Math. Soc., 54 (1993), Part 3, 667–672.

[To4] Ph. Tondeur, Riemannian foliations and the heat equation, Geometry and its Applications (Yokohama, 1991), World Sci. Publishing, 1993, 233–239.

[To5] Ph. Tondeur, Geometry of Riemannian foliations, Seminar on Mathematical Sciences 20, Keio University, Yokohama, 1994.

[To6] Ph. Tondeur, A characterization of Riemannian flows, preprint.

[To-T] Ph. Tondeur and G. Toth, On transversal infinitesimal automorphisms for harmonic foliations, Geom. Dedicata 24 (1987), 229–236.

[To-V1] Ph. Tondeur and L. Vanhecke, Reflections in submanifolds, Geom. Dedicata 28 (1988), 77–85.

[To-V2] Ph. Tondeur and L. Vanhecke, Isometric reflections with respect to submanifolds, Simon Stevin 63 (1989), 107–116.

[To-V3] Ph. Tondeur and L. Vanhecke, Isometric reflections with respect to submanifolds and the Ricci operator of geodesic spheres, Monatsh. Math. 108 (1989), 211–217.

[To-V4] Ph. Tondeur and L. Vanhecke, Transversally symmetric Riemannian foliations, Tôhoku Math. J. 42 (1990), 307–317.

[To-V5] Ph. Tondeur and L. Vanhecke, Characterizing special Riemannian foliations, Simon Stevin 67 (1993), 227–234.

[To-V6] Ph. Tondeur and L. Vanhecke, A characterization of Riemannian foliations and totally umbilical submanifolds, Bull. Austral. Math. Soc. 48 (1993), 101–108.

[To-V7] Ph. Tondeur and L. Vanhecke, Jacobi fields, Riccati equation and Riemannian foliations, Illinois J. Math., to appear.

[Ts] T. Tshikuna-Matamba, Submersions métriques presque de contact dont l'espace total est une variété de Kenmotsu, An. Stiint. Univ. "Al. I. Cuza" Iasi, Sect. I a Mat. 37 (1991), 197–206.

[U] K. Ueno, On foliations associated with differential equations of conformal type, Publ. Res. Inst. Math. Sci. 22 (1986), 177–207.

[V1] I. Vaisman, Variétés riemanniennes feuilletées, Czechoslovak Math. J. 21 (1971), 46–75.

[V2] I. Vaisman, Cohomology and differential forms, Marcel Dekker Inc., New York, 1973.

[V3] I. Vaisman, Conformal foliations, Kodai Math. J. 2 (1979), 26–37.

[Vi] E. Vidal, Sur les variétés à structure presque-produit complexe avec métrique presque-feuilletée, C. R. Acad. Sci. Paris 273 (1971), 1152–1155.

[Wad] A. Wadsley, Geodesic foliations by circles, J. Differential Geom. 10 (1975), 541–549.

[Walc1] P. Walczak, On foliations with leaves satisfying some geometrical conditions, Dissertationes Math. (Rozprawy Mat.) 226 (1983), 1–47.

[Walc2] P. Walczak, Mean curvature functions for codimension one foliations with all the leaves compact, Czechoslovak Math. J. 34 (1984), 146–155.

[Walc3] P. Walczak, Dynamics of the geodesic flow of a foliation, Ergodic Theory Dynamical Systems 8 (1988), 637–650.

[Walc4] P. Walczak, An estimate for the second fundamental tensor of a foliation, Proceedings of the Sixth International Colloquium on Differential Geometry (Santiago de Compostela, 1988), Cursos Congr. Univ. Santiago de Compostela 61, Univ. Santiago de Compostela, 1989, 247–252.

[Walc5] P. Walczak, An integral formula for a Riemannian manifold with two orthogonal complementary distributions, Colloq. Math. 58 (1990), 243–252.

[Walc6] P. Walczak, On quasi-Riemannian foliations, Ann. Global Anal. Geom. 9 (1991), 83–95; Erratum, ibid. 325.

[Walc7] P. Walczak, A finiteness theorem for Riemannian submersions, Ann. Polon. Math. 57 (1992), 283–290; Erratum, ibid. 58 (1993), 319.

[Walc8] P. Walczak, Foliations invariant under the mean curvature flow, Illinois J. Math. 37 (1993), 609–623.

[Walc9] P. Walczak, Jacobi operator for leaf geodesics, Colloq. Math. 65 (1993), 213–226.

[Walc10] P. Walczak, Existence of smooth invariant measures for geodesic flows of foliations of Riemannian manifolds, Proc. Amer. Math. Soc. 120 (1994), 903–906.

[Wals1] G. Walschap, Metric foliations and curvature, J. Geom. Anal. 2 (1992), 373–381.

[Wals2] G. Walschap, Foliations of symmetric spaces, Amer. J. Math. 115 (1993), 1189–1195.

[Wals3] G. Walschap, Some rigidity aspects of Riemannian fibrations, Proc. Sympos. Pure Math., Amer. Math. Soc., 54 (1993), Part 3, 679–683.

[Wals4] G. Walschap, Soul-preserving submersions, Michigan Math. J. 41 (1994), 609–617.

[Wals5] G. Walschap, On the metric structure of nonnegatively curved manifolds, Geometry and topology of submanifolds, VII (Belgium, 1994), World Sci. Publishing, 1995, 268–270.

[Wals6] G. Walschap, On measure-invariant flows in constant curvature, preprint.

[Wals7] G. Walschap, A nonexistence theorem for Riemannian foliations, preprint.

[Wals8] G. Walschap, Umbilic foliations and curvature, preprint.

[Wat] B. Watson, Almost Hermitian submersions, J. Differential Geom. 11 (1976), 147–165.

[Wi1] H. Winkelnkemper, The graph of a foliation, Ann. Global Anal. Geom. 1 (1983), 51–75.

[Wi2] H. Winkelnkemper, Infinitesimal obstructions to weakly mixing, Ann. Global Anal. Geom. 10 (1992), 209–218.

[Wo1] R. Wolak, Foliated G-structures and Riemannian foliations, Manuscripta Math. 66 (1989), 45–59.

[Wo2] R. Wolak, Pierrot's theorem for singular Riemannian foliations, Publ. Mat. 38 (1994), 433–439.

[Wo3] R. Wolak, Compact leaves in minimal Riemannian foliations, Indag. Math. (N.S.) 5 (1994), 375–379.

[Wo4] R. Wolak, Maximal subalgebras in the algebra of foliated vector fields of a Riemannian foliation, Comm. Math. Helv., to appear.

[Wl] J. A. Wolf, Growth of finitely generated solvable groups and curvature of Riemannian manifolds, J. Differential Geom. 2 (1968), 421–446.

[Woo] C. M. Wood, A class of harmonic almost-product structures, J. Geom. Phys. 14 (1994), 25–42.

[Wu] H. Wu, A remark on the Bochner technique in differential geometry, Proc. Amer. Math. Soc. 78 (1980), 403–408.

[Y] K. Yamato, Exotic characteristic classes of compact Riemannian foliations, Foliations (Tokyo, 1983), Adv. Stud. Pure Math. 5, North-Holland, Amsterdam, 1985, 211–227.

[Yoo] H. Yoo, Existence of complete metrics of a Riemannian foliation, Math. J. Toyama Univ. 15 (1992), 35–38.

[Yo1] S. Yorozu, Behaviour of geodesics in foliated manifolds with bundle-like metrics, J. Math. Soc. Japan 35 (1983), 251–272.

[Yo2] S. Yorozu, A_ν-operator on complete foliated Riemannian manifolds, Israel J. Math. 56 (1986), 349–354.

[Ze] A. Zeggar, Nombre de Lefschetz basique pour un feuilletage riemannien, Ann. Fac. Sci. Toulouse Math. 1 (1992), 105–131.

[Zeb1] A. Zeghib, Feuilletages géodésiques des variétés localement symétriques, Ph. D. Thesis, Université de Dijon, Dijon, 1985.

[Zeb2] A. Zeghib, Laminations et hypersurfaces géodésiques des variétés hyperboliques, Ann. Sci. Ecole Norm. Sup. (4) 24 (1991), 171–188.

[Zeb3] A. Zeghib, Sur les feuilletages géodesiques continus des variétés hyperboliques, Invent. Math. 114 (1993), 193–206.

[Zh1] N. Zhukova, On the stability of leaves of Riemannian foliations, Ann. Global Anal. Geom. 5 (1987), 261–271.

[Zh2] N. Zhukova, Local stability of leaves of Riemannian foliations with singularities, Methods in qualitative theory and bifurcation theory, Nizhegorod. Gos. Univ., Nizhnii Novgorod, 1990, 93–105.

[Zh3] N. Zhukova, A criterion for the stability of leaves of Riemannian foliations (Russian), Izv. Vyssh. Uchebn. Zaved. Mat. 1992, 88–91.

[Z1] R. Zimmer, Curvature of leaves in amenable foliations, Amer. J. Math. 105 (1983), 1011–1022.

[Z2] R. Zimmer, Ergodic theory and semisimple groups, Birkhäuser, Boston, 1984.

[Z3] R. Zimmer, Amenable actions and dense subgroups of Lie groups, J. Funct. Anal. 72 (1987), 58–64.

[Z4] R. Zimmer, Arithmeticity of holonomy groups of Lie foliations, J. Amer. Math. Soc. 1 (1988), 35–58.

Appendix A
Books and Surveys on Particular Aspects of Foliations

Alvarez López, J. A.

1987 Sucesion espectral asociada a foliaciones riemannianas, Publ. 72 del Departamento de Geometria y Topologia, Universidad de Santiago de Compostela.

Bott, R.

1972 Lectures on characteristic classes and foliations, Springer Lecture Notes in Math. 279, 1–94.

1973 Gelfand-Fuks cohomology and foliations, Proc. Symp. New Mexico State University.

1976 On characteristic classes in the framework of Gelfand-Fuks cohomology, Astérisque 32–33, 113–139.

Camacho, C. and Lins Neto, A. L.

1979 Geometric theory of foliations, I.M.P.A. Rio de Janeiro [Portuguese]. Engl. Translation: Birkhäuser, Boston (1985).

Conlon, L.

1985 Foliations and exotic classes, Lectures at the Universidad de Extremadura, Jaranville de la Vera (Caceres).

Connes, A.

1980 Feuilletages et algèbres d'opérateurs, Séminaire Bourbaki 1979–80, exp 551.

1982 A survey of foliations and operator algebras, Proc. Symp. Pure Math. 38, Part 1, 521–628.

1985 Noncommutative differential geometry, Publ. Math. IHES 62, 41–144.

1990 Géométrie non commutative, Inter Editions, Paris.

Ehresmann, Ch.

1961 Structures feuilletées, Proc. Fifth Canadian Math. Congress.

Fuks, D. B.

1978 Cohomology of infinite-dimensional Lie algebras and characteristic classes of foliations, Itogi Nauki-Seriya "Matematika" 10, 179–286 [Russian]. Translation: J. Soviet Math 11 (1979), 922–980.

1981 Foliations, Itogi Nauki-Seriya Algebra, Topologiya, Geometriya 18, 151–213 [Russian]. Translation: J. Soviet Math. 18 (1982), 255–291.

Ghys, E.

1989 L'invariant de Godbillon-Vey, Séminaire Bourbaki 1988–89, exp 706, Astérisque 177–178, 155–181.

Godbillon, C.

1971 Problèmes d'existence et d'homotopie dans les feuilletages, Séminaire Bourbaki 1970–71, exp 390, Springer Lecture Notes in Math. 244, 167–181.

1972 Cohomologies d'algèbres de Lie de champs de vecteurs formels, Séminaire Bourbaki 1972–73, exp 421, Springer Lecture Notes in Math. 383.

1979 Systèmes dynamiques sur les surfaces, Strasbourg Lecture Notes.

1983 Dynamical systems on surfaces, Universitext, Springer Verlag, New York.

1985 Feuilletages, Etudes géométriques I.

1986 Feuilletages, Etudes géométriques II. Institut de Recherche Mathématique Avancée, Université Louis Pasteur, Strasbourg.

1991 Feuilletages, Etudes géométriques, Progress in Mathematics Vol. 98, Birkhäuser, Boston.

Haefliger, A.

1958 Structures feuilletées et cohomologie à valeur dans un faisceau de groupöides, Comment. Math. Helv. 32, 248–329.

1967 Travaux de Novikov sur les feuilletages, Séminaire Bourbaki 1967–68, exp 339, 12 pp.

1972 Sur les classes caractéristiques des feuilletages, Séminaire Bourbaki 1971–72, exp 412, Springer Lecture Notes in Math. 317.

1976 Differential cohomology, C.I.M.E. Lectures, Varenna.

1989 Feuilletages riemanniens, Séminaire Bourbaki 1988–89, exp 707, 15 pp.

Hector, G.

1993 Groupöides, feuilletages et C^*-algèbres (quelques aspects de la conjecture de Baum-Connes), Geometric Study of Foliations, Tokyo 1993, 3–34, World Scientific, 1994.

Hector, G. and Hirsch, U.

1981 Introduction to the geometry of foliations, Part A, Vieweg Verlag, Braunschweig.

1983 Introduction to the geometry of foliations, Part B, Vieweg Verlag, Braunschweig.

Hurder, S.

1993 Coarse geometry of foliations, Geometric Study of Foliations, Tokyo 1993, 35–96, World Scientific, 1994.

Kamber, F. W. and Tondeur, Ph.

1975 Foliated bundles and characteristic classes, Springer Lecture Notes in Math. 493.

1978 G-foliations and their characteristic classes, Bull. Amer. Math. Soc. 84, 1086–1124.

Lawson, H. B.

1974 Foliations, Bull. Amer. Math. Soc. 80, 369–418.

1977 Lectures on the quantitative theory of foliations, CBMS Regional Conf. Series, Vol. 27.

Molino, P.

1988 Riemannian foliations, Progress in Mathematics Vol. 73, Birkhäuser, Boston.

1993 Orbit-like foliations, Geometric Study of Foliations, Tokyo 1993, 97–119, World Scientific, 1994.

Moore, C. C. and Schochet, C.

1988 Global analysis on foliated spaces, Mathematical Sciences Research Institute Publications Vol. 9, Springer Verlag.

Pittie, H.

1976 Characteristic classes of foliations, Pitman Research Notes in Math., Vol. 10.

Poénaru, V.

1978 Travaux de Thurston sur les difféomorphismes des surfaces, Séminaire Bourbaki 1978–79, exp 529, Springer Lecture Notes in Math. 770.

Reeb, G.

1952 Sur certaines propriétés topologiques des variétés feuilletées, Actualités Sci. Indust., Hermann, Paris.

1959 Sur les feuilletages analytiques, Séminaire Bourbaki, exp 192.

1972 Feuilletages: résultats anciens et nouveaux (Painlevé, Hector et Martinet), Presses Univ. Montréal 1974.

1978 Structures feuilletées, Springer Lecture Notes in Math. 652, 104–113.

Reinhart, B. L.

1983 Differential geometry of foliations, Ergeb. Math. 99, Springer Verlag, New York.

Remsing, C.

1984 Introducere in Teoria Geometrica a Foliatiilor, Monografii Matematice, Universitatea din Timisoara.

Rosenberg, H.

1971 Feuilletages sur des sphères (d'après H. B. Lawson), Séminaire Bourbaki, exp 393, Springer Lecture Notes in Math. 244, 221–232.

Roussarie, R.

1978 Constructions de feuilletages (d'après W. Thurston), Séminaire Bourbaki, exp 499, Springer Lecture Notes in Math 677, 138–154.

Tamura, I.

1976 Topology of foliations, Iwanami shoten, Publishers, Tokyo (Japanese): Engl. Translation: Amer. Math. Soc., Translations of Math. Monographs 97 (1992).

Thurston, W.

1988 On the geometry and dynamics of diffeomorphisms of surfaces, I, Bull. Amer. Math. Soc. 19 (1988), 417–431.

Tondeur, Ph.

1988 Foliations on Riemannian manifolds, Universitext, Springer Verlag, New York.

1994 Geometry of Riemannian foliations, Seminar on Mathematical Sciences 20, Keio University, Yokohama.

1996 Survey of Riemannian foliations, 24th Romanian National Conference on Geometry and Topology, Univ. of Timisoara, Timisoara.

Tsuboi, T.

1990 Foliations and homology of the group of diffeomorphisms, Sugaku 3, 145–181.

1993 A characterization of the Godbillon-Vey invariant, Sugaku 45, 128–140.

Vaisman, I.

1973 Cohomology and differential forms, Dekker, New York.

Ver Eecke, P.

1982 Introduction à la théorie des variétés feuilletées, Esquisses Mathématiques vol. 31.

Viviente, J.

1984 Una excursion por la teoria de foliaciones, Academia de Ciencias Exactas, Fisicas, Quimicas y Naturales de Zaragoza.

Appendix B
Proceedings of Conferences and Symposia devoted to Foliations

1963	U of Grenoble	Structures Feuilletées Ann. Institut Fourier 14, no. 1, 1964
1971	Bahia de Salvador	Dynamical Systems Academic Press, 1973
1971	IPN, Mexico City	Lectures on Algebraic and Differential Topology Eds: Bott-Gitler-James Springer Lecture Notes in Math. 279, 1972
1972	U of Santiago de Compostela	Géométrie Différentielle Ed: Vidal Springer Lecture Notes in Math. 392, 1974
1973	U of Tokyo	Manifolds U of Tokyo Press, 1975
1973	Stanford U	Differential Geometry Eds: Chern-Osserman Proc. Symp. Pure Math. AMS 27, 1975
1974	U of Warwick	Dynamical Systems Ed: Manning Springer Lecture Notes in Math. 468, 1975
1974	U of Dijon	Differential Geometry and Topology Eds: Joubert-Moussu-Roussarie Springer Lecture Notes in Math. 484, 1975
1975	Washington U	Lectures on the Quantitative Theory of Foliations Lawson CBMS 27, 1977
1976	Rio de Janeiro	Geometry and Topology Eds: Palis-Do Carmo Springer Lecture Notes in Math. 597, 1977

1976	Pontifica U Catolica do Rio de Janeiro	Differential Topology, Foliations and Gelfand-Fuks cohomology Ed: Schweitzer Springer Lecture Notes in Math. 652, 1978
1976	C.I.M.E.	Differential Cohomology C.I.M.E. Lectures, Varenna Liguori 1979
1979	U of Orsay	Travaux de Thurston sur les surfaces, Séminaire Orsay 1979, Eds: Fahti-Laudenbach-Poénaru Astérisque 66–67.
1981	Rio de Janeiro	Geometric Dynamics Springer Lecture Notes in Math. 1007, 1983
1981 –82	U of Maryland	Differential Geometry Proceedings of the Special Year 1981-82 Eds: Brooks-Gray-Reinhart Progress in Math. 32 Birkhäuser, Basel, 1983
1982	Schnepfenried	Feuilletages Géométrie Symplectique et de Contact Analyse Non Standard et Applications Vol. 1, Astérisque 107–108, 1983 Vol. 2, Astérisque 109–110, 1983
1982	U of Toulouse	Structure Transverse des Feuilletages Ed: Pradines Astérisque 116, 1984
1983	U of Lyon	Feuilletages et Quantification Géométrique Eds: Dazord-Desolneux Moulis Travaux en Cours 6 Hermann, Paris, 1984.
1983	U of Tokyo	Foliations Ed: Tamura Adv. in Pure Math. 5 North-Holland, Amsterdam, 1985
1983	Kyoto	Geometric Methods in Operator Algebras Res. Notes Math. Ser. 123 Pitman, London, 1986
1984	U of Santiago de Compostela	Differential Geometry Ed: Cordero Res. Notes Math. Ser. 131 Pitman, London, 1985

1984	Mexico City	The Lefschetz Centennial Conference Contemp. Math. 58, 1987
1985	Rio de Janeiro	Dynamical Systems and Bifurcation Theory Res. Notes Math. Ser. 160 Pitman, London, 1986
1986	New Orleans	Index Theory of Elliptic Operators, Foliations and Operator Algebras Contemp. Math. 70, 1988
1986	U of Lyon	Feuilletages Riemanniens, Quantification Géométrique et Mécanique, Eds: Dazord-Desolneux Moulis-Morvan, Travaux en Cours 26 Hermann, Paris, 1988
1986	Mexico City	Holomorphic Dynamics Eds: Gomez Mont-Seade-Verjovsky Springer Lecture Notes in Math. 1345, 1988
1988	U of Lyon and U of Barcelona	Mois de Juin Feuilleté Semester on Differential Geometry Publicacions Matematiques Univ. Autonoma de Barcelona Vol. 33, no. 3, 1989
1988	U of Santiago de Compostela	Differential Geometry Eds: Cordero Proceedings of the 6th Int. Colloq., 1989
1988	U of Tokyo	A Fête in Topology Papers dedicated to I. Tamura Academic Press, 1988
1989	U of Maryland	Bruce Reinhart Memorial Conference on Foliations Annals of Global Anal. and Geom. 9, no. 1, 1991.
1992	Rio de Janeiro	Complex Analytic Methods in Dynamical Systems Eds: Camacho-Lins Neto-Monsow-Sad Astérisque 222, 1994
1992	Rio de Janeiro	Differential Topology, Foliations and Group Actions Eds: Schweitzer-Hurder-dos Santos-Arraut Contemp. Math. 161, 1994.
1993	Tokyo Chuo U and Tokyo Institute of Technology	Geometric Study of Foliations Eds: Inaba-Masuda-Matsumoto -Mitsumatsu-Mizutani-Tsuboi World Scientific, 1994

1994 U of Santiago Analysis and Geometry in Foliated Manifolds
de Compostela Eds: Alvarez-Macias-Masa
Proceedings of the 7th Int. Colloq.,
Santiago de Compostela, 1994
World Scientific 1995

Appendix C
Bibliography on Foliations

K. Abe, Characterization of totally geodesic submanifolds of S^N and CP^N by an inequality, Tôhoku Math. J. 23 (1971), 219–244.

K. Abe, Applications of a Ricatti type differential equation to Riemannian manifolds with totally geodesic distributions, Tôhoku Math. J. 25 (1973), 425–444.

N. Abe, On foliations and exotic characteristic classes, Kodai Math. Sem. Rep. 28 (1977), 324–341.

N. Abe, Exotic characteristic classes of certain γ-foliations, Kodai Math. J. 2 (1979), 254–271.

Y. Abe, On Levi foliations, Memoirs of the Faculty of Science, Kyushu Univ., Ser. A 38 (1984), 169–176.

N. A'Campo, Un feuilletage de S^5, C. R. Acad. Sci. Paris 272 (1971), 1504–1506.

N. A'Campo, Feuilletages de codimension 1 sur des variétés de dimension 5, C. R. Acad. Sci. Paris 273 (1971), 603–604.

N. A'Campo, Feuilletages de codimension 1 sur les variétés simplement connexes de dimension 5, Comment. Math. Helv. 47 (1972), 514–525.

N. A'Campo and D. Kotschick, Contact structures, foliations, and the fundamental group, Bull. London Math. Soc. 26 (1994), 102–106.

M. Adachi, A note on γ_n^c-structures, J. Math. Kyoto Univ. 31 (1991), 583–591.

S. Adams, Superharmonic functions on foliations, Trans. Amer. Math. Soc. 330 (1992), 625–635.

S. Adams, Rank rigidity of foliations by manifolds of nonpositive curvature, Diff. Geom. and its Appl. 3 (1993), 47–70.

S. Adams and A. Freire, Nonnegatively curved leaves in foliations, J. Diff. Geom. 34 (1991), 681–700.

S. Adams and L. Hernández, A foliated metric rigidity theorem for higher rank irreducible symmetric spaces, Geom. and Funct. Anal. 4 (1994), 483–521.

S. Adams and G. Stuck, Splitting of non-negatively curved leaves in minimal sets of foliations, Duke Math. J. 71 (1993), 71–92.

Y. Agaoka, Geometric invariants associated with flat projective structures, J. Math. Kyoto Univ. 22 (1983), 701–718.

C. Albert, Feuilletages invariants et pseudoalgebres de Lie lisses, Cah. Topol. Geom. Diff. 13 (1972), 309–323.

C. Albert, Invariants riemanniens des champs de plans, C. R. Acad. Sci. Paris 296 (1983), 329–332.

C. Albert, Sur le feuilletage caractéristique des groupoïdes de Poisson, C. R. Acad. Sci. Paris 312 (1991), 529–531.

C. Albert and D. Lehmann, Une algèbre graduée universelle pour les connexions sans torsion, Math. Z. 159 (1978), 133–142.

I. Albu and D. Opris, The geometry of the vector bundle associated to a foliated manifold, the background of some applications in theoretical physics, Analele Universitatii din Timisoara 28 (1990), 103–119.

F. Alcalde Cuesta, Groupoïde d'homotopie d'un feuilletage riemannien et réalisation symplectique de certaines variétés de Poisson, Publicacions Mathematiques 33 (1989), 395–410.

F. Alcalde Cuesta, Integración simpléctica de las variedades de Poisson riemannianas, *Publ. Dpto. Xeometria e Topoloxia*, Santiago de Compostela 79 (1991).

F. Alcalde Cuesta, Intégration symplectique des variétés sans cycle evanouissant, Thèse, Univ. Clande Bernard-Lyon 1, 1993.

F. Alcalde Cuesta and G. Hector, Feuilletages en surfaces, cycles évanouissants et variétés de Poisson, preprint.

F. Alcalde Cuesta and G. Hector, Intégration symplectique des variétés de Poisson régulierès, Israel J. of Math., to appear.

D. Alekseevsky and P. Michor, differential geometry of $\mathfrak{g}$-manifolds, Diff. Geom. and its Appl. 5 (1995), 371–403.

J. Alexander and A. Verjovsky, First integrals for singular holomorphic foliations with leaves of bounded volume, Springer Lecture Notes in Math. 1345 (1988), 1–10

R. Almeida and P. Molino, Suites d'Atiyah et feuilletages transversalement complets, C. R. Acad. Sci. Paris 300 (1985), 13–15.

R. Almeida and P. Molino, Suites d'Atiyah, feuilletages et quantification, Sém. Gaston Darboux de Géométrie et Topologie Différentielle, Univ. de Montpellier, 1984–1985.

R. Almeida and P. Molino, Flots riemanniens sur les 4-variétés compactes, Tôhoku Math. J. 38 (1986), 313–326.

A. Alta'ai and J. Pradines, Caractérisation universelle du groupe de Haefliger-van Est d'un espace de feuilles ou d'orbites, et théorème de van Kampen, C. R. Acad. Sci. Paris 309(1989), 503–506.

J. Alvarez López, Sucesion espectral asociada a foliaciones riemannianas, Publicaciones del Departamento de Geometria y Topologia 72 (1987), Universidad de Santiago de Compostela.

J. Alvarez López, A finiteness theorem for the spectral sequence of a Riemannian foliation, Ill. J. of Math. 33 (1989), 79–92.

J. Alvarez López, Duality in the spectral sequence of Riemannian foliations, American J. of Math. 111 (1989), 905–926.

J. Alvarez López, on Riemannian foliations with minimal leaves, Ann. Inst. Fourier 40 (1990), 163–176.

J. Alvarez López, A decomposition theorem for the spectral sequence of Lie foliations, Trans. Amer. Math. Soc. 329 (1992), 173–184.

J. Alvarez López, A type of nonequivalent pseudogroups. Application to foliations. Ann. Polon. Math. 56 (1992), 187–194.

J. Alvarez López, the basic component of the mean curvature of Riemannian foliations, Ann. Global Anal. Geom. 10 (1992), 179–194.

J. Alvarez López, Morse inequalities for pseudogroups of local isometries, J. Diff. Geom. 37 (1993), 603–638.

J. Alvarez López, Modified Laplacians in foliated manifolds, to appear.

J. Alvarez López, An index formula for transversally elliptic operators with respect to pseudogroups of local isometries, to appear.

J. Alvarez López and G. Hector, The dimension of the leafwise reduced cohomology, preprint, 1994.

J. Alvarez López and G. Hector, Leafwise homologies, and subfoliations, Analysis and Geometry in Foliated Manifolds, Proc. VII Int. Colloq. on Diff. Geom., Santiago de Compostela 1994, World Scientific 1995.

J. Alvarez López and S. Hurder, Pure-point spectrum for foliation geometric operators, preprint, 1994.

J. Alvarez López and Ph. Tondeur, The heat flow along the leaves of a Riemannian foliation, Geometric and topological invariants of elliptic operators, Contemporary Mathematics Vol. . (1990), 271–280.

J. Alvarez López and Ph. Tondeur, Hodge decomposition along the leaves of a Riemannian foliation, J. Funct. Anal. 99 (1991), 443–458.

K. Amur and R. Venkataraman, On a foliated and parametric minimal hypersurface in Euclidean space, J. Ramanujan Math. Soc. 6 (1991), 9–27.

Y. Ando, An existence theorem of foliations with singularities A_k, D_k and E_k, Hokkaido Math. J. 20 (1991), 571–578.

O. Andrade and M. Soares, Chern numbers of a Kupka component, Ann. Inst. Fourier 44 (1994), 1237–1242.

P. Andrade, Fluxos que admitem folheacoes transversas, Thesis, PUC-Rio, Rio de Janeiro, 1985.

P. Andrade, On homology directions for flows, Japan J. Math. 17 (1991).

P. Andrade, The set of vector fields with transverse foliations, J. Math. Soc. Japan 45 (1993), 21–35.

P. Andrade and M. Pereira, on the cohomology of one-dimensional foliated manifolds, Bol. Soc. Brasil. Mat. 21 (1990), 79–89.

G. Andrzejczak, Characteristic classes of foliations preserved by a transverse k-field, Differential Geometry (Warsaw, 1979), 23–27, Banach Center Publ. 12, PWN, Warsaw, 1984.

G. Andrzejczak, Some characteristic invariants of foliated bundles, Dissertationes Math. (rozprawy mat.) 222 (1984), 67 pp.

G. Andrzejczak, More characteristic invariants of foliated bundles, Diff. Geom., Banach Center Publ. 12 (1984), 9–22.

G. Andrzejczak, Characteristic homomorphism for transversely holomorphic foliations via the Cauchy–Riemann equations, Deformation of mathematical structures (Lódz/Lublin, 1985/87), 55–63, Kluwer Acad. Publ., Dordrecht, 1989.

G. Andrzejczak, A semi-simplicial approach to foliations and their transverse structure, Dissertationes (Rozprawy Mat.) 314 (1991), 97 pp.

M. Anona, Sur la d_l-cohomologie, C. R. Acad. Sci. Paris 290 (1980), 649–651.

D. Anosov, Roughness of geodesic flows on compact Riemannian manifolds of negative curvature, Dokl. Akad. Nauk SSSR 145, 707–709 [Russian]. Translation: Soviet Math. Dokl. 3 (1962), 1068–1069.

D. Anosov, ergodic properties of geodesic flows on closed Riemannian manifolds of negative curvature, Dokl. Akad. Nauk SSSR 151, 1250–1252 [Russian]. Translation: Soviet Math. Dokl. 4 (1963), 1153–1156.

D. Anosov, Geodesic flows on closed Riemannian manifolds with negative curvature, Trudy Mat. Inst. Steklov 90 (1967) [Russian]. Translation: Proc. Steklov Inst. Math. 90, AMS 1969.

D. Anosov, Flows on surfaces, Proc. Steklov Inst. of Math. 3 (1993), 7–11.

M. Aof, Topological aspects of holonomy groupoids, Univ. College of North Wales Pure Maths Preprint, 147 pp., 88.10, Bangor, 1987.

T. Aoki, N. Matsuoka and S. Yorozu, Notes on vector fields and transverse fields on foliated Riemannian manifolds, Ann. Sci. Kanazawa Univ. 26 (1989), 1–6.

T. Aoki and S. Yorozu, l^2-transverse conformal and killing fields on complete foliated Riemannian manifolds, Yokohama Math. J. 36 (1988), 27–41.

L. Apostolova, Nearly Kähler manifolds are holomorphic foliations, C. R. Acad. Bulgare Sci. 38 (1985), 977–979.

C. Apreutesei, Quelques classes caractéristiques et g_t-structures, C. R. Acad. Sci. Paris 280 (1975), 41–44.

C. Apreutesei, Algèbres HC (,), HP (,) et obstructions à l'intégrabilité, Att. Accad. Naz. Lincei. Rend. Cl. Sci. Fis. Mat. nat. 62 (1977), 17–25.

S. Aranson, Topology of vector fields, foliations with singularities, and homeomorphisms with invariant foliations on closed surfaces, Proc. Steklov Inst. of Math. 3 (1993), 13–18.

S. Aranson, V. Mamaev and E. Zhuzhoma, Asymptotial properties of codimension one foliations and Anosov-Weyl problem, Geometrical Study of Foliations, Tokyo 1993, 145–151, World Scientific, 1994.

S. Aranson, T. Medvedev and E. Zhuzhoma, Cherry foliations and Cherry flows on the sphere, Sel. Math. 13, No. 4, 283–303 (1994).

S. Aranson and E. Zhuzhoma, Classification of transitive foliations on a sphere with four singularities of "needle" type, Methods on the qualitative theory of differential equations, 3–10, 197, Gorky Univ. Publ. Gorki, 1984.

S. Aranson and E. Zhuzhoma, On the trajectories of coverings of flows in the case of coverings of a sphere and a projective plane, Math. Notes 53 (1993), 463–468.

S. Aranson and E. Zhuzhoma, Quasiminimal sets of foliations, and one-dimensional basic sets of a-diffeomorphisms of surfaces (Russian), Dokl. Akad. Nauk 330 (1993), 280–281.

S. Aranson and E. Zhuzhoma, On the structure of quasiminimal sets of foliations on surfaces, to appear.

G. Arca, Espaces-temps symplectiques admettant un double feuilletage lagrangien transverse, Tensor 47 (1988), 255–259.

V. Arnold, Topological and ergodic properties of closed 1-forms with incommensurable periods, Funct. Anal. Appl. 25 (1991), 81–90.

J. Arraut, Note on foliations, Dynamical Systems, Bahia, 1971, 1–6, Academic Press, 1973.

J. Arraut, A two-dimensional foliation on s^7, Topology 3 (1973), 243–245.

J. Arraut and M. Craizer, Foliations of M^3 defined by $\mathbb{R}^2$-actions, Ann. Inst. Fourier 45 (1995), 1091–1118.

J. Arraut and N. dos Santos, Actions of $\mathbb{R}^p$ on closed manifolds, Topology and its Appl. 29 (1988), 41–54.

J. Arraut and N. dos Santos, Differentiable conjugation of actions of $\mathbb{R}^p$, Bol. Soc. Bras. Mat. 19 (1988), 1–19.

J. Arraut and N. dos Santos, Linear foliations of t^n, Bol. Soc. Bras. Mat. 21 (1991), 189–204.

J. Arraut and N. dos Santos, The characteristic mapping of a foliated bundle, Topology 31 (1992), 545–557.

D. Arrowsmith, Anosov flows and parallel foliations, J. London Math. Soc. 10 (1975), 48–52.

D. Asimov, Round handles and homotopy of nonsingular vector fields, Bull. Amer. Math. Soc. 81 (1975), 417–419.

D. Asimov, On the average Gaussian curvature of leaves of foliations, Bull. Amer. Math. Soc. 84 (1978), 131–133.

D. Asimov and H. Gluck, Morse-Smale fields of geodesics, Lecture Notes in Math. 819, Springer-Verlag (1980), 1–17.

K. Aso and S. Yorozu, A generalization of Clairaut's theorem and umbilic foliations, Nihonkai Math. J. 2 (1991), 139–153.

M. Atiyah, Complex analytic connections in fibre bundles, Trans. Amer. Math. Soc. 85 (1957), 181–207.

M. Atiyah, Vector fields on manifolds, Arbeitsgemeinschaft für Forschung des Landes Nordrhein-Westfalen, vol. 200, Westdeutscher Verlag (1970).

M. Atiyah, Differential geometry, foliations and characteristic classes, Canberra Notes, 1972.

M. Atiyah, Elliptic operators and compact groups, Springer Lecture Notes in Math. 401 (1974).

O. Attie and S. Hurder, Manifolds which cannot be leaves of foliations, Topology 35, to appear.

P. Baird and J. Wood, Harmonic morphisms and conformal foliations by geodesics of three-dimensional space forms, J. Austral. Math. Soc. 41 (1991), 118–153.

P. Baird and J. Wood, Harmonic morphisms, seifert fibre spaces and conformal foliations, Proc. London Math. Soc. 64 (1992), 170–196.

P. Baird and J. C. Wood, The geometry of a pair of Riemannian foliations by geodesics and associated harmonic morphisms, Bull. Soc. Math. Belg. 44 (1992), 115–139.

D. Baker, On a class of foliations and the evaluation of their characteristic classes, Bull. Amer. Math. Soc. 83 (1977), 394–396.

D. Baker, On a class of foliations and the evaluation of their characteristic classes, Comment. Math. Helv. 53 (1978), 334–363.

D. Baker, Some cohomology invariants for deformations of foliations, Ill. J. Math. 25 (1981), 169–189.

R. Balan, A note about integrability of distributions with singularities, Boll. Un. Mat. Ital. A (7) 8 (1994), 335–344.

E. Ballico, A splitting theorem for the kupka component of a foliation of $\mathbb{C}P^n$, $n \geq 6$. Addendum to a paper of O. Calvo-Andrade and N. Soares, Ann. Inst. Fourier 45 (1995), 1119–1121.

W. Ballman, M. Brin and K. Burns, On the differentiability of horocycles and horocycle foliations, J. Diff. Geom. 26 (1987), 337–347.

S. Bando, M. Masuda, and H. Sato, Topological Blaschke conjecture for cohomological projective spaces, to appear.

V. Bangert, The existence of gaps in minimal foliations, Aequationes Math. 34 (1987), 153–166.

V. Bangert, On minimal laminations of the torus, Ann. Inst. H. Poincaré Anal. Non Linéaire 6 (1989), 95–138.

V. Bangert, Laminations of 3-tori by least area surfaces, Analysis, et cetera, 85–114, Academic Press, Boston, 1990.

V. Bangert, Minimal foliations and laminations, Proc. Int. Congress of Math., Zürich, 1994, 453–464; Birkhäuser, Basel, 1995.

L. Banghe and A. Haefliger, Currents on a circle invariant by a Fuchsian group, Springer Lecture Notes in Math. 1007(1983), 369–378.

A. Banyaga and P. Rukimbira, Weak stability of almost regular contact foliations, J. Geom. 50 (1994), 16–27.

A. Banyaga and P. Rukimbira, On characteristics of circle invariant presymplectic forms, Proc. Ams 123 (1995), 3901–3906.

J. Barbosa, J. Gomes and A. Silveira, Foliations of 3-dimensional space forms by surfaces with constant mean curvature, Bol. Soc. Bras. Math. 18 (1987), 1–12.

J. Barbosa, K. Kenmotsu and G. Oshikiri, Foliations by hypersurfaces with constant mean curvature, Math. Z. 207 (1991), 97–108.

T. Barbot, Géométrie transverse des flots d'Anosov, Thesis, Ecole Norm. Sup., Lyon, 1992.

T. Barbot, Caractérization des flots d'Anosov en dimension 3 par leurs feuilletages faibles, Erg. Th. Dyn. Syst. 15 (1995), 247–270.

T. Barbot, Flots d'Anosov sur les variétés graphées au sens de Waldhausen I: Morceaux fibrés de feuilletages d'Anosov, to appear.

T. Barbot, Mise en position optimale d'un tore par rapport à un flot d'Anosov, to appear.

D. Barlet, L'espace des feuilletages d'un espace analytique compact, Ann. Inst. Fourier 37 (1987), 117–130.

E. Barletta and S. Dragomir, Transversally CR foliations, to appear.

R. Barre, De quelques aspects de la théorie des Q-variétés différentielles et analytiques, Ann. Inst. Fourier 23 (1973), 227–312.

R. Barre, Quelques problèmes liés à la théorie des Q-variétés différentielles et analytiques, Astérisque 116 (1984), 15–24.

R. Barre, Fermeture de l'espace des divergences et séparation de l'espace des feuilles, Ann. Fac. Sci. Toulouse Math. 8 (1986/87), 121–130.

D. Barrett, Complex analytic realization of Reeb's foliation of S^3, Math. Z. 203 (1990), 355–361.

D. Barrett and J. Fornaess, On the smoothness of Levi-foliations, Publ. Math. 32 (1988), 171–177.

A. Basmajian and G. Walschap, Metric flows in space forms of nonpositive curvature, Proc. AMS 123 (1995), 3177–3181.

M. Bauer, Connexions L-equivalentes associées à un feuilletage, C. R. Acad. Sci. Paris 278 (1974), 1633–1636.

M. Bauer, Almost regular foliations, C. R. Acad. Sci. Paris 299 (1984), 387–390.

M. Bauer, Codimension one, almost regular foliations, C. R. Acad. Sci. Paris 299 (1984), 819–822.

M. Bauer, Feuilletage singulier défini par une distribution presque régulière, Collect. Math. 37 (1986), 189–209.

P. Baum, Structure of foliation singularities, Adv. Math. 15 (1975), 361–374.

P. Baum and R. Bott, On the zeros of meromorphic vectorfields, Essays Topol. Relat. Topics, Berlin et al. (1970), 29–47.

P. Baum and R. Bott, Singularities of holomorphic foliations, J. Diff. Geometry 7 (1972), 279–342.

P. Baum and A. Connes, Geometric K-theory for lie groups and foliations, IHES, preprint, 1982.

P. Baum and A. Connes, Leafwise homotopy equivalence and rational Pontrjagin classes, Foliations (Tokyo, 1983), 1–14, Adv. Stud. Pure Math. 5, North-Holland, Amsterdam, 1985.

E. Bedford and M. Kalka, Foliations and Monge-Ampère equations, Comm. Pure Appl. Math. 30 (1977), 543–571.

A. Bejancu and K. Duggal, Gauge theory on foliated manifolds, Rend. Sem. Mat. Messina Ser. II 14 (1991), no. 1, 31–68.

I. Belko, The Atiyah-Molino class of foliated Lie algebroid, Dokl. Akad. Nauk. Belarusi 37 (1993), no.5, 16–18, 121 (1994).

S. Benenti and W. Tulczyjew, Sur un feuilletage coisotrope du fibré cotangent d'un groupe de Lie, C. R. Acad. Sci. Paris 300 (1985), 119–122.

Y. Benoist, P. Foulon and F. Labourie, Flots d'Anosov à distributions stable et instable différentiables, C. R. Acad. Sci. Paris. 311 (1990), 351–354.

Y. Benoist, P. Foulon and F. Labourie, Sur les difféomorphismes d'Anosov á feuilletages stable et instable différentiables, c. r. acad. sci. paris 313 (1991), 45–47.

Y. Benoist, P. Foulon and F. Labourie, Anosov flows with stable and unstable differentiable distributions, J. Amer. Math. Soc. 5 (1992), 33–74.

Y. Benoist and F. Labourie, Sur les difféomorphismes d'Anosov affines à feuilletages stable et instable différentiables, Invent. Math. 111 (1993), 285–308.

C. Benson, Characteristic classes for symplectic foliations, Mich. Math. J. 33 (1986), 105–118.

C. Benson and R. Ellis, Characteristic classes of transversely homogeneous foliations, Trans. Amer. Math. Soc. 289 (1985), 849–859.

V. Berestovskii, Compact homogeneous manifolds with integrable invariant distributions, Izv. Vyss. Uchebn. Zaved. Mat. 6 (1992), 42–48.

M. Berger and D. Ebin, Some decompositions of the space of symmetric tensors on a Riemannian manifold, J. Diff. Geom. 3 (1969), 376–392.

M. Berger, P. Gauduchon and E. Mazet, Le spectre d'une variété riemannienne, Lecture Notes in Math. 194, Springer-Verlag (1971).

I. Bernshtein and B. Rozenfeld, On characteristic classes of foliations, Funkcional. Anal. i Prilozen 6, 68–69 [Russian]. Translation: Functional Anal. Apppl. 6 (1972), 60–61.

I. Bernshtein and B. Rozenfeld, Homogeneous spaces of infinite-dimensional Lie algebras and characteristic classes of foliations, Uspehi Mat. Nauk. 28 (2973), 103–138 [Russian]. Translation: Russian Math. Surveys 28 (1973), 107–142.

A. Besse, Einstein manifolds, Ergeb. Math. 3. Folge, Band 10, Springer, New York 1987.

G. Besson and M. Bordoni, On the spectrum of Riemannian submersions with totally geodesic fibers, Atti Acad. Naz. Lincei Cl. Sci. Fis. Mat. Natur. Rend (9) Mat. Appl. 1 (1990), 335–340.

I. Bielko, Affine transformations of a transversal projectable connection on a manifold with a foliation, Mat. Sbor. 117 (1982), 181–195: Translations: Math. USSR Sbornik 45 (1983), 191–204.

I. Bielko, On the structural function of foliated $G_\mathcal{K}$-structures, Mat. Zametki 40 (1986), 662–670. Translation: Math. Notes 40 (1986), 875–879.

B. Bigonnet and J. Pradines, Graphe d'un feuilletage singulier, C. R. Acad. Sci. Paris 300 (1985), 439–442.

A. Biś, Entropy of transverse foliations, Acta Univ. Lodz. Folia Math. 4 (1991), 9–17.

R. Bishop, Clairaut submersions, Diff. Geom. in honor of K. Yano, Kinokuniya, Tokyo (1972), 21–31.

R. Bishop and B. O'Neill, Manifolds of negative curvature, Trans. Amer. Math. Soc. 145 (1969), 1–49.

I. Bivens, Orthogonal geodesics and minimal distributions, Trans. Amer. Math. Soc. 275 (1983), 397–408.

I. Bivens, When do orthogonal families of curves possess a complex potential? Math. Mag. 65 (1992), 226–235.

D. Blair and J. Vanstone, A generalization of the helicoid, minimal submanifolds and geodesics, Proc. Japan-Us Seminar Tokyo 1977, 13–16.

S. Blank and F. Laudenbach, Isotopie des formes fermées en dimension 3, C. R. Acad. Sci. Paris 285 (1977), 215–218.

S. Blank and F. Laudenbach, Isotopies des formes fermées en dimension trois, Invent. Math. 54 (1979), 103–177.

J. Block and E. Getzler, Quantization of foliations, in: Catto, Sultan (ed.) et al., Differential geometric methods in theoretical physics. Proc. of the 20th International Conference, June 3–7, 1991, New York City, Vol. 2, 471–487. Singapore: World Scientific, 1992.

R. Blumenthal, Transversely homogeneous foliations, Ann. Institut Fourier 29 (1979), 143–158.

R. Blumenthal, The base-like cohomology of a class of transversely homogeneous foliations, Bull. Des Sciences Math. 104 (1980), 301–303.

R. Blumenthal, Riemannian homogeneous foliations without holonomy, Nagoya Math. J. 83 (1981), 197–207.

R. Blumenthal, Foliated manifolds with flat basic connection, J. Diff. Geom. 16 (1981), 401–406.

R. Blumenthal, Basic connections with vanishing curvature and parallel torsion, Bull. des Sciences Math. 106 (1982), 393–400.

R. Blumenthal, Riemannian foliations with parallel curvature, Nagoya Math. J. 90 (1983), 145–153.

R. Blumenthal, Transverse curvature of foliated manifolds, Astérisque 116 (1984), 25–30.

R. Blumenthal, Foliations with locally reductive normal bundle, Ill. J. Math. 28 (1984), 691–702.

R. Blumenthal, Stability theorems for conformal foliations, Proc. Amer. Math. Soc. 91 (1984), 485–491.

R. Blumenthal, Cartan connections in foliated bundles, Mich. Math. J. 31 (1984), 55–63.

R. Blumenthal, Local isomorphisms of projective and conformal structures, Geometriae Dedicata 16 (1984), 73–78.

R. Blumenthal, Affine submersions, and foliations of affinely connected manifolds, C. R. Acad. Sci. Paris 299 (1984), 1013–1015.

R. Blumenthal, Connections on foliated manifolds, Springer Lecture Notes 1165 (1985), 30–35.

R. Blumenthal, Sprays, fiber spaces, and product decompositions, Proceedings of the fifth international colloquium on differential geometry, santiago de Compostela (1984), pitman research notes 131 (1985), 156–161.

R. Blumenthal, Affine submersions, Ann. Global Anal. Geom. 3 (1985), 275–285.

R. Blumenthal, Les applications de cartan et les feuilletages à modèle transverse d'un espace à connexion de Cartan, C. R. Acad. Sci. Paris 301 (1985), 919–922.

R. Blumenthal, Mappings between manifolds with cartan connections, Springer Lecture Notes 1209 (1986), 94–99.

R. Blumenthal, Cartan submersions and cartan foliations, Ill. J. Math. 31 (1987), 327–343.

R. Blumenthal and J. Hebda, De Rham decomposition theorems for foliated manifolds, Ann. Inst. Fourier 33 (1983), 183–198.

R. Blumenthal and J. Hebda, Complementary distributions which preserve the leaf geometry and applications to totally geodesic foliations, Quarterly J. Math. Oxford 35 (1984), 383–392.

R. Blumenthal and J. Hebda, Ehresmann connections for foliations, Indiana Univ. Math. J. 33 (1984), 597–611.

R. Blumenthal and J. Hebda, Un analogue de la nappe d'holonomie pour une variété feuilletée, C. R. Acad. Sci. Paris 303 (1986), 931–934.

R. Blumenthal and J. Hebda, A sufficient condition for the leaves of a totally umbilic foliation to be conformally complete, Ann. Global Anal. Geom. 6 (1988), 165–175.

R. Blumenthal and J. Hebda, An analogue of the holonomy bundle for a foliated manifold, Tôhoku Math. J. 40 (1988), 189–197.

N. Blyakhman and N. Zhukova, Foliations that are compatible with second-order differential equations, methods of the qualitative theory of differential equations, 31–42, Gor'kov. Gos. Univ., Gorki, 1987.

J. Bolton, Transnormal hypersurfaces, Proc. Cambridge Phil. Soc. 74 (1973), 43–48.

J. Bolton, Transnormal systems, Quart. J. Math. Oxford 24 (1973), 385–395.

F. Bonahon, Earthquakes on Riemann surfaces and on measured geodesic laminations, Trans. Amer. Math. Soc. 330 (1992), 69–95.

C. Bonatti, Existence of codimension one singular foliations with dense leaves on closed manifolds, C. R. Acad. Sci. Paris 300 (1985), 493–496.

C. Bonatti, Sur les feuilletages singuliers stables des variétés de dimension trois, Comment. Math. Helv. 60 (1985), 429–444.

C. Bonatti, Existence de feuilles compactes pour les feuilles proches d'une fibration, C. R. Acad. Sci. Paris 305 (1987), 199–202.

C. Bonatti, Stabilité de feuilles compactes pour les feuilletages définis par des fibrations, Topology 29 (1990), 231–245.

C. Bonatti and S. Firmo, Feuilles compactes d'un feuilletage générique en codimension un, Ann. Ec. Norm. Sup. 27 (1994), 407–462.

C. Bonatti and X. Gómez-Mont, The index of holomorphic vector fields on singular varieties I, Astérisque 222 (1994), 9–35.

C. Bonatti and A. Haefliger, Déformations de feuilletages, Topology 29 (1990), 205–229.

C. Bonatti and R. Langevin, Un exemple de flot d'Anosov transitif transverse à un tore et non conjugué à une suspension, Ergodic Theory Dyn. Systems 14 (1994), 633–643.

A. Bonome and L. A. Cordero, The gla-cohomology of vector-valued differential forms on foliated manifolds, Boletin Acad. Galega de Ciencias, vol. I (1982), 53–65.

W. Boothby, Transversely complete e-foliations of codimension one and accessibility properties of non-linear systems, Lie groups: History, Frontiers and Appl. Vol. VII, Math. Sci. Press, Brookline (1977), 361–385.

A. Borisenko and A. Yampolski, Riemannian geometry of bundles, Russian Math Surveys 46 (1991), 55–106.

R. Bott, Vector fields and characteristic numbers, Mich. Math. J. 14 (1967), 231–244.

R. Bott, On a topological obstruction to integrability. In: Global Analysis, Proceedings of Symposia in Pure Math., vol. 16 (1970), 127–131.

R. Bott, On topological obstructions to integrability, Actes Congr. Int. Mathematiciens 1970, 1, Paris (1971), 27–36.

R. Bott, Lectures on characteristic classes and foliations, Springer Lecture Notes in Math. 279 (1972), 1–94.

R. Bott, The Lefschetz formula and exotic characteristic classes, Geometria Differenziale, Roma, Symp. Math. 10 (1972), 95–105.

R. Bott, Gelfand-Fuks cohomology and foliations, Proc. Symp. New Mexico State University (1973).

R. Bott, On the Chern-Weil homomorphism and the continuous cohomology of Lie Groups., Adv. in Math. 11 (1973), 289–303.

R. Bott, Some aspects of invariant theory in differential geometry. In: Differential Operators on Manifolds, Varenna, C.I.M.E. Lectures, 3 Ciclo 1975, 49–145.

R. Bott, Some remarks on continuous cohomology, Manifolds-Tokyo 1973, Proc. Internat. Conf. on Manifolds and Related Topics in Topology, 161–170, Univ. Tokyo Press, 1975.

R. Bott, On characteristic classes in the framework of Gelfand-Fuks cohomology, Astérisque 32–33 (1976), 113–139.

R. Bott, On the characteristic classes of groups of diffeomorphisms, Enseign. Math. 239 (1977), 209–220.

R. Bott, On some formulas for the characteristic classes of group actions, Springer Lecture Notes in Math. 652 (1978), 25–61.

R. Bott and A. Haefliger, On characteristic classes of Γ-foliations, Bull. Amer. Math. Soc. 78 (1972), 1039–1044.

R. Bott and J. Heitsch, A remark on the integral cohomology of B_q, Topology 11 (1972), 141–146.

R. Bott and g. segal, the cohomology of the vector fields on a manifold, topology 16 (1977), 285–298.

R. Bott, H. Shulman and J. Stasheff, On the De Rham theory of certain classifying spaces, Adv. Math. 20 (1976), 43–56.

H. Boualem, Feuilletages riemanniens singuliers transversalement intégrables, C. R. Acad. Sci. Paris 314 (1992), 547–550.

H. Boualem, Feuilletages riemanniens singuliers transversalement intégrables, Thèse, Montpellier 1993.

H. Boualem, Théorème de décomposition d'un feuilletage riemannien singulier transversalement intégrable, C. R. Acad. Sci. Paris 316 (1993), 59–62.

H. Boualem, théorème de décomposition d'un feuilletage riemannien transversalement intégrable, C. R. Acad. Sci. Paris 316 (1993), 59–62.

H. Boualem, Feuilletages riemanniens singuliers transversalement intégrables, Compositio Math. 95(1995), 101–125.

H. Boualem and p. molino, modèles locaux saturés de feuilletages riemanniens singuliers, C. R. Acad. Sci. Paris 316 (1993), 913–916.

L. Bouma and W. van Est, Manifold schemes and foliations on the 2-torus and the klein bottle, I, Proc. K. Ned. Akad. Wet., Ser. A 81 (1978), 313–325.

L. Bouma and W. van Est, Manifold schemes and foliations on the 2-torus and the Klein bottle, II, Proc. K. Ned. Akad. Wet., Ser. A 81 (1978), 326–338.

L. Bouma and W. van Est, Manifold schemes and foliations on the 2-torus and the Klein bottle, III, Proc. K. Ned. Akad. Wet., Ser. A 81 (1978), 339–347.

D. Boutat, Feuilletages isodrastiques de Weinstein et phase de Berry-Weinstein pour les mouvements des sous-variétés lagrangiennes: Cas des surfaces symplectiques, Thèse, Lyon-I, 1993.

D. Boutat, Sur les feuilletages isodrastiques de Weinstein, C. R. Acad. Sci. Paris 318 (1994), 477–480.

R. Bowen, Unique ergodicity for foliations, Astérisque 40 (1976), 11–16.

R. Bowen, Anosov foliations are hyperfinite, Ann. of Math. 106 (1977), 549–565.

R. Bowen and B. Marcus, Unique ergodicity for horocycle foliations, Israel J. Math. 26 (1977), 43–67.

J. Bracho, Strong classificiation of Haefliger structures, Some geometry of BG, Algebraic Topology Oaxtepec 1981, Contemp. Math. 12 (1982), 61–72.

J. Bracho, Haefliger structures and linear homotopy, Trans. Amer. Math. Soc. 282 (1984), 341–350.

A. Brakhman, Foliations without limit cycles, Mat. Zametki 9 (1971), 181–191.

D. Brill and F. Flaherty, Isolated maximal surfaces in space-time, Commun. Math. Phys. 50 (1976), 157–165.

D. Brill and F. Flaherty, Maximizing properties of extremal surfaces in general relativity, Ann. Inst. Henri Poincaré, a, 28 (1978), 335–347.

F. Brito, Une obstruction géométrique à l'existence de feuilletages de codimension 1 totalement géodésiques, J. Diff. Geom. 16 (1981), 675–684.

F. Brito, A remark on minimal foliations of codimension two, Tôhoku Math. J. 36 (1984), 341–350.

F. Brito, R. Langevin and H. Rosenberg, Intégrales de courbure sur une variété feuilletée, C. R. Acad. Sci. Paris 285 (1977), 533–536.

F. Brito, R. Langevin and H. Rosenberg, Intégrales de courbure sur une variété feuilletée, J. Diff. Geom. 16 (1981), 19–50.

F. Brito and P. Walczak, Totally geodesic foliations with integrable normal bundles, Bol. Soc. Bras. Mat. 17 (1986), 41–46.

F. Brito and P. Walczak, Total curvature of orthogonal vector fields on three-manifolds, Bull. Polish Acad. Sc. Math. 35 (1987), 553–556.

M. Brittenham, Essential laminations in non-Haken-3-manifolds, Topology Appl. 53 (1993), 317–324.

M. Brittenham, Essential laminations in Seifert fibered spaces, Topology 32 (1993), 61–85.

M. Brittenham, Essential laminations and Haken normal form, Pac. J. Math. 168 (1995), 217–234.

M. Brittenham, Essential laminations and Haken normal form: laminations with no holonomy, Comm. in Anal. and Geom. 3 (1995), 465–477.

M. Brittenham, Essential laminations and Haken normal form: regular cell decompositions, preprint.

I. Bronstein and A. Kopanskii, Smooth invariant manifolds and normal forms, World Scientific, 1994.

R. Brooks, Volumes and characteristic classes of foliations, Topology 18 (1979), 295–304.

R. Brooks, Some riemannian and dynamical invariants of foliations, Proc. of the 1981–82 year in Differential Geometry, Univ. of Maryland, Birkhäuser, Progress in Math. 32 (1983), 56–72.

R. Brooks, The spectral geometry of foliations, Amer. J. Math. 106 (1984), 1001–1012.

R. Brooks and W. Goldman, The godbillon-vey invariant of a transversely homogeneous foliation, Trans. Amer. Math. Soc. 286 (1984), 651–664.

H. Browne, Codimension 1 totally geodesic foliations of H^n, Tôhoku Math. J. 36 (1984), 315–340.

M. Brunella, Expansive flows on Seifert manifolds and on torus bundles, Bol. Soc. Brasil. Mat. 24 (1993), 89–104.

M. Brunella, Foliations on the complex projective plane with many parabolic leaves, Ann. Inst. Fourier 44 (1994), 1237–1242.

M. Brunella, On the discrete Godbillon-Vey invariant and Dehn surgery on geodesic flows, Ann. de la Fac. Sc. Toulouse 3 (1994), 335–344.

M. Brunella, Remarks on structurally stable proper foliations, Math. Proc. Cambridge Philos. Soc. 115 (1994), 111–120.

M. Brunella, Surfaces of section for expansive flows on three-manifolds, J. Math. Soc. Japan 47 (1995), 491–501.

M. Brunella, Une remarque sur des champs de vecteurs holomorphes transverses au bord d'un convexe, preprint.

M. Brunella and E. Ghys, Umbilical foliations and transversely holomorphic flows, J. Diff. Geom. 41 (1995), 1–19.

M. Brunella and P. Sad, Holomorphic foliations in certain holomorphically convex domains of $\mathbb{C}^2$, Bull. Soc. Math. France 123 (1995), 535–546.

J. Brüning and F. Kamber, On the spectrum and index of transversal Dirac operators associated to Riemannian foliations, to appear.

J. L. Brylinski, Noncommutative Ruelle-Sullivan type currents, the Grothendieck Festschrift i, 477–498, Progr. in Math. 86, Birkhäuser, 1990.

K. Buchner and R. Rosca, Invariant submanifolds and proper CR foliations on a paraco-Kählerian manifold with concircular structure vector field, Rend. Circ. Mat. Palermo 37 (1988), 161–173.

A. Bucki, Geometry of leaves of some distributions on almost r-paracontact riemannian manifolds, Analysis and Geometry in Foliated Manifolds, Proc. VII Int. Colloq. on Diff. Geom., Santiago de Compostela 1994, World Scientific 1995.

R. Buemi, An obstruction to certain non-integrable 2-plane fields, Topology 16 (1977), 173–176.

J.-P. Buffet and J.-C. Lor, Une construction d'un universel pour une classe assez large de Γ-structures, C. R. Acad. Sci. Paris 270 (1970), 640–642.

K. Bugajska, Structure of a leaf of some codimension one Riemannian foliation, Ann. Inst. Fourier 38 (1988), 169–174.

D. Burns, Curvatures of monge-ampère foliations, Ann. of Math. 115 (1982), 349–373.

R. Caddeo, On the torsional cohomology of Molino for an almost complex manifold, Rend. Sem. Fac. Sci. Univ. Cagliari 50 (1980), 765–777.

G. Cairns, Feuilletages riemanniens et classes caractéristiques fines et exotiques, Thèse 3e cycle, Montpellier (1982).

G. Cairns, Géométrie globale des feuilletages totalement géodésiques, C. R. Acad. Sci. Paris 297 (1983), 525–527.

G. Cairns, Feuilletages totalement géodésiques de dimension 1, 2 ou 3, C. R. Acad. Sci. Paris 298 (1984), 341–344.

G. Cairns, Feuilletages totalement géodésiques, Séminaire de géométrie différentielle 1983–84, Montpellier.

G. Cairns, Une remarque sur la cohomologie basique d'un feuilletage riemannien, Séminaire de géométrie différentielle 1984–1985, 1–7, Montpellier, 1985.

G. Cairns, Aspects cohomologiques des feuilletages totalement géodésiques, C. R. Acad. Sci. Paris 299 (1984), 1017–1019.

G. Cairns, A general description of totally geodesic foliations, Tôhoku Math. J. 38 (1986), 37–55.

G. Cairns, Some properties of a cohomology group associated to a totally geodesic foliation, Math. Z. 192 (1986), 391–403.

G. Cairns, Feuilletages géodésibles, Thèse, Univ. des Sciences et Techniques du Languedoc, Montpellier, 1987.

G. Cairns, Feuilletages géodésibles sur les variétés simplement connexes, in Séminaire Sud-Rhodanien de Géométrie VII, Vol. II, ed. N. Desolneux-Moulis, P. Dazord, Travaux en Cours, Herman, Paris, 1987.

G. Cairns, Feuilletages totalement géodésiques sur les variétés simplement connexes, Feuilletages riemanniens, quantification géométrique et mécanique (Lyon, 1986), 1–14, Travaux en Cours, 26, Hermann, Paris, 1988.

G. Cairns, The duality between Riemannian foliations and geodesible foliations, in P. Molino, Riemannian foliations, Birkhäuser (1988), 249–263.

G. Cairns, Totally umbilic Riemannian foliations, Mich. Math. J. 37 (1990), 145–159.

G. Cairns, Compact 4-manifolds that admit totally umbilic metric foliations, Diff. Geom. and its Appl., Proc. Conf. 1989, Brno, World Scientific 1990, 9–16.

G. Cairns and R. Escobales, Bundle-like flows on curved manifolds, preprint.

G. Cairns and R. Escobales, Further geometry of the mean curvature one-form and the normal plane field one-form on a foliated Riemannian manifold, preprint.

G. Cairns and E. Ghys, Totally geodesic foliations on 4-manifolds, J. Diff. Geom. 23 (1986), 241–254.

E. Calabi, An intrinsic characterization of harmonic 1-forms, Global Analysis, Papers in honor of K. Kodaira, Ed. by D. C. Spencer and S. Iyanaga, Princeton Math. Series 29 (1969), 101–117.

M. Calapso and R. Rosca, Cosymplectic quasi-Sasakian pseudo-Riemannian manifolds and coisotropic foliations, Rend. Circ. Mat. Palermo 36 (1987), 407–422.

B. Callenaere and D. Lehmann, Classes exotiques universelles, Ann. Inst. Fourier 24 (1974), 301–306.

O. Calvo, Persistencia de folheacoes definidas por formas logaritmicas, Thesis, IMPA, 1990.

O. Calvo-Andrade, Deformations of holomorphic foliations, Contemp. Math. 161 (1994), 21–28.

C. Camacho, Structural stability of foliations with singularities, Springer Lecture Notes in Math. 652 (1978), 128–137.

C. Camacho, Structural stability theorems for integrable differential forms on 3-manifolds, Topology 17 (1978), 143–155.

C. Camacho, Singularities of holomorphic differential equations, Singularities and dynamical systems, Iraklion, 1983; North-Holland Math. Stud. 103 (1985), 137–159.

C. Camacho, Quadratic forms and holomorphic foliations on singular surfaces, Math. Ann. 282 (1988), 177–184.

C. Camacho, Problems on limit sets of foliations on complex projective spaces, Proc. Int. Congr. Math., Kyoto 1990, Vol. II (1991), 1235–1239.

C. Camacho, F. Cano and P. Sad, Absolutely isolated singularities of holomorphic vector fields, Invent. Math. 98 (1989), 351–369.

C. Camacho, N. Kuiper and J. Palis, La topologie du feuilletage d'un champ de vecteurs holomorphes près d'une singularité, C. R. Acad. Sci. Paris 282 (1976), 959–961.

C. Camacho, N. Kuiper and J. Palis, The topology of holomorphic flows with singularities, Publ. Math. Inst. Hautes Etudes Sci. 48 (1978), 5–38.

C. Camacho and A. Lins Neto, Orbit preserving diffeomorphisms and the stability of Lie group actions and singular foliations, Geometry and Topology, Rio de Janeiro 1976, Springer Lecture Notes in Math. 597 (1977), 82–103.

C. Camacho and A. Lins Neto, Stabilité des feuilletages définis par une forme différentielle intégrable au voisinage d'une singularité, C. R. Acad. Sci. Paris 290 (1980), 423–425.

C. Camacho and A. Lins Neto, Geometric Theory of Foliations, IMPA 1979 (Portuguese) and Birkhäuser, Boston, 1985 (English).

C. Camacho and A. Lins Neto, Minimal sets of foliations on complex projective spaces, Inst. Hantes Etudes Sci. Publ. Math. 68 (1988), 187–203.

C. Camacho, A. Lins Neto and P. Sad, Foliations with algebraic limit set, Ann. of Math. 136 (1992), 429–446.

M. Camacho and C. Palmeira, Polynomial foliations of degree 3 in the plane, Pitman Res. Notes Math. Ser 160 (1987), 27–58.

F. Campana, Algébricité de l'espace des feuilletages d'un espace analytique compact, Math. Ann. 281 (1988), 387–391.

A. Candel, Uniformization of surface laminations, Ann. Sci. Ec. Norm. Sup. 26 (1993), 489–516.

A. Candel and X. Gómez-Mont, Uniformization of the leaves of a rational vector field, Ann. Inst. Fourier 45 (1995), 1123–1133.

F. Cano, Desingularization strategies for three-dimensional vector fields, Springer Lecture Notes in Math. 1259 (1987), 194 pp.

F. Cano, Réduction des singularités des feuilletages holomorphes, C. R. Acad. Sci. Paris 307 (1988), 795–798.

F. Cano, Dicritical singular foliations, Mem. Real Acad. Cienc. Exact. Fis. Natur. Madrid 24 (1989), V1+ 154 pp.

J. Cantwell and L. Conlon, Open leaves in closed 3-manifolds, Bull. Amer. Math. Soc. 82 (1976), 256–258.

J. Cantwell and L. Conlon, Closed transversals and the genus of closed leaves in foliated 3-manifolds, J. Math. Anal. Appl. 55 (1976), 653–657.

J. Cantwell and L. Conlon, Leaves with isolated ends in foliated 3-manifolds, Topology 16 (1977), 311–322.

J. Cantwell and L. Conlon, Growth of leaves, Comment. Math. Helv. 53 (1978), 93–111.

J. Cantwell and L. Conlon, Leaf prescriptions for closed 3-manifolds, Trans. Amer. Math. Soc. 236 (1978), 239–261.

J. Cantwell and L. Conlon, Endsets of leaves, Topology 21 (1980), 333–352.

J. Cantwell and L. Conlon, Poincaré-Bendixson theory for leaves of codimension one, Trans. Amer. Soc. 265 (1981), 181–209.

J. Cantwell and L. Conlon, Reeb stability for noncompact leaves in foliated 3-manifolds, Proc. Amer. Math. Soc. 83 (1981), 113–135.

J. Cantwell and L. Conlon, Smoothing fractional growth, Tôhoku Math. J. 33 (1981), 249–262.

J. Cantwell and L. Conlon, Tischler fibrations of open foliated sets, Ann. Inst. Fourier 31 (1981), 113–135.

J. Cantwell and L. Conlon, Nonexponential leaves at finite level, Trans. Amer. Math. Soc. 269 (1982), 637–661.

J. Cantwell and L. Conlon, Analytic foliations and the theory of levels, Math. Ann. 265 (1983), 253–261.

J. Cantwell and L. Conlon, The dynamics of open, foliated manifolds and a vanishing theorem for the Godbillon-Vey class, Advances in Math. 53 (1984), 1–27.

J. Cantwell and L. Conlon, Every surface is a leaf, Topology 26 (1987), 265–285.

J. Cantwell and L. Conlon, Foliations and subshifts, Tôhoku Math. J. 40 (1988), 165–187.

J. Cantwell and L. Conlon, Smoothability of proper foliations, Ann. Inst. Fourier 38 (1988), 219–244.

J. Cantwell and L. Conlon, The theory of levels, Contemp. Math. 70 (1988), 1–10.

J. Cantwell and L. Conlon, Leaves of Markov local minimal sets in foliations of codimension one, Publicacions Matemàtiques 33 (1989), 461–484.

J. Cantwell and L. Conlon, Leafwise hyperbolicity of proper foliations, Comment. Math. Helv. 64 (1989), 329–337; a correction, ibid. 66 (1991), 319–321.

J. Cantwell and L. Conlon, Depth of knots, Topology and its Appl. 42 (1991), 277–289.

J. Cantwell and L. Conlon, Markov minimal sets have hyperbolic leaves, Ann. Global Anal. Geom. 9 (1991), 13–25.

J. Cantwell and L. Conlon, Foliations of $E(5_2)$ and related knot complements, Proc. Amer. Math. Soc. 118 (1993), 953–962.

J. Cantwell and L. Conlon, Surgery and foliations of knot complements, Journal of Knot Theory and its Ramifications 2 (1993), 369–397.

J. Cantwell and L. Conlon, Endsets of exceptional leaves; a theorem of G. Duminy, (an account prepared for informal circulation only).

J. Cantwell and L. Conlon, Isotopy of depth one foliations, Geometric study of Foliations, Tokyo 1993, 153–173, World Scientific, 1994.

J. Cantwell and L. Conlon, Topological obstructions to smoothing proper foliations, Contemp. Math. 161 (1994), 1–20.

J. Cantwell and L. Conlon, General position in tautly foliated, sutured manifolds, to appear.

J. Cantwell and L. Conlon, Isotopies of foliated 3-manifolds without holonomy, to appear.

J. Carballés, Characteristic homomorphism for (F_1, F_2)-foliated bundles over subfoliated manifolds, Ann. Inst. Fourier 34 (1984), 219–245.

A. Carfagna D'Andrea, A characterization of the tangent bundle of a foliation, C. R. Acad. Sci. Paris 301 (1985), 77–80.

M. Carfora, Zero-lapse loci in asymptotically flat maximally foliated spacetime manifolds, Phys. Lett. A 84 (1981), 53–55.

J. Carinena and L. Ibort, On Lax equations arising from Lagrangian foliations, Lett. Math. Phys. 8 (1984), 21–26.

M. Carnicer, Cremona transformations and foliations on the complex projective plane, Symp. on Singularity Theory, ICTP, Aug 19–Sept 6, 1991, World Scientific (1995), 153–172.

P. Caron, Flots transversalement de Lie. Thèse de 3ème cycle, Université de Lille (1980).

P. Caron and Y. Carrière, Flots transversalements de Lie $\mathbb{R}^n$, flots de Lie minimaux, C. R. Acad. Sci. Paris 280 (1980), 477–478.

F. Carreras, Linear invariants of Riemannian almost product manifolds, Math. Proc. Camb. Phil. Soc. 91 (1982), 99–106.

F. Carreras and A. Naveira, On the Pontrjagin algebra of a certain class of flags of foliations, Canad. Math. Bull. 28 (1985), 77–83.

Y. Carrière, Flots riemanniens et feuilletages géodésibles de codimension un, Thèse Université des Sciences et Techniques de Lille I (1981).

Y. Carrière, Flots riemanniens, Astérisque 116 (1984), 31–52.

Y. Carrière, Les propriétés topologiques des flots riemanniens retrouvées à l'aide du théorème des variétés presque plates, Math. Z. 186 (1984), 393–400.

Y. Carrière, Sur la croissance des feuilletages de Lie, prépublication IRMA, Lille, Vol. VI, fasc. 3, 1984.

Y. Carrière, Feuilletages riemanniens à croissance polynomiale, Comment. Math. Helv. 63 (1988), 1–20.

Y. Carrière, Variations on Riemannian flows, Appendix A, pp. 217–234 in Riemannian foliations, by P. Molino, Progress in Math 73 (1988), Birkhäuser.

Y. Carrière and E. Ghys, Feuilletages totalement géodésiques, Anais da Acad. Bras. de Ciencias 53 (1981), 427–432.

Y. Carrière and E. Ghys, Relations d'equivalence moyennables sur les groupes de Lie, C.R. Acad. Sci. Paris 300 (1985), 677–680.

E. Cartan, Sur certaines expressions différentielles et le problème de Pfaff, Ann. Sci. Ecole Norm. Sup. 16 (1899). Oeuvres II, 303–396.

E. Cartan, Sur l'intégration des systèmes d'équations aux différentielles totales, Ann. Sci. Ecole Norm. Sup. 18 (1901), 241–311, Oeuvres II, 411–481.

H. Cartan, Cohomologie réelle d'un espace fibré principal différentiable. In: Sém. Cartan 1949/50, exp. 19–20. Paris: Ecole Norm. Sup.

S. Carter and Z. Sentürk, The space of immersions parallel to a given immersion, J. London Math. Soc. 50 (1994), 404–416.

S. Carter and A. West, Isoparametric systems and transnormality, Proc. London Math. Soc. 51 (1985), 520–542.

D. Cass, Minimal leaves in foliations, Trans. Amer. Math. Soc. 287 (1985), 201–213.

I. Cattaneo-Gasparini, Connessioni adattate ad una struttura quasi prodotto, Ann. Mat. Pura ed Appl. 63 (1963), 133–150.

I. Cattaneo-Gasparini, Global reduction of a dynamical system on a foliated manifold, J. Math. Phys. 25 (1984), 2918–2921.

I. Cattaneo-Gasparini, Global reduction of a dynamical system on a foliated manifold and controlled projectability, Dynamical systems and microphysics, Academic Press, 1984, 183–205.

V. Cavalier, Feuilletages transversalment holomorphes, quasi-transversalment parallelisables, Thèse, Academie Montpellier, Université des Sciences et Techniques du Languedoc, Montpellier, 1987.

H. Cendra, E. Lacomba and A. Verdiell, A new proof of Frobenius theorem and applications, Z. Angew. Math. Phys. 44 (1993), 266–281.

B. Cenkl, On the de Rham complex of $B\Gamma$, Proc. Symp. Pure Math. 27 (1973), Part 1, 265–274.

B. Cenkl, Residues of singularities of holomorphic foliations, J. Diff. Geom. 13 (1978), 11–23.

B. Cenkl, Formulas for the characteristic classes of groups of diffeomorphisms, Rend. Mat. (7) 1 (1981), 443–462.

J. Cerf, 1-forms fermées non singulières sur les variétés compactes de dimension 3, Séminaire Bourbaki, exp. 574, Springer Lecture Notes in Math. 901 (1981), 205–219.

D. Cerveau, Distributions involutives singulières, Ann. Inst. Fourier 29 (1979), 261–294.

D. Cerveau, Integrating agents and the cobordism problem of germs of one-codimensional singular holomorphic foliations, Hokkaido Math. J. 14 (1985), 21–32.

D. Cerveau, Feuilletages réglés, C. R. Acad. Sci. Paris 307 (1988), 33–36.

D. Cerveau, Minimaux des feuilletages algébriques de $\mathbb{C}P(n)$, Ann. Inst. Fourier 43 (1993), 1535–1543.

D. Cerveau, Théorèmes de type Fuchs pour les tissus feuilletés, Astérisque 222 (1994), 49–92.

D. Cerveau and A. Lins Neto, Holomorphic foliations in $\mathbb{C}P^2$ having an invariant algebraic curve, Ann. Inst. Fourier 41 (1991), 883–903.

D. Cerveau and A. Lins Neto, Codimension one foliations in $\mathbb{C}P^n$, $n \geq 3$, with Kupka components, Astérisque 222 (1994), 93–133.

D. Cerveau and F. Maghous, Algebraic foliations in C^n, C. R. Acad. Sci. Paris 303 (1986), 643–645.

D. Cerveau et J. Mattei, Formes intégrables holomorphes singulières, Astérisque 97 (1982).

D. Cerveau and T. Suwa, Determinary of complex analytic foliation germs without integrating factors, Proc. Amer. Math. Soc. 112 (1991), 989–997.

C. Charitos, Foliations with Singularities of Morse type on compact surface of genus zero, Math. Nachr. 169 (1994), 81–88.

G. Chatelet, Sur les feuilletages induits par l'action de groupes de Lie nilpotents, Ann. Inst. Fourier 27 (1977), 161–189.

G. Chatelet and H. Rosenberg, Un théorème de conjugaison des feuilletages, Ann. Inst. Fourier 21 (1971), 95–106.

G. Chatelet and H. Rosenberg, Manifolds which admit $\mathbb{R}^n$-actions, Publ. Math. Hautes Etudes Sci. 43 (1973), 245–260.

G. Chatelet, H. Rosenberg and D. Weil, A classification of the topological types of $\mathbb{R}^2$-actions on closed orientable 3-manifolds, Publ. Math. IHES 43 (1973), 261–272.

F. Chazal, Un théorème de fibration pour les feuilletages algébriques de codimension un de $\mathbb{R}^n$, C. R. Acad. Sci. Paris 321 (1995), 327–330.

B. Chen and P. Piccinni, The canonical foliation of a locally conformal Kähler manifold, Ann. Mat. Pura Appl. 141 (1985), 289–305.

S. Cheng and S. Yau, Maximal spacelike hypersurfaces in the Lorentz-Minkowski space, Ann. of Math. 104 (1976), 407–419.

S. Chern, The geometry of G-Structures, Bull. Amer. Math. Soc. 72 (1966), 167–219.

S. Chern and J. Simons, Characteristic forms and geometric invariants, Ann. of Math. 99 (1974), 48–69.

S. Chern and K. Tenenblat, Foliations on a surface of constant curvature and the modified Korteweg-de Vries equation, J. Diff. Geom. 16 (1981), 347–349.

P. R. Chernoff, Essential self-adjointness of powers of generators of hyperbolic equations, J. Funct. Anal. 12 (1983), 401–404.

T. Cherry, Analytic quasi-periodic curves of discontinuous type on a torus, Proc. London Math. Soc. 44 (1938), 175–215.

D. Chinea, M. de León and J. Marrero, Symplectic and cosymplectic foliations on cosymplectic manifolds, Publ. Inst. Math. (Beograd), Nouv. Ser. 50 (64), (1991), 163–169.

D. Chinea, M. de León and J. Marrero, Stability of invariant foliations on almost contact manifolds, Publ. Math. Debrecen 43 (1993), 41–52.

K. Cho, J. Kwan and J. Pak, Transverse conformal mappings of complete foliated Riemannian manifolds with harmonic foliation, Math. J. Toyama Univ. 15 (1992), 43–58.

K. Cho, J. Pak and W. Sohn, A note on spectral characterizations of Sasakian foliations, Kyungpook Math. J. 34 (1994), 283–291.

S. Chow, X. Lin, and K. Lu, Smooth invariant foliations in infinite-dimensional spaces, J. Differential Equations 94 (1991), 266–291.

W. Chow, Über Systeme von linearen partiellen Differentialgleichungen erster Ordnung, Math. Ann. 117 (1940/41), 98–105.

D. Christodoulou and M. Francaviglia, Some dynamical properties of Einstein spacetimes admitting a Gaussian foliation, Gen. Relativity Gravitation 10 (1979), 455–459.

J. Christy, Intransitive Anosov flows on 3-manifolds, to appear.

P. Chrusciel, Sur les feuilletages conformément minimaux des variétés riemanniennes de dimension trois, C. R. Acad. Sci. Paris 301 (1985), 609–612.

W. Cieslak and W. Mozgawa, Euclidean plane foliations, Ann. Univ. Mariae Curie-Sklodowska Sect. A 41 (1987), 1–7.

W. Cieslak and W. Mozgawa, Quelques remarques sur les flots riemanniens singuliers, An. Univ. Timisoara Ser. Stiint, Mat. 25 (1987), 15–20.

W. Claus, Essential laminations in closed Seifert fibered spaces, Thesis, Univ. of Texas, Austin, 1991.

A. Clebsch, Über die simultane Integration linearer partieller Differentialgleichungen, J. Reine Angew. Math. 65 (1866), 257–268.

Y. Clifton and J. Smith, The Enter class as an obstruction in the theory of foliations, Proc. Nat. Acad. Sci. USA 50 (1963), 949–954.

F. Cohen and L. Taylor, Computations of Gelfand-Fuks cohomology, the cohomology of function spaces, and the cohomology of configuration spaces, Springer Lecture Notes in Math. 657 (1978), 106–143.

M. Cohen, Approximation of foliations, Can. Math. Bull. 14 (1971), 311–314.

M. Cohen, Smoothing one-dimensional foliations on $S^1 \times S^1$, Can. Math. Bull. 16 (1973), 43–44.

M. Cohen, Foliations on 3-manifolds, Amer. Math. Monthly 81 (1974), 462–473.

L. Conlon, Lectures on Foliations and Characteristic Classes by Raoul Bott, (Notes by Lawrence Conlon), Springer Lectures in Math. 279 (1972), 1–80.

L. Conlon, Foliations and locally free transformation groups of codimension two, Mich. Math. J. 21 (1974), 87–96.

L. Conlon, Locally free Lie transformation groups of codimension two, Proc. Symp. Pure Math., Amer. Math. Soc. 27 (1975), 275–276.

L. Conlon, Erratum to "Transversally parallelizable foliations of codimension two," Trans. Amer. Math. Soc. 207 (1975), 406.

L. Conlon, Foliations and exotic classes, Lectures at the Universidad de Extramadura, Jarandilla de la Vera (Caceres), 1985.

L. Conlon and S. Goodman, Opening closed leaves in foliated 3-manifolds, Topology 14 (1975), 59–61.

L. Conlon and S. Goodman, The closed leaf index of foliated manifolds, Trans. Amer. Math. Soc. 233 (1977), 205–221.

A. Connes, The von Neumann algebra of a foliation, Springer Lecture Notes in Phys. 80 (1978), 145–151.

A. Connes, Sur la théorie noncommutative de l'intégration. In: Algèbres d'opérateurs, pp. 19–143, Springer Lecture Notes in Math. 725 (1979).

A. Connes, C^*-algèbres et géométrie différentielle, C. R. Acad. Sci. Paris 290 (1980), 599–604.

A. Connes, Feuilletages et algèbres d'opérateurs. In: Sém. Bourbaki 1979–80, exp. 551, Springer Lecture Notes in Math. 842 (1980).

A. Connes, A survey of foliations and operator algebras. Proc. Symp. Pure Math Amer. Math. Soc., 38, Part 1 (1982), 521–628.

A. Connes, Non commutative differential geometry, Publ. Math. IHES 62 (1985), 41–144.

A. Connes, Non commutative differential geometry, Part I, The Chern character in K-homology, Publ. Math. IHES 62 (1985), 49–93.

A. Connes, Non commutative differential geometry, Part II, De Rham homology and non commutative algebra, Publ. Math. IHES 62 (1985), 94–144.

A. Connes, Cyclic cohomology and the transverse fundamental class of a foliation, Geometric methods in operator algebras (Kyoto, 1983), Pitman Res. Notes in Math. 123 (1986), 52–144.

A. Connes, Cyclic cohomology and noncommutative differential geometry, Proc. Int. Congress of Mathematicians (Berkeley, 1986), Vol. 1, 2 (1987), 879–889.

A. Connes, Introduction à la géométric non-commutative, Proc. Symp. Pure Math. Amer. Math. Soc. 50 (1990), 91–118.

A. Connes, Noncommutative geometry, Academic Press, 1994.

A. Connes and G. Skandalis, Théorème de l'indice pour les feuilletages, C. R. Acad. Sci. Paris 292 (1981), 871–876.

A. Connes and G. Skandalis, The longitudinal index theorem for foliations, Publ. Res. Inst. Math. Sci., Kyoto Univ. 20 (1984), 1139–1183.

D. Cooper, D. Long and A. Reid, Bundles and finite foliations, Invent. Math. 118 (1994), 255–283.

L. Cordero, Nota sobre una decomposicion del operador diferencial exterior en una estructura casi-producto, Actas de la undécime reunion annal de matematicos Espanoles (1971), 124–129.

L. Cordero, Sur une théorie de cohomologie associée aux feuilletages, C. R. Acad. Sci. Paris 272 (1971), 1056–1057.

L. Cordero, P-normal almost-product structure, Tensor 28 (1974), 229–238.

L. Cordero, The extension of G-foliations to tangent bundles of higher order, Nagoya Math. J. 56 (1975), 29–44.

L. Cordero, The horizontal lift of a foliation and its exotic classes, Springer Lecture Notes in Math. 484 (1975), 192–200.

L. Cordero, Special connections on almost-multifoliate Riemannian manifolds, Math. Ann. 216 (1975), 209–215.

L. Cordero, Sheaves and cohomologies associated to subfoliations, Resultate Math. 8 (1985), 9–20.

L. Cordero (Ed), Differential Geometry, Proc. Colloq. Santiago de Compostela 1984, Pitman Research Notes 131 (1985).

L. Cordero and A. de Prada, Foliacion en el fibrado tangente a una variedad foliada y la obstruccion de Bott a la integrabilidad, Actas Prim. J. Mat. Luso-Es. Publs. Inst. 'Jorge Juan' Mat., Madrid (1973), 269–274.

L. Cordero and P. Gadea, Exotic characteristic classes and subfoliations, Ann. Inst. Fourier 26 (1976), 225–237.

L. Cordero and X. Masa, Characteristic classes of subfoliations, Ann. Inst. Fourier 31 (1981), 61–86.

L. Cordero and M. Prada, Sur un feuilletage dans le fibré tangent à une variété feuilletée, C. R. Acad. Sci. Paris 275 (1972), 831–833.

L. Cordero and M. Prada, Sobre las cohomologias y metricas del fibrado tangente a una variedad foliada, Actas Prim. J. Mat. Luso-Esp. Publs. Inst. 'Jorge Juan' Mat., Madrid (1973), 255–259.

L. Cordero and R. Wolak, Examples of foliations with foliated geometric structures, Pacific J. Math. 142 (1990), 265–276.

L. Cordero and R. Wolak, Properties of the basic cohomology of transversely Kähler foliations, Rendiconti del Circolo Matematico di Palermo, Serie II, 40 (1991), 177–188.

M. Craioveanu, Sur un théorème de De Rham pour les fibrés integrables, An. Univ. Bucaresti 19 (1970), 15–19.

M. Craioveanu, Sur les sous-feuilletages d'une structure feuilletée, C. R. Acad. Sci. Paris 272 (1971), 731–733.

M. Craioveanu, Variétés banachiques feuilletées, I, An. Univ. Timisoara, Ser. Sti. Mat. 9 (1971), 35–48.

M. Craioveanu, Variétés banachiques feuilletées, II, An. Univ. Timisoara, Ser. Sti. Math. 13 (1975), 11–12.

M. Craioveanu and M. Puta, DeRham-type currents on Riemannian foliated manifolds, Colloquia Math. Soc. Janos Bolyai, Budapest (1979), 159–165.

M. Craioveanu and M. Puta, Cohomology on a Riemannian foliated manifold with coefficients in the sheaf of germs of foliated currents, Math. Nachr. 99 (1980), 43–53.

M. Craioveanu and M. Puta, Cohomology classes and foliated manifolds, in Nonlinear Analysis, ed. Th. Rassias, World Scientific Publ. Co., Singapore, 1987.

M. Craioveanu and M. Puta, On the basic Laplacian of a Riemannian foliation, The XVIIIth National Conference on Geometry and Topology (Oradea, 1987), 49–52.

M. Craioveanu and M. Puta, Some remarks concerning the basic Laplacian, An. Univ. Timisoara, Seria st. matematice 25 (1987), 3–13.

M. Craioveanu and M. Puta, Asymptotic properties of eigenvalues of the basic Laplacian associated to certain Riemannian foliations, Bull. Math. Soc. Sci. Math. Roum. Nouv. Sér. 35 (83) (1991), 61–65.

P. Crawford and P. Vargas Moniz, Bianchi type III foliations of the de Sitter Space, Int. J. Theor. Phys. 32 (1993), 841–848.

R. Crew and D. Fried, Nonsingular holomorphic flows, Topology 25 (1986), 471–473.

C. Cumenge, Sheaves and cohomology of leaf spaces of foliations, C. R. Acad. Sci. Paris 297 (1983), 195–198.

C. Cumenge, Cohomologie et classes caractéristiques transverses des feuilletages, C. R. Acad. Sci. Paris 305 (1987), 27–30.

C. Curras-Bosch, The geometry of totally geodesic foliations admitting Killing field, Tôhoku Math. J. 40 (1988), 535–548.

C. Curras-Bosch, Codimension-one foliations with one compact leaf, Geometriae Dedicata 32 (1989), 329–340.

C. Curras-Bosch, Killing fields preserving minimal foliations, Yokohama Math. J. 37 (1989), 1–4.

C. Curras-Bosch, On codimension-one foliations, Springer Lecture Notes in Math 1410 (1989), 100–107.

C. Curras-Bosch, Sur les feuilletages lagrangiens à holonomie linéaire, C. R. Acad. Sci. Paris 317 (1993), 605–608.

C. Curras-Bosch and P. Molino, Voisinage d'une feuille compacte dans un feuilletage lagrangien: le problème de linéarisation symplectique, Hokkaido Math. J. 23 (1994), 355–360.

C. Curras-Bosch and P. Molino, Réduction symplectique d'un feuilletage lagrangien an voisinage d'une feuille compacte, C. R. Acad. Sci. Paris 318 (1994), 661–664.

S. Dal Jung and J. Pak, A transversal Dirac operator and some vanishing theorems on a complete foliated Riemannian manifold, Math. J. Toyama Univ. 16 (1993), 97–108.

C. Danthony, Feuilletages orientés des surfaces: le problème de la section globale, Ann. Inst. Fourier 38 (1988), 201–227

C. Danthony, The connectedness of the space of foliations by ties, J. London Math. Soc. 38 (1988), 166–178.

C. Danthony and A. Nogueira, Involutions linéaires et feuilletages mesurés, C. R. Acad. Sci. Paris 307 (1988), 409–412.

C. Danthony and A. Nogueira, Measured foliations on nonorientable surfaces, Ann. Sci. Ecole Norm. Sup. 23 (1990), 469–494.

A. Davis and F. Wilson, Vector fields tangent to foliations, I, Reeb foliations, J. Diff. Eq. 11 (1972), 491–498.

P. Dazord, Sur le géométrie des sous-fibrés et des feuilletages lagrangiens, Ann. Sci. Ec. Norm. Sup. 14 (1981), 465-480; erratum: Ann. Sci. Ec. Norm. Sup 18 (1985), 685.

P. Dazord, Feuilletages en géométrie symplectique, C. R. Acad. Sci. Paris 294 (1982), 489–491.

P. Dazord, Feuilletages et mécanique Hamiltonienne, interventions de la géométrie en analyse et en physique mathématique, Université C. Bernard (1982).

P. Dazord, Sur l'existence de feuilles sphériques, C. R. Acad. Sci. Paris 296 (1983), 77–79.

P. Dazord, Holonomie des feuilletages singuliers, C. R. Acad. Sci. Paris 298 (1984), 27–30.

P. Dazord, Feuilletages à singularités, Ned. Akad. van Wet. Indag. Math. 47 (1985), 21–39.

P. Dazord and G. Hector, Intégration symplectique des variétés de Poisson totalement asphériques, Séminaire Sud-Rhodanien de Géométrie à Berkeley, MSRI Publ. 20 (1991), 37–72.

P. Dazord and P. Molino, Γ-structures Poissoniennes et feuilletages de Libermann, Publ. Dept. Math. Lyon, 1/B (1988), 69–89.

F. Deahna, Über die Bedingungen der Integrabilität linearer Differentialgleichungen erster Ordnung zwischen einer beliebigen Anzahl veränderlicher Grössen, J. reine angew. Math. 20 (1840), 340–349.

K. Decesare and T. Nagano, On compact foliations, Proc. Symp. Pure Math., 27 (1975), Part 1, 277–281.

T. Delzant, Foliations of symplectic manifolds, C. R. Acad. Sci. Paris 300 (1985), 201–204.

A. Denjoy, Sur les courbes définis par les équations différentielles à la surface du tore, J. Math. Pures Appl. 11 (1932), 333–375.

S. Deshmukh, Reduction in codimension of mixed foliate CR-submanifolds of a Kähler manifold, Kodai Math. J. 11 (1988), 155–158.

S. Deshmukh, Mixed foliate CR-submanifolds of a Kaehler manifold, Math. Chronicle 19 (1990), 23–25.

N. Desolneux-Moulis, Sur certaines familles à un paramètre de $T^2 \times S^2$, C. R. Acad. Sci. Paris 287 (1978), 1043–1046.

N. Desolneux-Moulis, Familles à un paramètre de feuilletages proches d'une fibration, Astérisque 80 (1980), 77–84.

N. Desolneux-Moulis, Singular Lagrangian foliation associated to an integrable Hamiltonian vector field, Symplectic geometry, groupoids, and integrable systems (Berkeley, 1989), 129–136, Math. Sci. Res. Inst. Publ. 20, Springer, 1991.

M. Diener, Sur un feuilletage de $\mathbb{R}^3$, Enseign. Math. 22 (1976), 35–40.

M. Diener, Feuilletages de Briot et Bouquet, Enseign. Math. 28 (1977), 101–114.

C. Diop, Sur les feuilletages singuliers presque isométriques, Thèse, Dakar-Montpellier (1993).

C. Diop and P. Molino, Une observation sur les feuilletages presque isométriques, Sémin. Gaston Darboux Géom. Topologie Diff. 1990–1991 (1992), 45–53.

P. Dippolito, Codimension one foliations of closed manifolds, Ann. of Math. 107 (1978), 403–453; erratum: Ann. of Math. 110 (1979), 203.

P. Djoharian, Modèles de Segal pour les structures multifeuilletées, Manuscripta Math. 53 (1985), 107–144.

P. Dombrowski, Jacobi fields, totally geodesic foliations and geodesic differential forms, Resultate der Math. 1 (1978), 156–194.

P. Dombrowski, Classification up to diffeomorphism of measure preserving foliations of the torus $S^2 \times S^2$, following S. Sternberg, Singularities, foliations and Hamiltonian mechanics (Balaruc 1985), Travaux en cours, Hermann (1985), 1–19.

D. Domínguez, Sur les classes caractéristiques des sous-feuilletages, Publ. Res. Inst. Math. Sci. 23 (1987), 813–840.

D. Domínguez, Classes caractéristiques non triviales des feuilles de sous-feuilletages localement homogènes, Geom. Dedicata 28 (1988), 229–249.

D. Domínguez, Classes caractéristiques non triviales des feuilles de sous-feuilletages localement homogènes, Kodai Math. J. 11 (1988), 177–204.

D. Domínguez, On the linear independence of certain cohomology classes in the classifying space of subfoliations, Trans. Amer. Math. Soc. 329 (1992), 221–232.

D. Domínguez, Deformations of secondary classes for subfoliations, Canad. Math. Bull. 35 (1992), 167–173.

D. Domínguez, Residues and secondary characteristic classes for subfoliations, Diff. Geom. and its Applications 4 (1994), 25–44.

D. Domínguez, A tenseness theorem for Riemannian foliations, C. R. Acad. Sci. Paris 320 (1995), 1331–1335.

D. Domínguez, Finiteness and tenseness theorems for Riemannian foliations, to appear.

R. Douglas, J. Glazebrook, F. Kamber and G. Yu, Index formulas for geometric Dirac operators in Riemannian foliations, K-Theory 9 (1995), 407–441.

R. Douglas, S. Hurder and J. Kaminker, Cyclic cocycles, renormalization and eta-invariants, Invent. Math. 103 (1991), 101–179.

S. Druck, Stabilité des feuilles compactes dans les feuilletages donnés par des fibrés, C. R. Acad. Sci. Paris 303 (1986), 471–474.

S. Druck and S. Firmo, Stability of compact leaves close to invariant fibered manifolds, J. Fac. Sc. Univ. Tokyo Sect IA Math. 40 (1993), 285–306.

T. Duchamp, Characteristic invariants of G-foliations, Ph.D. Thesis, University of Illinois, Urbana, Illinois (1976).

T. Duchamp and M. Kalka, Holomorphic foliations and the Kobayashi metric, Proc. Amer. Math. Soc. 67 (1977), 117–122.

T. Duchamp and M. Kalka, Deformation theory for holomorphic foliations, J. Diff. Geom. 14 (1979), 317–337.

T. Duchamp and M. Kalka, Stability theorems for holomorphic foliations, Trans. Amer. Math. Soc. 260 (1980), 255–266.

T. Duchamp and M. Kalka, Holomorphic foliations and deformations of the Hopf foliation, Pac. J. of Math. 112 (1984), 69–81.

T. Duchamp and M. Kalka, Invariants of complex foliations and the Monge-Ampère equation, Michigan Math. J. 35 (1988), 91–115.

T. Duchamp and M. Kalka, Complex foliations, Contemp. Math. 101 (1989), 323–338.

T. Duchamp and M. Kalka, The equivalence problem for complex foliations of complex surfaces, Illinois J. of Math. 34 (1990), 59–77.

J. Dufour (Ed), Singularités, feuilletages et mécanique hamiltonienne, Séminaire Sud-Rhodanien (1985).

G. Duminy, L'invariant de Godbillon-Vey d'un feuilletage se localise dans les feuilles ressort, preprint Lille (1982).

G. Duminy, Sur les cycles feuilletés de codimension un, preprint Lille (1982).

G. Duminy and V. Sergiescu, Sur la nullité de l'invariant de Godbillon-Vey, C. R. Acad. Sci. Paris 292 (1981), 821–824.

T. Dumitru, On the Godbillon-Vey characteristic class, Stud. Cerc. Mat. 37 (1985), 472–475.

J. Dupont, The dilogarithm as a characteristic class for flat bundles, J. Pure Appl. Algebra 44 (1987), 137–164.

J. Dupont, Dilogarithm identities and characteristic classes for flat bundles, to appear.

J. Dupont and F. Kamber, On a generalization of Cheeger-Chern-Simons classes, Ill. J. of Math. 34 (1990), 221–255.

J. Dupont and F. Kamber, Cheeger-Chern-Simons classes of transversally symmetric foliations: dependence relations and eta-invariant, Math. Ann. 295 (1993), 449–468.

A. Durfee, Foliations of odd-dimensional spheres, Ann. of Math. 96 (1972), 407–411; erratum, Ann. of Math. 97 (1973), 187.

A. Durfee and H. Lawson, Fibered knots and foliations of highly connected manifolds, Invent. Math. 17 (1972), 203–215.

S. Duzhin, A spectral sequence related to a foliation and the cohomology of certain Lie algebras of vector fields, Uspekhi Mat. Nauk 39 (1984), 135–136; Translation: Russian Math. Surveys 39 (1984), 147–148.

W. Dwyer, R. Ellis and R. Szczarba, Foliations with nonorientable leaves, Proc. Amer. Math. Soc. 89 (1983), 733–738.

I. Dynnikov, Proof of the Novikov conjecture for the case of small perturbations of rational magnetic fields, Uspekhi Math. Nauk 47 (1992), 161–162.

I. Dynnikov, Proof of the Novikov conjecture about semi classical electron motion, Matematicheskyie Zametki 53 (1993), 57–68; Translation: Math. Notes 53 (1993), 495–501.

C. Earle and J. Eells, Foliations and fibrations, J. Diff. Geom. 1 (1967), 33–41.

R. Edwards, A question concerning compact foliations, Springer Lecture Notes in Math. 468 (1975), 2–4.

R. Edwards, K. Millett, and D. Sullivan, Foliations with all leaves compact, Topology 16 (1977), 13–32.

J. Eells, Elliptic operators on manifolds, complex analysis and its appl. (Trieste), I (1975), 95–152.

J. Eells and L. Lemaire, A report on harmonic maps, Bull. London Math. Soc. 10 (1978), 1–68.

J. Eells and J. Sampson, Harmonic mappings of Riemannian manifolds, Amer. J. Math. 86 (1964), 109–160.

S. Egashira, Expansion growth of foliations, Ann. Fac. Sci. Toulouse 2 (1993), 15–52.

S. Egashira, Expansion growth of horospherical foliations, J. Fac. Sci. Univ. Tokyo 40 (1993), 663–682.

S. Egashira, Expansion growth of smooth codimension one foliations, to appear.

C. Ehresmann, Sur les espaces fibrés différentiables, C. R. Acad. Sci. Paris 222 (1947), 1611–1612.

C. Ehresmann, Sur les variétés plongées dans une variété différentiable, C. R. Acad. Sci. Paris 226 (1948), 1879–1880.

C. Ehresmann, Les connexions infinitésimales dans un espace fibré différentiable, Colloque de Topologie, Bruxelles, 1950, 29–55.

C. Ehresmann, Les prolongements d'une variété différentiable. I. Calcul des jets. II. L'espace des jets d'ordre r de V_n dans V_m. III. Transitivité des prolongements, C. R. Acad. Sci. Paris 233 (1951), 598–600, 777–779, 1081–1083.

C. Ehresmann, Sur la théorie des variétés feuilletées, Rendiconti di Matematica e delle sue applicazioni, serie V. vol. X (1951), 64–82.

C. Ehresmann, Structures locales et structures infinitésimales, C. R. Acad. Sci. Paris 243 (1952), 587–589.

C. Ehresmann, Les prolongements d'une variété différentiable. IV. Eléments de contact et éléments d'envelope. V. Covariants différentiels et prolongements d'une structure infinitésimale, C. R. Acad. Sci. Paris 234 (1952), 1028–1030, 1424–1425.

C. Ehresmann, Structures feuilletées, Proc. Fifth Canad. Math. Congress, Montréal (1961), 109–172.

C. Ehresmann and G. Reeb, Sur les champs d'éléments de contact de dimension p complètement intégrables dans une variété continuement différentiable V_n, C.R. Acad. Sci. Paris 216 (1944), 628–630.

C. Ehresmann and G. Reeb, Sur les champs d'éléments de contact de dimension p complètement intégrables dans une variété continuement différentiable, C. R. Acad. Sci. Paris 218 (1944), 955–957.

C. Ehresmann and Shi-Weishu, Sur les espaces feuilletées: théorème de stabilité, C. R. Acad. Sci. Paris 243 (1956), 344–346.

D. Eisenbud, U. Hirsch and W. Neumann, Transverse foliations of Seifert bundles and self homeomorphisms of the circle, Comment. Math. Helv. 56 (1981), 638–660.

T. Ekedahl, Foliations and inseparable morphisms, Proc. Symp. Pure Math. AMS 46 (1987), Part 2, 139–149.

Y. Eliashberg and N. Mishachev, Surgery of singularities of foliations, Funkts. Anal. Ego Prilozhen 11 (1977), 43–53. Translation Funct. Anal. Appl. 11 (1977), 197–206.

A. El Kacimi, Sur la cohomologie feuilletée, Comp. Math. 49 (1983), 195–215.

A. El Kacimi, Equation de la chaleur sur les espaces singuliers, C. R. Acad. Sc. Paris 303 (1986), 243–246.

A. El Kacimi, Dualité pour les feuilletages transversalement holomorphes, Manuscripta Math. 58 (1987), 417–433.

A. El Kacimi, Stabilité des V-variétés kählériennes, Holomorphic dynamics, Proc. of a 1986 Conf. in Mexico City, Springer Lecture Notes in Math. 1345 (1988), 111–123.

A. El Kacimi, Opérateurs transversalement elliptiques sur un feuilletage riemannien et applications, Compositio Math. 73 (1990), 57–106.

A. El Kacimi, Examples of foliations and problems in transverse complex analysis, Functional analytic methods in complex analysis and applications to partial differential equations (Trieste, 1988), 341–364, World Sci. Publ., 1990.

A. El Kacimi and E. Gallego Gomez, Applications harmoniques feuilletées, Ill. J. of Math., to appear.

A. El Kacimi and G. Hector, Décomposition de Hodge sur l'espace des feuilles d'un feuilletage riemannien, C. R. Acad. Sci. Paris 298 (1984), 289–292.

A. El Kacimi and G. Hector, Décomposition de Hodge basique pour un feuilletage riemannien, Ann. Inst. Fourier 36 (1986), 207–227.

A. El Kacimi, G. Hector and V. Sergiescu, La cohomologie basique d'un feuilletage riemannien est de dimension finie, Math. Z. 188 (1985), 593–599.

A. El Kacimi and M. Nicolau, Déformations des feuilletages transversalement holomorphes à type différentiable fixe, Publicaciones del Depto de Geometria y Topologia, Universidad de Santiago de Compostela, 57 (1982), 485–500.

A. El Kacimi and M. Nicolau, Structures géométriques invariantes et feuilletages de Lie, Indag. Math., N.S. 1 (1990), 323–334.

A. El Kacimi and M. Nicolau, A class of C^∞-stable foliaions, Ergodic Theory and Dynam. Systems 13 (1993), 697–704.

A. El Kacimi and M. Nicolau, On the topological invariance of the basic cohomology, Math. Ann. 295 (1993), 627–634.

A. El Kacimi and A. Tihami, Bigraded cohomology of certain foliations, Bull. Soc. Math. Belg. Sér. B 38 (1986), 144–156; Erratum ibid. B 39 (1987), 379.

C. Ennis, Sufficient conditions for smoothing codimension one foliations, Thesis, Univ. of California, Berkeley, 1980.

C. Ennis, Sufficient conditions for smoothing codimension one foliations, Trans. Amer. Math. Soc. 276 (1983), 311–322.

C. Ennis, M. Hirsch and C. Pugh, Foliations that are not approximable by smoother ones, Springer Lecture Notes in Math. 1007 (1983), 146–176.

B. Enriquez, Sur le théorème de Gauss-Bonnet pour les feuilletages avec mesure harmonique, C. R. Acad. Sci. Paris 309 (1989), 733–736.

C. Epstein, Foliations by geodesic circles, Appendix A of A. Besse, Manifolds all of whose geodesics are closed, Ergebuisse der Mathematik und ihrer Grenzgebiete vol. 93 (1978), 214–224.

C. Epstein, Pointwise homeomorphisms, Proc. London Math. Soc. 42 (1981), 415–460.

C. Epstein, Orthogonally integrable line fields in H^3, Comm. Pure Appl. Math. 38 (1985), 593–608.

D. Epstein, The simplicity of certain groups of homeomorphisms, Compos. Math. 22 (1970), 165–173.

D. Epstein, Periodic flows on three-dimensional manifolds, Ann. of Math. 95 (1972), 66–82.

D. Epstein, Foliations with all leaves compact, Springer Lecture Notes in Math. 468 (1975), 1–2.

D. Epstein, Foliations with all leaves compact, Ann. Inst. Fourier 26 (1976), 265–282.

D. Epstein, A topology for the space of foliations, Springer Lecture Notes in Math. 597 (1977), 132–150.

D. Epstein, Transversely hyperbolic 1-dimensional foliations, Astérisque 116 (1984), 53–69.

D. Epstein, K. Millet and D. Tischler, Leaves without holonomy, J. London Math. Soc. 16 (1977), 548–552.

D. Epstein and H. Rosenberg, Stability of compact foliations, Springer Lecture Notes in Math. 597 (1978), 151–160.

D. Epstein and E. Vogt, A counterexample to the periodic orbit conjecture in codimension 3, Ann. Math. 108 (1978), 539–552.

R. H. Escobales, Riemannian submersions with totally geodesic fibers, J. Diff. Geom. 10 (1975), 253–276.

R. Escobales, Riemannian submersions from complex projective space, J. Diff. Geom. 13 (1978), 93–107.

R. Escobales, Sufficient conditions for a bundle-like foliation to admit a Riemannian submersion onto its leaf space, Proc. Amer. Math. Soc. 84 (1982), 280–284.

R. Escobales, The integrability tensor for bundle-like foliations, Trans. Amer. Math. Soc. 270 (1982), 333–339.

R. Escobales, Bundle-like foliations with Kählerian leaves, Trans. Amer. Math. Soc. 276 (1983), 853–859.

R. Escobales, Riemannian foliations of the rank one symmetric spaces, Proc. Amer. Math. Soc. 95 (1985), 495–498.

R. Escobales, The mean curvature cohomology class for foliations and the infinitesimal geometry of the leaves, Diff. Geom. and its Appl. 2 (1992), 167–178.

R. Escobales and Ph. Parker, Geometric consequences of the normal curvature cohomology class in umbilic foliations, Indiana Univ. Math J. 37 (1988), 389–408.

W. van Est, Group cohomology and Lie algebra cohomology in Lie groups. Nederl. Akad. Wetensch. Indag. Math. 15 (1953), 484–492, 493–504.

W. van Est, On the algebraic cohomology concepts in Lie groups. Nederl. Akad. Wetensch. Indag. Math. 17 (1955), 225–233, 286–294.

W. van Est, Une application d'une méthode de Cartan-Leray, Nederl. Akad. Wetensch. Indag. Math. 17 (1955), 542–544.

W. van Est, A generalization of the Cartan-Leray spectral sequence, Nederl. Akad. Wetensch. Indag. Math. 20 (1958), 399–413.

W. van Est, Fundamental group of manifold schemes, Topological structures, II, Part 1, Math. Centre Tracts Amsterdam 115 (1979), 79–90.

W. van Est, Sur le groupe fondamental des schémas analytiques de variété à une dimension, Ann. Inst. Fourier 30 (1980), 45–77; erratum: ibid 30–4 (1980).

W. van Est, Quelques questions de géométrie en rétrospective, Astérisque 107–108 (1983), 13–30.

W. van Est, Rapport sur les S-atlas, Astérisque 116 (1984), 235–292.

W. van Est, Structures transverses et intégrales premières. Remarques, Indag. Mathem., N.S. 4 (1993), 439–445.

T. Fack, Quelques remarques sur le spectre des C^*-algébres de feuilletages, Bull. Soc. Math. Belg. Ser. B 36 (1984), 113–129.

T. Fack and G. Skandalis, Structure des ideaux de la C^*-algèbre associeé à un feuilletage, C. R. Acad. Sci. Paris 290 (1980), 1057–1059.

T. Fack and G. Skandalis, Sur les représentations et ideaux de la C^*-algèbre d'un feuilletage, J. Operator Theory 8 (1982), 95–129.

T. Fack and G. Skandalis, Some properties of the C^*-algebra associated with a foliation, Proc. Symp. Pure Math. 38 (1982), Part 1, 629–632.

T. Fack and X. Wang, The C^*-algebras of Reeb foliations are not AF-embeddable, Proc. Amer. Math. Soc. 108 (1990), 941–946.

A. Fahti, F. Laudenbach and V. Poénaru, Travaux de Thurston sur les surfaces, Séminaire Orsay 1979, Astérisque 66–67.

J. Faran, Local invariants of foliations by real hypersurfaces, Mich. Math. J. 35 (1988), 395–407.

H. Farran, G-structures on manifolds with parallel foliations, J. Univ. Kuwait Sci. 7 (1980), 59–67.

H. Farran, On Anosov foliations, Math. Rep. Toyama Univ. 7 (1984), 1–11.

F. Farrell and L. Jones, H-cobordism with foliated control, Bull. Amer. Math. Soc. 15 (1986), 69–72; erratum, ibid. 16 (1987), 177.

F. Farrell and L. Jones, Foliated control with hyperbolic leaves, K-theory 1 (1987), 337–359.

F. Farrell and L. Jones, Foliated control theory. I, II. K-theory 2 (1988), 357–430.

F. Farrell and L. Jones, Foliated control without radius of injectivity restrictions, Topology 30 (1991), 117–142.

A. Fathi, The Poisson bracket on the space of measured foliations on a surface, Duke Math. J. 55 (1987), 693–697.

L. Favaro, S-transversality, Proc. Amer. Math. Soc. 53 (1975), 481–488.

L. Favaro, Differentiable mappings between foliated manifolds, Bol. Soc. Brasil. Mat. 8 (1977), 39–46.

E. Fedida, Sur les feuilletages de Lie, C. R. Acad. Sci. Paris 272 (1971), 999–1002.

E. Fedida, Structures différentiables sur le branchement simple et équations différentielles dans le plan, C. R. Acad. Sci. Paris 276 (1973), 1657–1659.

E. Fedida, Sur l'existence des feuilletages de Lie, C. R. Acad. Sci. Paris 278 (1974), 835–837.

E. Fedida, Sur la théorie des feuilletages associées au repère mobile: cas des feuilletages de Lie, Springer Lecture Notes in Math. 652 (1978), 183–195.

E. Fedida, Feuilletages du plan, Feuilletages de Lie, Thèse, Strasbourg (1983).

E. Fedida and P. Furness, Feuilletages transversalement affine de codimension 1, C. R. Acad. Sci. Paris 282 (1976), 825–827.

E. Fedida and P. Furness, Transversally affine foliations, Glasgow Math. J. 17 (1976), 106–111.

E. Fedida, C. Hyjazi and F. Pluvinage, Sur les feuilletages transverses du plan, Afrika Mat. 7 (1985), 63–87.

E. Fedida and F. Pluvinage, Sur les structures feuilletées déterminées par des equations polynomiales, C. R. Acad. Sci. Paris 267 (1968), 101–104.

B. Feigin, Characteristic classes of flags of foliations, Funkts. Anal. Ego Prilozhen. 9 (1975), 49–56. Translation: Funct. Anal. Appl. 9 (1975), 312–317.

B. Feigin and D. Fuks, Homology of the Lie algebra of vector fields on the line, Funct. Anal. Appl. 14 (1980), 201–212.

B. Feigin and D. Fuks, Stable cohomology of the algebra W_n and relations in the algebra L_1, Funktsional. Anal. i Prilozen 18 (1984), 94–95.

S. Fenley, Depth one foliations in hyperbolic 3-manifolds, Ph.D. thesis, Princeton, 1990.

S. Fenley, Quasi-isometric foliations, Topology 31 (1992), 667–676.

S. Fenley, Asymptotic properties of depth one foliations in hyperbolic 3-manifolds, J. Diff. Geom. 36 (1992), 269–312.

S. Fenley, Anosov flows in 3-manifolds, Annals of Math. 139 (1994), 79–115.

S. Fenley, One sided branching in Anosov foliations, Comment. Math. Helvetici 70 (1995), 248–266.

S. Fenley, Quasigeodesic Anosov flows and homotopic properties of flow lines, J. Diff. Geom. 41 (1995), 479–514.

S. Fenley, Topological and homotopic properties of Anosov flows in 3-manifolds, Foliations Tokyo 1993, to appear.

R. Feres, Geodesic flows on manifolds of negative curvature with smooth horospheric foliations, Ergod. Theory and Dynam. Sys. 11 (1991), 653–686.

R. Feres, The center foliation of an affine diffeomorphism, Geom. Dedicata 46 (1993), 233–238.

R. Feres, Hyperbolic dynamical systems, invariant geometric structures and rigidity, Math. Res. Letters 1 (1994), 11–26.

R. Feres and A. Katok, Invariant tensor fields of dynamical systems with pinched Lyapounov exponents and rigidity of geodesic flows, Ergodic Theory and Dynam. Sys. 9 (1989), 427–432.

R. Feres and A. Katok, Anosov flows with smooth foliations and rigidity of geodesic flows on three-dimensional manifolds of negative curvature, Ergodic Theory and Dynam. Sys. 10 (1990), 657–670.

S. Ferry and A. Wasserman, Morse theory for codimension-one foliations, Trans. Amer. Math. Soc. 298 (1986), 227–240.

D. Ferus, Totally geodesic foliations, Math. Ann. 188 (1970), 313–316.

D. Ferus, On the completeness of nullity foliations, Mich. Math. J. 18 (1971), 61–64.

B. Fine, P. Kirk and E. Klassen, A local analytic splitting of the holonomy map on flat connections, Math. Ann. 299 (1994), 171–189.

M. Fliess, Cascade decompositions of control systems and invariant foliations, Bull. Soc. Math. France 113 (1985), 285–293.

M. Fliess, Cascade decomposition of nonlinear systems, foliations and ideals of transitive Lie algebras, Systems Control Lett. 5 (1985), 263–265.

R. Foote, Stein manifolds that admit Monge-Ampère foliations, Complex analysis and applications (Varna, 1985), 220–227, Bulgar. Acad. Sci., Sofia, 1986.

R. Foote, Differential geometry of real Monge-Ampère foliations, Math. Z. 1994 (1987), 331–350.

R. Forman, Adiabatic limits, small eigenvalues and spectral sequences, XXth Int. Conference on Differential Geometric Methods in Theoretical Physics, World Scientific Press (1992), 306–315.

R. Forman, Hodge theory and spectral sequences, Topology 33 (1994), 591–611.

R. Forman, Spectral sequences and adiabatic limits, to appear.

P. Foulon, Feuilletages des sphères et dynamiques Nord-Sud, C. R. Acad. Sci. Paris 318 (1994), 1041–1042.

J. Franks, Two foliations in the plane, Proc. Amer. Math. Soc. 58 (1976), 262–264.

J. Franks, Holonomy invariant cochains for foliations, Proc. Amer. Math. Soc. 62 (1977), 161–164.

J. Franks and R. Williams, Anomalous Anosov flows, Springer Lectrue Notes 819 (1980).

M. Freeman, Local complex foliation of real submanifolds, Math. Ann. 209 (1974) 1–30.

M. Freeman, The Levi form and local complex foliations, Proc. Amer. Math. Soc. 57 (1976), 369–370.

D. Fried, Fibrations over S^1 with pseudo-Anosov monodromy, Astérisque 66–67 (1979), 251–266.

D. Fried, The geometry of cross sections to flows, Topology 21 (1982), 353–372.

D. Fried, Transitive Anosov flows and pseudo-Anosov maps, Topology 22 (1983), 299–303.

D. Fried, Anosov foliations and cohomology, Ergodic Theory Dynamical Systems 6 (1986), 9–16; Erratum, ibid. 8 (1988), 491–492.

H. Frings, Generalized entropy for foliations, Thesis Univ. Bielefeld, 1991.

F. Frobenius, Über das Pfaffsche Problem, J. reine Angew. Math. 82 (1877), 267–282. Ges. Abh. I, 286–301.

D. Fuks, Characteristic classes of foliations, Usp. Mat. Nauk. 28 (1973), 3–17; Russ. Math. Surveys 28 (1973), 1–16.

D. Fuks, Finite-dimensional Lie algebras of formal vector fields and characteristic classes of homogeneous foliations, Izv. Akad. Nauk SSSR, Ser. Mat. 40 (1976), 57–6. Translation: Math. USSR Izv. 10 (1976), 55–62.

D. Fuks, Cohomology of infinite-dimensional Lie algebras and characteristic classes of foliations, Itogi Nauki-Seriya "Matematika" 10 (1976), 179–286 [Russian]. Translation: J. Soviet Math. 11 (1979), 922–980.

D. Fuks, Non-trivialité des classes caractéristiques des g-structures, Applications aux classes caractéristiques des feuilletages, C. R. Acad. Sci. Paris 284 (1977), 1017–1019.

D. Fuks, Non-trivialité des classes caractéristiques des g-structures, Applications aux variations des classes caractéristiques de feuilletages, C. R. Acad. Sci. Paris 284 (1977), 1105–1107.

D. Fuks, Foliations, Itogi Nauki-Seriya Algebra, Topologiya, Geometriya 18 (1981), 151–213 [Russian]. Translation: J. Soviet Math. 18 (1982), 255–291.

D. Fuks, Cohomology of infinite dimensional Lie algebras, Nauka, Moscow, 1984.

D. Fuks, A. Gabrielov and I. Gel'fand, The Gauss-Bonnet theorem and the Atiyah-Patodi-Singer functionals for the characteristic classes of foliations, Topology 15 (1976), 165–188.

K. Fukui, Codimension 1 foliations on simply connected 5-manifolds, Proc. Japan Acad. 49 (1973), 432–434.

K. Fukui, An application of the Morse theory to foliated manifolds, Nagoya Math. J. 54 (1974), 165–178.

K. Fukui, On the foliated cobordisms of codimension-one foliated 3-manifolds, Acta Hum. Sci. Univ., Sangio Kiotiensis Nat. Sci. Ser. 7 (1978), 42–49.

K. Fukui, A remark on the foliated cobordism of codimension-one foliated 3-manifolds, J. Math. Kyoto Univ. 18 (1978), 189–197.

K. Fukui, Perturbations of compact foliations, Proc. Japan Acad. Ser. A. 58 (1982), 341–344.

K. Fukui, Stability and instability of certain foliations of 4-manifolds by closed orientable surfaces, Publ. RIMS, Kyoto Univ. 22 (1986), 1155–1171.

K. Fukui, Stability of foliations of 3-manifolds by circles, J. Math. Soc. Japan 39 (1987), 117–126.

K. Fukui, Perturbations of compact foliations, Foliations (Tokyo, 1983), Adv. Stud. Pure Math. 5, North-Holland, Amsterdam, 1985.

K. Fukui, Stability of a compact leaf homeomorphic to the Klein bottle and its applications, J. Math. Kyoto Univ. 29 (1989), 257–265.

K. Fukui, Stability of Hausdorff foliations of 4-manifolds by circles, Math. Japonica. 34 (1989), 925–932.

K. Fukui, Stability of foliations of 4-manifolds by Klein bottles, J. Math. Kyoto Univ. 31 (1991), 133–137.

K. Fukui, Instability of certain foliations of 4-manifolds by Klein bottles, Acta Hum. Sci. Univ. Samgio Kyotiensis 22, Nat. Sci. Ser. I (1992), 1–10.

K. Fukui, Stability of Hausdorff foliation of 5-manifolds by Klein bottles, J. Math. Kyoto Univ. 34 (1994), 251–261.

K. Fukui, Remark on the actions of $\mathbb{R}^p$ on foliated manifolds, Geometric Study of Foliations, Tokyo 1993, 193–199, World Scientific, 1994.

K. Fukui, Remark on the actions of $\mathbb{R}^p$ on foliated manifolds II, Proc. VII Colloq. on Diff. Geom., Santiago de Compostela 1994, World Scientific, 1995.

K. Fukui and N. Tomita, Lie algebra of foliation preserving vector fields, J. Math. Kyoto Univ. 22 (1983), 685–699.

K. Fukui and S. Ushiki, On the homotopy type of $F\text{Diff}(S^3, \mathcal{F}_R)$, J. Math. Kyoto Univ. 15 (1975), 201–210.

R. Fulp and J. Marlin, Integrals of foliations on manifolds with a generalized symplectic structure, Pacific J. Math. 67 (1976), 373–387.

P. Furness, Affine foliations of codimension one, Q. J. Math. No. 98, 25 (1974), 151–161.

P. Furness and S. Robertson, Parallel framings and foliations on pseudoriemannian manifolds, J. Diff. Geom. 9 (1974), 409–422.

A. Futaki and S. Morita, Invariant polynomials on compact complex manifolds, Proc. Japan Acad. Ser. A Math. Sci. 60 (1984), 369–372.

D. Gabai, Foliations and the topology of 3-manifolds, Bull. Amer. Math. Soc. 8 (1983), 77–80.

D. Gabai, Foliations and the topology of 3-manifolds, J. Diff. Geom. 18 (1983), 445–503.

D. Gabai, Foliations and the genera of links, Topology 23 (1984), 381–394.

D. Gabai, Foliations and the topology of 3-manifolds, II, J. Diff. Geom. 26 (1987), 461–478.

D. Gabai, Foliations and the topology of 3-manifolds, III, J. Diff. Geom. 26 (1987), 479–536.

D. Gabai, Foliations and 3-manifolds, Proc. Int. Congr. Math., Kyoto 1990, vol. I (1991), 609–619.

D. Gabai, Taut foliations of 3-manifolds and suspensions of S^1, Ann. Inst. Fourier 42 (1992), 193–208.

D. Gabai and U. Oertel, Essential laminations in 3-manifolds, Ann. of Math. 130 (1989), 41–73.

E. Gallego, L. Gualandri, G. Hector and A. Reventós, Groupoides riemanniens, Publ. Math. 33 (1989), 417–422.

E. Gallego and A. Reventós, Curvature and plane fields, C. R. Acad. Sci. Paris 306 (1988), 675–679.

E. Gallego and A. Reventós, Lie flows of codimension 3, Trans. Amer. Math. Soc. 326 (1991), 529–541.

S. Gallot and D. Meyer, Opérateurs de courbure et Laplacien des formes différentielles d'une variété riemannienne, J. Math. Pures et Appl. 54 (1975), 259–284.

M. Garançon, Le rang de certaines variétés closes, Ann. Inst. Fourier 20 (1970), 1–19.

M. Garançon, Homotopie et holonomie de certains feuilletages de codimension 1, Ann. Inst. Fourier 22 (1972), 61–71.

M. Garançon, Feuilletages transversalement analytiques de codimension 1 admettant une transversale fermée qui coupe toutes les feuilles, Ann. Inst. Fourier 22 (1972), 271–287.

J. Garcia, Multiplicity of a foliation on projective spaces along an integral curve, Rev. Mat. Univ. Complutense Madrid 6 (1993), 207–217.

J. Garcia and A. Naveira, Two remarks about foliations and minimal foliations of codimension greater than two, Analysis and Geometry in Foliated Manifolds, Proc. VII Colloq. on Diff. Geom., Santiago de Compostela 1994, World Scientific 1995, 29–38.

J. Garcia and A. Naveira, Some remarks about foliations and totally geodesic foliations of codimension greater than one, to appear.

R. Gardner, Differential geometry and foliations: the Godbillon-Vey invariant and the Bott-Pasternack vanishing theorems, Springer Lecture Notes in Math. 652 (1978), 75–94.

L. Garnett, Statistical properties of foliations, Springer Lecture Notes in Math. 1007 (1983), 294–299.

L. Garnett, Foliations, the ergodic theorem and Brownian motion, J. Funct. Anal. 51 (1983), 285–311.

H. Gauchman, An integral inequality for normal contact Riemannian manifolds and its applications, Geom. Dedicata 23 (1987), 53–58.

H. Gauchman, Basic cohomology classes of compact Sasakian manifolds, Acta Sci. Math. (Szeged) 56 (1992), 269–285.

D. Gauld, Submersions and foliations of topological manifolds, Math. Chron. 1 (1971), 139–146.

D. Gauld, Foliations on topological manifolds, Math. Chron. 2 (1972), 29–41.

B. Gaveau, Equations for hulls of holomorphy and foliations, Ennio De Giorgi colloquium (Paris, 1983), Pitman Res. Notes Math. Ser. 125 (1985), 50–61.

M. Gazolaz (del Carmen), Fibrés de Seifert: classification et existence de feuilletages, C. R. Acad. Sci. Paris 295 (1982), 677–679.

I. Gelfand and D. Fuks, Cohomologies of the Lie algebra of tangent vector fields of a smooth manifold. I, Funkts. Anal. Appl. 3 (1969), 194–210; II, 4 (1970), 110–116.

I. Gelfand and D. Fuks, The Cohomologies of the Lie algebra of formal vector fields, Izv. Akad. Nauk SSSR, Ser. Mat. 34 (1970), 322–337.

I. Gelfand and D. Fuks, PL-foliations, Funkcional. Anal. i Ego Prilozen. 7 (1973), 29–37. Translation: Funct. Anal. Appl. 7 (1973), 278–284.

I. Gelfand and D. Fuks, PL-foliations, II, Funkcional. Anal. i Ego Prilozen 8 (1974), 7–11. Translation: Funct. Anal. Appl. 8 (1974), 197–200.

I. Gelfand, B. Feigin and D. Fuks, Cohomologies of the Lie algebra of formal vector fields with coefficients in the dual space and variations of characteristic classes of foliation, Funkcional. Anal. i Ego Prilozen 8 (1974), 13–29. Translation: Funct. Anal. Appl. 8 (1974), 99–112.

I. Gelfand, D. Fuks and D. Kalinin, On the cohomology groups of the Lie algebra of Hamiltonian formal vector fields. Funkcional. Anal. i Ego Prilozen 6, 25–29[Russian]. Translation: Functional Anal. Appl. 6 (1972), 193–196.

I. Gelfand, D. Fuks and D. Kazhdan, The actions of infinite dimensional Lie algebras, Funkcional. Anal. i Ego Prilozen 6, 10–15 [Russian]. Translation: Functional Anal. Appl. 6 (1972), 9–13.

R. Gerard and J. P. Jouanolou, Etude de l'existence de variétés intégrales compactes de certains systèmes de Pfaff, C. R. Acad. Sci. Paris 277 (1973), 167–169.

R. Gerard and J. P. Jouanolou, Etude de l'existence de feuilles compactes de certains feuilletages analytiques complexes, C. R. Acad. Sci. Paris 277 (1973), 311–314.

R. Gerard and G. Reeb, Le théorème de Floquet et la théorie de de Rham (pour les formes de degré 1) comme cas particulier d'un théorème d'Ehresmann sur les structures feuilletées, Ann. Scuola Norm. Sup. Pisa 21 (1967), 93–98.

R. Gerard and A. Sec, Feuilletages de Painlevé et équations de Pfaff, C. R. Acad. Sci. Paris 270 (1970), 1166–1169.

R. Gerard and A. Sec, Feuilletages de Painlevé, Bull. Soc. Math. France 100 (1972), 47–72.

C. Gerhardt, H-surfaces in Lorentzian manifolds, Commun. Math. Phys. 89 (1983), 523–553.

E. Ghys, Classification des feuilletages totalement géodésiques de codimension un, Comment. Math. Helv. 58 (1983), 543–572.

E. Ghys, Feuilletages riemanniens sur les variétés simplement connexes, Ann. Inst. Fourier 34 (1984), 203–223.

E. Ghys, Actions localement libres du groupe affine, Invent. Math. 82 (1985), 479–526.

E. Ghys, Flots d'Anosov sur les 3-variétés fibrés en cercles, Ergod. Th. & Dynam. Syst. 4 (1984), 67–80.

E. Ghys, Groupes d'holonomie des feuilletages de Lie, Proc. Kon. Nederl. Acad. Sci. A 88 (1985), 173–182.

E. Ghys, Un feuilletage analytique dont la cohomologie basique est de dimension infinie, Publ. IRMA Lille 7 (1985).

E. Ghys, Une variété qui n'est pas une feuille, Topology 24 (1985), 67–73.

E. Ghys, Classe d'Euler et minimal exceptionnel, Topology 26 (1987), 93–105.

E. Ghys, Flots d'Anosov dont les feuilletages stables sont différentiables, Ann. Sci. Ecole Norm. Sup (4) 20 (1987), 251–270.

E. Ghys, Groupes d'homéomorphismes du cercle et cohomologie bornée, Contemp. Math. 58 (1987), 81–105.

E. Ghys, Sur l'invariance topologique de la classe de Godbillon-Vey, Ann. Inst. Fourier 37 (1987), 59–76.

E. Ghys, Codimension one Anosov flows and suspensions, Springer Lecture Notes in Math. 1331 (1988), 59–72.

E. Ghys, Gauss-Bonnet theorem for 2-dimensional foliations, J. Funct. Anal. 77 (1988), 51–59.

E. Ghys, Riemannian foliations: examples and problems, in P. Molino, Riemannian foliations, Birkhäuser (1988), 297–314.

E. Ghys, L'invariant de Godbillon-Vey, Séminaire Bourbaki 1988/89, Astérisque 177–178 (1989), exp 706, 155–181.

E. Ghys, Flots transversalement affines et tissus feuilletés, Mém. Soc. Math. France (N.S.) 46 (1991), 123–150.

E. Ghys, Le cercle à l'infini des surfaces à courbure négative, Proc. Int. Congress Math., Kyoto 1990, Math. Soc. Japan (1991), 501–509.

E. Ghys, Déformations de flots d'Anosov et groupes fuchsiens, Ann. Inst. Fourier 42 (1992), 209–247.

E. Ghys, Rigidité différentiable des groupes fuchsiens, Publ. Math. IHES 78 (1993), 163–185.

E. Ghys, Construction de champs de vecteurs sans orbite periodique, Séminaire Bourbaki 1993–94, exp. 785, Astérisque 227 (1995), 283–307.

E. Ghys, Holomorphic Anosov systems, Invent. Math. 119 (1995), 585–614.

E. Ghys, Topologie des feuilles génériques, Ann. of Math. 141 (1995), 387–422.

E. Ghys, G. Hector and Y. Moriyama, On codimension one nilfoliations and a theorem of Malcev, Topology 28 (1989), 197–210.

E. Ghys and X. Gómez-Mont, The space of wandering leaves of holomorphic foliations, to appear.

E. Ghys, R. Langevin and P. Walczak, Entropie mesurée et partitions de l'unité, C. R. Acad. Sci. Paris 303 (1986), 251–254.

E. Ghys, R. Langevin and P. Walczak, Entropie géometrique des feuilletages, Acta Math. 160 (1988), 105–142.

E. Ghys and V. Sergiescu, Stabilité et conjugaison différentiable pour certains feuilletages, Topology 19 (1980), 179–197.

E. Ghys and V. Sergiescu, Sur un groupe remarquable de difféomorphismes du cercle, Comment. Math. Helv. 62 (1987), 185–239.

E. Ghys and T. Tsuboi, Différentiabilité des conjugaisons entre systèmes dynamiques de dimension 1, Ann. Inst. Fourier 38 (1988), 209–247.

E. Ghys and A. Verjovsky, Locally free holomorphic actions of the complex plane, Geometric Study of Foliations, Tokyo 1993, 201–217, World Scientific, 1994.

G. Gigante, Positive bundle over foliations with complex leaves, Riv. Mat. Univ. Parma (5) 3 (1994), 169–176.

G. Gigante and G. Tomassini, Foliations with complex leaves, Diff. Geom. and its Appl. 5 (1995), 33–49.

P. Gilkey and J. H. Park, Eigenvalues of the Laplacian and riemannian submersions, Yokohama Math. J. 43 (1995), 7–11.

P. Gilkey and J. H. Park, Riemannian submersions which preserve the eigenforms of the Laplacian, Ill. J. of Math., to appear.

O. Gil-Medrano and A. Naveira, The Gauss-Bonnet integrand for a class of Riemannian manifolds admitting two orthogonal complementary foliations, Canad. Math. Bull. 26 (1983), 358–364.

O. Gil-Medrano and A. Naveira, Some remarks about the Riemannian curvature operator of a Riemannian almost-product manifold, Rev. Roumaine Math. Pures Appl. 30 (1985), 647–658.

G. Giraud, Connexions subordonnées et classes caractéristiques, Revue Roumaine Math. Pures Appl. 20 (1975), 917–925.

J. Girbau, Vanishing theorems for complex analytic foliated manifolds, Proc. IV. Int. Coll. Diff. Geom., Santiago de Compostela 1978, 131–140.

J. Girbau, Vanishing theorems and stability of complex analytic foliations, Springer Lecture Notes in Math. 792 (1980), 247–251.

J. Girbau, Sur le théorème de stabilité des feuilletages de Hamilton, Epstein et Rosenberg, C. R. Acad. Sci. Paris 291 (1980), 41–44.

J. Girbau, Vanishing cohomology theorems and stability of complex analytic foliations, Israel J. Math. 40 (1981), 235–254.

J. Girbau, Some examples of deformations of transversely holomorphic foliations, Differential Geometry, Penisola 1982, Springer Lecture Notes in Math. 1045 (1984), 53–62.

J. Girbau, On deformations of holomorphic and transversely holomorphic foliations, Géométrie différentielle (Paris, 1986), 51–63, Travaux en Cours, 33, Hermann, Paris, 1988.

J. Girbau, A versality theorem for transversely holomorphic foliations of fixed differentiable type, Ill. J. Math. 36 (1992), 428–446.

J. Girbau and G. Guasp, Deformations of transversely symplectic and transversely contact foliations, Tsukuba J. of Math. 15 (1991), 479–508.

J. Girbau, A. Haefliger and A. Sundararaman, On deformations of transversely holomorphic foliations, J. für die reine und angew. Math. Crelle J. 345 (1983), 122–147.

J. Girbau and M. Nicolau, Pseudodifferential operators on V-manifolds and foliations, I, Collect. Math. 30 (1979), 247–265.

J. Girbau and M. Nicolau, Pseudodifferential operators on V-manifolds and foliations, II, Collect. Math. 31 (1980), 63–95.

J. Girbau and M. Nicolau, Deformations of holomorphic foliations and transversely holomorphic foliations, Pitman Research Notes in Math. 131 (1985), 162–173.

J. Girbau and M. Nicolau, On deformations of holomorphic foliations, Ann. Inst. Fourier 39 (1989), 417–449.

E. Giroux, Une structure de contact, même tendue, est plus ou moins tordue, Ann. Sci. Ecole Norm. Sup. 27 (1994), 697–705.

M. Giry, Study of nontransversally complete foliations in a regular case, Bull. Sci. Math. 106 (1982), 433–442.

J. Glazebrook, S. Hurder and F. Kamber, Higher transverse index theory for Riemannian foliations, to appear.

J. Glazebrook and F. Kamber, Transversal Dirac families in Riemannian foliations, Comm. Math. Phys. 140 (1991), 217–240.

J. Glazebrook and F. Kamber, On spectral flow of transversal Dirac operators and a theorem of Vava-Witten, Ann. Global Anal. Geom. 9 (1991), 27–35.

J. Glazebrook and F. Kamber, Determinant line bundles for Hermitian foliations and a generalized Quillen metric, AMS Symp. Pure Math. 52 (1991), Part 2, 225–232.

J. Glazebrook and F. Kamber, Secondary invariants and chiral anomalies of basic Dirac families, Differ. Geom. Appl. 3 (1993), 285–299.

J. Glazebrook and F. Kamber, Chiral anomalies and Dirac families in Riemannian foliations, to appear.

J. Glazebrook, F. Kamber, H. Pedersen and A. Swann, Foliation reduction and self-duality, Geometric Study of Foliations, Tokyo 1993, 219–249, World Scientific, 1994.

J. Glazebrook and D. Sundararaman, Deformation of foliated quatornionic structures and the twistor correspondence, Bol. Soc. Mat. Mexicana (2) 37 (1992), 177–187.

H. Gluck, Can space be filled by geodesics, and if so, how?, open letter (1979).

H. Gluck, Dynamical behavior of geodesic fields, Springer Lecture Notes in Math. 819 (1980), 190–215.

H. Gluck and F. Warner, Great circle fibrations of the three sphere, Duke Math. J. 50 (1983). . . .

H. Gluck and W. Ziller, On the volume of a unit vector field on the three-sphere, Comment. Math. Helv. 61 (1986), 177–192.

H. Goda, Depth of foliations on tunnel number one genus one knot complements, Analysis and Geometry of Foliated Manifolds, Proc. VII Colloq. on Diff. Geom., Santiago de Compostella 1994, World Scientific 1995, 39–53.

C. Godbillon, Feuilletages ayant la propriété du prolongement des homotopies, Ann. Inst. Fourier 17 (1967), 219–260.

C. Godbillon, Holonomie transversale, C. R. Acad. Sci. Paris 264 (1967), 1050–1052.

C. Godbillon, Problèmes d'existence et d'homotopie dans les feuilletages, Springer Lecture Notes in Math. 244 (1971), 167–181.

C. Godbillon, Cohomologies d'algèbres de Lie de champs de vecteurs formels. In: Sém. Bourbaki 1972/73 exp. 421, Springer Lecture Notes in Math. 383 (1972).

C. Godbillon, Fibrés en droites et feuilletages du plan, Enseign. Math. 18 (1972), 213–224.

C. Godbillon, Cohomologies d'algèbres de Lie de champs de vecteurs formels, Springer Lecture Notes in Math. 383 (1974), 69–87.

C. Godbillon, Feuilletages de Lie, Springer Lecture Notes in Math. 392 (1974), 10–13.

C. Godbillon, Invariants of foliations, Global Anal. Appl. Lect. Int. Semin. Course Trieste, Vienna 2 (1974), 215–219.

C. Godbillon, Dynamical systems on surfaces, Strasbourg 1979 (French) and Springer 1983 (English).

C. Godbillon, Une construction nonvelle des classes caractéristiques des feuilletages, Ann. Inst. Fourier 40 (1990), 709–721.

C. Godbillon, Feuilletages: Etudes géométriques I, II, Publ. IRMA Strasbourg (1985–86);Progress in Math. vol 98 (1991), Birkhäuser.

C. Godbillon and G. Reeb, Fibrés sur le branchement simple, Enseign. Math. 12 (1966), 277–287.

C. Godbillon and J. Vey, Un invariant des feuilletages de codimension un, C. R. Acad. Sci. Paris 273 (1971), 92–95.

A. Goddard, Foliations of space-time by space-like hypersurfaces of constant mean curvature, Commun. Math. Phys. 54 (1977), 279–282.

B. Golbus, On the singularities of foliations and of vector bundle maps, Bol. Soc. Brasil. Mat. 7 (1976), 11–35.

B. Golbus, On extending local foliations, Quart. J. Math. 28 (1977), 163–176.

L. Goldberg, k-flat structures and exotic characteristic classes, Trans. Amer. Math. Soc. 306 (1988), 433–453.

V. Goldberg and R. Rosca, Foliate conformal Kählerian manifolds, Rend. Sem. Mat. Messina Ser. II 14 (1991), 105–122.

R. Goldman, The holonomy ring on the leaves of foliated manifolds, J. Diff. Geom. 11 (1976), 411–449.

W. Goldmann, M. Hirsch and G. Levitt, Invariant measures for affine foliations, Proc. Amer. Math. Soc. 86 (1982), 511–518.

X. Gómez-Mont, Transversal holomorphic structures, J. Diff. Geom. 15 (1980), 161–185.

X. Gómez-Mont, Unfoldings of holomorphic foliations, Publicaciones del Depto de Geometria y Topologia, Universidad de Santiago de Compostela, 57 (1982), 510–515.

X. Gómez-Mont, On families of rational vector fields, Coll. on Dyn. Systems, Guanajuato, 1983, Aportaciones Mat., Soc. Mat. Mexicana 1985, 36–65.

X. Gómez-Mont, Transverse deformations of holomorphic foliations, Contemp. Math. 58 (1986), 127–139.

X. Gómez-Mont, Foliations by curves of complex analytic spaces, Contemp. Math. 58 (1987), 123–141.

X. Gómez-Mont, Universal families of foliations by curves, Astérisque 150-151 (1987), 109–129.

X. Gómez-Mont, The transverse dynamics of a holomorphic flow, Ann. of Math. 127 (1988), 49–92.

X. Gómez-Mont, Holomorphic foliations in ruled surfaces, Trans. Amer. Math. Soc. 312 (1989), 179–201.

X. Gómez-Mont, Integrals for holomorphic foliations with singularities having all leaves compact, Ann. Inst. Fourier 39 (1989), 451–458.

X. Gómez-Mont, On closed leaves of holomorphic foliations by curves, Springer Lecture Notes in Math. 1414 (1989), 61–98.

X. Gómez-Mont, Unfolding of holomorphic foliations, Publicacions Matemàtiques U.A.B. 33 (1989), 501–515.

X. Gómez-Mont, On the spaces of polynomial vector fields modulo projectivities, Proc. Cong. Dyn. Syst., Trieste 1988, Pitman (1990), 112–127.

X. Gómez-Mont, On foliations in $\mathbb{C}P^2$ tangent to an algebraic curve, Proc. Cong. Alg. Geom., Cimat, 1989; Aportaciones Matematicas, Investigacion 5 (1992), 87–99.

X. Gómez-Mont, An algebraic formula for the index of a vector field on a variety with an isolated singularity, to appear.

X. Gómez-Mont and G. Kempf, Stability of meromorphic vector fields in projective spaces, Comment. Math. Helv. 64 (1989), 462–473.

X. Gómez-Mont and G. Kempf, A law of conservation of number for local Euler characteristics, to appear.

X. Gómez-Mont and A. Lins Neto, Structural stability of singular holomorphic foliations having a meromorphic first integral, Topology 30 (1991), 315–334.

X. Gómez-Mont and I. Luengo, Germs of holomorphic vector fields in $\mathbb{C}^3$ without a separatrix, Inventiones Math. 109 (1992), 211–219.

X. Gómez-Mont and J. Mucino, Persistent cycles for holomorphic foliations having a meromorphic first integral, Springer Lecture Notes in Math. 1345 (1988), 129–162.

X. Gómez-Mont and L. Ortiz, Sistemas dinamicos holomorfos en superficies, Aportaciones Matematicas, Sociedad Matematica Mexicana 1989, 207 pp.

X. Gómez-Mont, J. Seade and A. Verjovsky, Holomorphic dynamics, Springer Lectures Notes 1345, 1988.

X. Gómez-Mont, J. Seade and A. Verjovsky, The index of a holomorphic flow with an isolated singularity, Math. Ann. 291 (1991), 737–751.

X. Gómez-Mont, J. Seade and A. Verjovsky, On the topology of a holomorphic vector field around an isolated singularity, Funct. Anal. and its Applications (1994).

X. Gómez-Mont and D. Sundararaman, Remarks on the versal families of deformations of holomorphic and transversely holomorphic foliations, Deformations of Mathematical Structures (Lódz/Lublin, 1985/87), 205–213, Kluwer Acad. Publ., Boston, 1989.

X. Gómez-Mont and B. Wirtz, Entropy of fixed points of conformal pseudogroups, to appear.

L. Goncharova, Cohomology of Lie algebra of formal vector fields on the line, Funct. Anal. Appl. 7 (1973), 6–14.

J. González-Dávila, Espacios transversalmente simetricos de tipo Killing, Ph.D. thesis, La Laguna (Tenerife), 1992.

J. González-Dávila, KTS-spaces and natural reductivity, Nihonkai Math. J. 5 (1994), 115–129.

J. González-Dávila and M. González-Dávila, Relations of curvatures in locally transversally Killing symmetric spaces, Geom. and Topol., Proc. 15th Port.-Span. Meet. Math. (Évora, 1990), Vol. III (1991), 155–160.

J. González-Dávila and M. González-Dávila, The Gelfand theorem in Killing transversally symmetric spaces, Geom. and Topol., Proc. 15th Port.-Span. Meet. Math. (Évora, 1990), Vol. III (1991), 185–190.

J. González-Dávila, M. González-Dávila and L. Vanhecke, The Gelfand theorem in flow geometry, C. R. Math. Rep. Acad. Sci. Canada 15 (1993), 281–285.

J. González-Dávila, M. González-Dávila and L. Vanhecke, Reflections and isometric flows, Kyungpook Math. J. 35 (1995), 113–144.

J. González-Dávila, M. González-Dávila and L. Vanhecke, Normal flow space forms and their classification, Publ. Math. Debrecen 48 (1996), 151–173.

J. González-Dávila, M. González-Dávila and L. Vanhecke, Invariant submanifolds in flow geometry, to appear.

J. González-Dávila, M. González-Dávila and L. Vanhecke, Classification of Killing-transversally symmetric spaces, Tsukuba J. Math., to appear.

J. González-Dávila and L. Vanhecke, Geodesic spheres and isometric flows, Colloq. Math. 67 (1994), 223–240.

J. González-Dávila and L. Vanhecke, Geometry of tubes and isometric flows, Math. J. Toyama Univ. 18 (1995), 9–35.

J. González-Dávila and L. Vanhecke, New examples of weakly symmetric spaces, to appear.

S. Goodman, Closed leaves in foliated 3-manifolds, Proc. Nat. Acad. Sci. 71 (1974), 4414–4415.

S. Goodman, Closed leaves in foliations of codimension one, Comment. Math. Helv. 50 (1975), 383–388.

S. Goodman, On the structure of foliated 3-manifolds separated by a compact leaf, Invent. Math. 39 (1977), 213–221.

S. Goodman, Dehn surgery and Anosov flows, Springer Lecture Notes 1007 (1983).

S. Goodman, Vector fields with transverse foliations, Topology 24 (1985), 333–340.

S. Goodman, Vector fields with transverse foliations II, Ergod. Th and Dyn. Syst. 6 (1986), 193–203.

S. Goodman and J. Plante, Holonomy and averaging in foliated sets, J. Diff. Geom. 14 (1979), 401–407.

V. Gordin, Complex foliations of submanifolds, Usp. Mat. Nauk. 27 (1972), 233–234.

T. Gotoh, Harmonic foliations on a complex projective space, Tsukuba J. Math. 14 (1990), 99–106.

T. Gotoh, A remark on foliations on a complex projective space with complex leaves, Tsukuba J. Math. 16 (1992), 169–172.

J. Grabowski, Lie algebra of vector fields and generalized foliations, Publicacions Matematiques 37 (1993), 359–367.

A. Gray, Pseudo-Riemannian almost product manifolds and submersions, J. Math. Mech. 16 (1967), 715–737.

A. Gray and L. Vanhecke, Riemannian geometry as determined by the volumes of small geodesic balls, Acta Math. 142 (1979), 157–198.

A. Gray and L. Vanhecke, The volume of tubes about curves in a Riemannian manifold, Proc. London Math. Soc. 44 (1982), 215–243.

P. Greenberg, Classifying spaces for foliations with isolated singularities, Trans. Amer. Math. Soc. 304 (1987), 417–429.

P. Greenberg, Pseudogroups from group actions, Amer. J. Math. 109 (1987), 893–906.

P. Greenberg, Pseudogroups of C^1 piecewise projective homeomorphisms, Pacific J. Math. 129 (1987), 67–75.

R. Grimaldi, The asymptotic geometry of the leaves of a foliation, Rend. Circ. Mat. Palermo 32 (1983), 199–207.

R. Grimaldi, Non-esistenza di cusps nella geometria asintotica delle foglie, Atti Accad. Naz. Lincei Rend. Cl. Sc. Fis. Mat. Natur. 80 (1986), 292–297.

R. Grimaldi, Géometrie asymptotique et noeds fibrés, C. R. Acad. Sci. Paris 309 (1989), 933–936.

R. Grimaldi, Géométrie asymptotique et fibrés à groupe structural discret, C. R. Acad. Sci. Paris 311 (1990), 347–349.

R. Grimaldi, Noeds et links fibrés et géométrie asymptotique de feuilles, Riv. Mat. Univ. Parma, IV Ser. 17 (1991), 149–158.

R. Grimaldi and G. Passante, The asymptotic geometry of the leaves of a foliation, Atti Acad. Sci. Torino Cl. Sci. Fis. Mat. Natur. 118 (1984), 97–100.

R. Grimaldi and G. Passante, Asymptotic geometry for Anosov foliations, C. R. Acad. Sci. Paris 300 (1985), 275–276.

R. Grimaldi and G. Passante, La geometria asintotica per le fogliazioni di Anosov, Boll. U.M.I. (6) 5-A (1986), 321–329.

R. Grimaldi and G. Passante, La géométrie asymptotique des feuilles exceptionelles des feuilletages d'Anosov, C. R. Acad. Sci. Paris 305 (1987), 85–87.

R. Grimaldi and G. Passante, La geometria asintotica delle foglie eccezionali delle foliazioni di Anosov, Ann. di Mat. Pura ed Appl. 152 (1988), 345–358.

D. Gromoll and K. Grove, One dimensional metric foliations in constant curvature spaces, Diff. Geom. and Complex Analysis, Rauch Memorial Volume 1985, 165–168.

D. Gromoll and K. Grove, The low-dimensional metric foliations of Euclidean spheres, J. Differential Geom. 28 (1988), 143–156.

M. Gromov, Transversal mappings of foliations, Soviet Math. Dokl. 9 (1968), 1126–1129.

M. Gromov, Stable mappings of foliations into manifolds, Izv. Akad. Nauk SSSR, Ser. Mat. 33 (1969), 707–734. Translation: Math. USSR Izv. 3 (1969), 671–694.

M. Gromov, Foliated Plateau Problem. I. Minimal varieties, Geom. Funct. Anal. 1 (1991), 14–79; II. Harmonic maps of foliations, Geom. Funct. Anal. 1 (1991), 253–320.

M. Gromov, Stability and pinching, Geometry Seminars. Sessions on Topology and Geometry of Manifolds (Bologna, 1990), 55–97, Univ. Stud. Bologna, Bologna, 1992.

S. Guelorget, Algèbre de Weil du groupe linéaire, Application aux classes caractéristiques d'un feuilletage, Springer Lecture Notes in Math. 484 (1975), 179–191.

S. Guelorget and G. Joubert, Algèbre de Weil et classes caractéristiques d'un feuilletage, C. R. Acad. Sci. Paris 277 (1973), 11–14.

V. Guillemin, Remarks on some results of Gelfand and Fuks, Bull. Amer. Math. Soc. 78 (1972), 539–540.

V. Guillemin, Cohomology of vector fields on a manifold, Adv. in Math. 10 (1973), 192–220.

V. Guinez, Positive quadratic differential forms and foliations with singularities on surfaces, Trans. Amer. Math. Soc. 309 (1988), 477–502.

V. Guinez, Nonorientable polynomial foliations on the plane, J. Differential Equations 87 (1990), 391–411.

B. Gurevic, The entropy of horocycle flows, Soviet Math. Dokl. 2 (1961), 134–138.

C. Gutierrez, Smoothing continuous flows and the converse of Denjoy–Schwartz theorem, An. Acad. Brasil. Ciens. 51 (1979), 581–589.

V. Gutiérrez, Foliations on surfaces having exceptional leaves, Springer Lecture Notes in Math. 1331 (1988), 73–85.

A. Haefliger, Sur les feuilletages des variétés de dimension n par des feuilles fermées de dimension $n-1$, Colloque de Topologie, Strasbourg, 1955.

A. Haefliger, Sur les feuilletages analytiques, C. R. Acad. Sci. Paris 242 (1956), 2908–2910.

A. Haefliger, Structures feuilletées et cohomologie à valeurs dans un faisceau de groupoides, Comment. Math. Helv. 32 (1958), 249–329.

A. Haefliger, Variétés feuilletées, Ann. Scuola Norm. Sup. Pisa 16 (1962), 367–397.

A. Haefliger, Travaux de Novikov sur les feuilletages. In: Sém. Bourbaki 1967/68, exp. 339, New York, W. A. Benjamin, 1968.

A. Haefliger, Feuilletages sur les variétés ouvertes, Topology 9 (1970), 183–194.

A. Haefliger, Homotopy and integrability, Springer Lecture Notes in Math. 197 (1971), 133–163.

A. Haefliger, Lectures on Gromov's theorem, Liverpool Singularities Symposium, Springer Lecture Notes in Math. 209 (1971), 128–141.

A. Haefliger, Teorema de Bott sobre una obstruccion topologica a la integrabilidad completa, Rev. Mat. Hisp. Amer. 32 (1972), 21–32.

A. Haefliger, Sur les classes caractéristiques des feuilletages, Sém. Bourbaki 1971–72, exp. 412, Springer Lecture Notes in Math. 317 (1973), 239–260.

A. Haefliger, Sur la cohomologie de Gelfand-Fuchs, Differential topology and geometry, Proc. Colloq. Dijon 1974, Springer Lecture Notes in Math. 484 (1975), 121–152.

A. Haefliger, Sur la cohomologie de l'algèbre de Lie des champs de vecteurs, Ann. Sci. Ecole Norm. Sup. 9 (1976), 503–532.

A. Haefliger, Whitehead products and differential forms, Springer Lecture Notes in Math. 652 (1978), 13–24.

A. Haefliger, On the Gelfand-Fuks cohomology, Enseign. Math. 24 (1978), 143–160.

A. Haefliger, Cohomology of Lie algebras and foliations, Springer Lecture Notes in Math. 652 (1978), 1–12.

A. Haefliger, Differentiable cohomology, C.I.M.E. Lectures 1976, Varenna; Liguori 1979, 19–70.

A. Haefliger, Feuilletages avec feuilles minimales, Proceedings of the IV International Colloqium on Differential Geometry, University of Santiago de Compostela, 1979, 275–284.

A. Haefliger, Some remarks on foliations with minimal leaves, J. Diff. Geom. 15 (1980), 269–284.

A. Haefliger, Groupoïdes d'holonomie et classifiants, Astérisque 116 (1984), 70–97.

A. Haefliger, Deformations of transversely holomorphic flows on spheres and deformations of Hopf manifolds, Compositio Math. 55 (1985), 241–251.

A. Haefliger, Pseudogroup of local isometries, Pitman Research Notes 131 (1985), 174–197.

A. Haefliger, Leaf closures in Riemannian foliations, A fête of topology, 3–32, Papers dedicated to I. Tamura, Academic Press, Boston, 1988.

A. Haefliger, Feuilletages riemanniens, Séminaire Bourbaki, 41 ème année, 1988–89, no. 707 (March 1989).

A. Haefliger and Quach Ngoc Du, Une présentation du groupe fondamental d'un orbifold, Astérisque 116 (1984), 98–107.

A. Haefliger and G. Reeb, Variétés (non séparées) à une dimension et structures feuilletées du plan, Enseign. Math. 3 (1957), 107–126.

A. Haefliger and E. Salem, Pseudogroupes d'holonomie des feuilletages riemanniens sur des variétés compactes 1-connexes, Géométrie différentielle (Paris, 1986), 141–160, Travaux en Cours, 33, Hermann, Paris, 1988.

A. Haefliger and E. Salem, Riemannian foliations on 1-connected manifolds and actions of tori on orbifolds, Illinois J. Math. 34 (1990), 706–730.

A. Haefliger and E. Salem, Actions of tori on orbifolds, Ann. Global Anal. Geom. 9 (1991), 37–59.

A. Haefliger and K. Sithanantham, A proof that $B\bar{\Gamma}_1^c$ is 2-connected, Algebraic Topology Oaxtepec 1981, Contemp. Math. 12 (1982), 129–139.

A. Haefliger and D. Sundararaman, Complexifications of transversely holomorphic foliations, Math. Ann. 272 (1985), 23–27.

A. Hamasaki, Continuous cohomologies of Lie algebras of formal G-invariant vector fields and obstructions to lifting foliations, Publ. RIMS, Kyoto Univ. 20 (1984), 401–429.

A. Hamasaki, Cohomologies of Lie algebras of formal vector fields preserving a foliation, Publ. Res. Inst. Math. Sci 24 (1988), 639–652.

U. Hamenstaedt, Harmonic measures for leafwise elliptic operators along foliations, First European Congress of Math., Vol. II (1994), 73–95.

U. Hamenstaedt, Invariant two-forms for geodesic flows, Math. Ann. 301 (1995), 677–698.

R. Hamilton, Deformation theory of foliations, preprint Cornell University (1978).

M. Handel, One dimensional minimal sets and the Seifert conjecture, Ann. of Math. 111 (1980), 35–66.

Th. Hangan, Totally geodesic distributions of the Riemannian nilpotent group H_{2p+1}, Rend. Sem. Fac. Sci. Univ. Cagliari 55 (1985), 31–227.

Th. Hangan, Accessibilité en géométrie riemannienne non-holonome, Advances in differential geometry and topology (1990), 39–45.

Th. Hangan and R. Lutz, Champs d'hyperplans totalement géodésiques sur les sphères, IIIe rencontre de géométrie du Schnepfenried, 1982, Astérisque 107–108 (1983), 189–200.

Y. Hantout, Déformations de connexions, résidus et classes caractéristiques des feuilletages, Publ. IRMA-Lille 8 III (1987).

Y. Hantout, Classes caractéristiques rigides de feuilletages, C. R. Acad. Sci. Paris 307 (1988), 263–265.

Y. Hantout, Classes caractéristiques de G (G, H)-structures et finitude de leur évaluation, Manuscripta Math. 62 (1988), 383–399.

D. Hardorp, All compact orientable three dimensional manifolds admit total foliations, Memoirs Amer. Math. Soc. 233 (1980).

J. Harrison, Unsmoothable diffeomorphisms, Ann. of Math. 102 (1975), 85–94.

J. Harrison, On unsmoothable diffeomorphisms, Bull. Amer. Math. Soc. 81 (1975), 746.

J. Harrison, Structure of a foliated neighborhood, Math. Proc. Camb. Phil. Soc. 79 (1976), 101–110.

J. Harrison, Unsmoothable diffeomorphisms on higher dimensional manifolds, Proc. Amer. Math. Soc. 73 (1979), 249–255.

J. Harrison, Opening closed leaves of foliations, Bull. London Math. Soc. 15 (1983), 218–220.

J. Harrison, C^2 counterexamples to the Seifert conjecture, Topology 27 (1988), 249–278.

J. Harrison and J. Yorke, Flows on S^3 and $\mathbb{R}^3$ without periodic orbits, Geometric Dynamics, Rio de Janeiro 1981, Springer Lecture Notes in Math. 1007 (1983), 401–407.

C. Hartzman, Non singular periodic flows on T^3 and periodic homeomorphisms of T^2, Canad. Math. Bull. 24 (1981), 23–28.

R. Harvey and H. Lawson, Calibrated foliations, Amer. J. of Math. 104 (1982), 607–633.

N. Hashiguchi, On the Anosov diffeomorphisms corresponding to geodesic flows on negatively curved closed surfaces, J. Fac. Sci. Tokyo Univ. 37 (1990), 485–494.

N. Hashiguchi, PL-representations of Anosov foliations, Ann. Inst. Fourier 42 (1992), 937–965.

N. Hashiguchi and H. Minakawa, Continuous variation of the discrete Godbillon-Vey invariant, J. Fac. Sci. Univ. Tokyo Sect IA Math 39 (1992), 271–278.

J. Hass, Minimal surfaces in foliated manifolds, Comment. Math. Helv. 61 (1986), 1–32.

B. Hasselblatt, Horospheric foliations and relative pinching, J. Differential Geom. 39 (1994), 57–63.

B. Hasselblatt, Periodic bunching and invariant foliations, Math. Res. Letters 1 (1994), 597–600.

B. Hasselblatt, Regularity of the Anosov splitting and of horospheric foliations, Ergodic Theory and Dynamical systems, 14 (1994), 645–666.

A. Hatcher, Some examples of essential laminations in 3-manifolds, Ann. Inst. Fourier 42 (1992), 313–325.

A. Hatcher, Measured laminations in 3-manifolds, preprint.

J. Hausmann, Extension d'une homotopie à un feuilletage topologique, Manifolds-Amsterdam 1970, Springer Lecture Notes in Math 197 (1971), 164–175.

J. Hebda, Curvature and focal points in Riemannian foliations, Indiana Univ. Math. J. 35 (1986), 321–331.

J. Hebda, An example relevant to curvature pinching theorems for Riemannian foliations, Proc. Amer. Math. Soc. 114 (1992), 195–199.

G. Hector, Théorème de Haefliger et feuilletages uniformes, C. R. Acad. Sci. Paris 265 (1967), 617–619.

G. Hector, Sur le type des feuilletages transverses de $\mathbb{R}^3$, C. R. Acad. Sci. Paris 273 (1971), 810–813.

G. Hector, Ouverts incompressibles et théorème de Denjoy-Poincaré pour les feuilletages, C. R. Acad. Sci. Paris 274 (1972), 159–162.

G. Hector, Ouverts incompressibles et théorème de Denjoy-Poincaré pour les feuilletages, Thesis, University of Strasbourg, 1972.

G. Hector, Ouverts incompressibles et structure des feuilletages de codimension 1, C. R. Acad. Sci. Paris 274 (1972), 741–744.

G. Hector, Sur les feuilletages presque sans holonomie, C. R. Acad. Sci. Paris 274 (1972), 1703–1706.

G. Hector, Actions de groupes de difféomorphismes de $[0,1]$, Springer Lecture Notes in Math. 392 (1974), 14–22.

G. Hector, Sur un théorème de structure des feuilletages de codimension un, in G. Reeb, Feuilletages: Résultats anciens et nouveaux (Painlevé, Hector et Martinet), Montréal 1972, Presses Univ. Montréal 1974, 48–54.

G. Hector, Quelques exemples de feuilletages, espèces rares, Ann. Inst. Fourier 26 (1976), 239–264.

G. Hector, Leaves whose growth is neither exponential nor polynomial, Topology 16 (1977), 451–459.

G. Hector, Feuilletages en cylindres. In: Geometry and Topology, Proc. III. Latin Amer. School of Math., Springer Lecture Notes in Math. 597 (1977), 252–270.

G. Hector, Germes de feuilletages, Proc. of the 10th Braz. Math. Colloq. 1975, Vol. II, Inst. Mat. Pura Apl. Pocos de Caldas 1978, 547–552.

G. Hector, Cohomology of Lie algebras and foliations, Springer Lecture Notes in Math. 652 (1978), 1–12.

G. Hector, Croissance des feuilletages presque sans holonomie, Springer Lecture Notes in Math. 652 (1978), 141–182.

G. Hector, Architecture des feuilletages de classe C^2, Astérisque 107–108 (1983), 243–258.

G. Hector, On the classification of manifolds foliated by the action of a nilpotent Lie group, Diff. Geom. and Math. Physics, Liège 1980, Math. Phys. Stud. 3 (1983), 31–35.

G. Hector, Cohomologies transversales des feuilletages riemanniens, Feuilletages riemanniens, quantification géometrique et mécanique, Travaux en Cours, 26, Hermann 1988, Paris.

G. Hector, Une nouvelle obstruction à l'intégrabilité des variétés de Poisson régulières, Hokkaido Math. J. 21 (1992), 159–185.

G. Hector, Groupoides, feuilletages et C^*-algebres, Geometric Study of Foliations, Tokyo 1993, 3–34, World Scientific, 1994.

G. Hector, Géométrie et topologie des espaces difféologiques, Analysis and Geometry in Foliated Manifolds, Proc. VII Colloq. on Diff. Geom., Santiago de Compostela 1994, World Scientific 1995, 55–80

G. Hector, Espaces difféologiques quotients de feuilletages et géométrie en dimension infinie, to appear.

G. Hector and W. Bouma, All open surfaces are leaves of simple foliations of $\mathbb{R}^3$, Nederl. Akad. Wetensch. Indag. Math. 45 (1983), 443–452.

G. Hector and U. Hirsch, Introduction to the Geometry of Foliations, Parts A and B, Vieweg, Braunschweig, 1981 and 1983.

G. Hector and E. Macias, Sur le théorème de DeRham pour les feuilletages de Lie, C.R. Acad. Sc. Paris 311 (1990), 633–636.

G. Hector, E. Macias and M. Saralegi, Lemme de Moser feuilleté et classification des variétés de Poisson regulierès, Publicaciones del Depto de Geometria y Topologia, Universidad de Santiago de Compostela, 57 (1982), 423–430; Publicacions Matemàtiques 33 (1989), 423–430.

G. Hector and S. Matsumoto, Bouts de feuilles et classification des feuilletage de Lie, Foliations Tokyo 1993, to appear.

G. Hector and M. Saralegi, Intersection cohomology of S^1-actions, to appear.

W. Heil and W. Whitten, The Seifert fiber space conjecture and torus theorem for nonorientable 3-manifolds, Can. Math. Bull. 37 (1994), 482–489.

J. Heitsch, The cohomologies of classifying spaces for foliations, Thesis, University of Chicago, Chicago, IL (1971).

J. Heitsch, Deformations of secondary characteristic classes, Topology 12 (1973), 381–388.

J. Heitsch, A cohomology for foliated manifolds, Bull. Amer. Math. Soc. 79 (1973), 1283–1285.

J. Heitsch, A cohomology for foliated manifolds, Comment. Math. Helv. 50 (1975), 197–218.

J. Heitsch, A remark on the residue theorem of Bott, Indiana Univ. Math. J. 25 (1976), 1139–1147.

J. Heitsch, Residues and characteristic classes of foliations, Bull. Amer. Math. Soc. 83 (1977), 397–399.

J. Heitsch, Independent variations of secondary classes, Ann. of Math. 108 (1978), 421–460.

J. Heitsch, Derivatives of secondary characteristic classes, J. Diff. Geom. 13 (1978), 311–339.

J. Heitsch, A residue formula for holomorphic foliations, Mich. Math. J. 27 (1980), 181–194.

J. Heitsch, Linearity and residues for foliations, Bol. Soc. Bras. Mat. 12 (1981), 87–94.

J. Heitsch, Flat bundles and residues for foliations, Invent. Math. 73 (1983), 271–285.

J. Heitsch, Secondary invariants of transversely homogeneous foliations, Mich. Math. J. 33 (1986) 47–57.

J. Heitsch, Some interesting group actions, Contemp. Math. 70 (1988), 113–123.

J. Heitsch and S. Hurder, Secondary classes, Weil measures and the geometry of foliations, J. Diff. Geom. 20 (1984), 291–309.

J. Heitsch and S. Hurder, Exotic cohomology for foliations, in preparation.

J. Heitsch and C. Lazarov, A Lefschetz theorem for foliated manifolds, Topology 29 (1990), 127–162.

J. Heitsch and C. Lazarov, The Lefschetz fixed point theorem for foliated manifolds, Contemp. Math. 105 (1990), 83–89.

J. Heitsch and C. Lazarov, A Lefschetz theorem on open manifolds, Contemp. Math. 105 (1990), 33–45.

J. Heitsch and C. Lazarov, Homotopy invariance of foliation Betti numbers, Invent. Math. 104 (1991), 321–347.

J. Heitsch and C. Lazarov, Rigidity theorems for foliations by surfaces and spin manifolds, Michigan Math. J. 38 (1991), 285–297.

J. Heitsch and C. Lazarov, Spectral asymptotics of foliated manifolds, Ill. J. of Math. 38 (1994), 653–678.

L. Helena and M. Vilas Bôas, Interval exchange transformation and foliation, Bol. Soc. Brasil Mat. 17 (1987), 57–74.

D. Henc, Transverse intersections and stability of foliations, Bol. Soc. Brasil Mat. 13 (1982), 1–18.

D. Henc, Ergodicity of foliations with singularities, J. Funct. Anal. 75 (1987), 349–361.

M. Hermann, Conjugaison C^∞ des difféomorphismes du cercle dont le nombre de rotation satisfait à une condition arithmétique, C. R. Acad. Sci. Paris 282 (1976), 503–506.

M. Hermann, Conjugaison C^∞ des difféomorphismes du cercle pour presque tout nombre de rotation, C. R. Acad. Sci. Paris (1976), 579–582.

M. Hermann, The Godbillon-Vey invariant of foliations by planes of T^3, Springer Lecture Notes in Math. 597 (1977), 294–307.

M. Hermann, Sur la conjugaison différentiable des difféomorphismes du cerle à des rotations, Inst. Hautes Etudes Sci. Publ. Math. 49 (1979), 5–233.

R. Hermann, A sufficient condition that a mapping of Riemannian manifolds be a fiber bundle. Proc. Amer. Math. Soc. 11 (1960), 236–242.

R. Hermann, The differential geometry of foliations, I, Ann. of Math. 72 (1960), 445–457.

R. Hermann, The differential geometry of foliations, II, J. Math. Mech. 11 (1962), 305–315.

R. Hermann, Totally geodesic orbits of groups of isometries, Indag. Math. 24 (1962), 291–298.

R. Hermann, The Born theory of rigid motions in relativity and the theory of Riemannian foliations. In: Development of Mathematics in the Nineteenth Century, Appendices, Kleinian Mathematics from an Advanced Standpoint, 549–573. Brookline, MA: Math Sci. (1979).

B. Herrera, Transverse structure of Lie foliations, Foliations Tokyo 1993, to appear.

B. Herrera, M. Llabres and A. Reventós, Transverse structure of Lie foliations, to appear.

S. Hiepko, Eine innere Kennzeichnung der verzerrten Produkte, Math. Ann. 241 (1979), 209–215.

S. Hiepko and H. Reckziegel, Über sphärische Blätterungen und die Vollständigkeit ihrer Blätter, Manuscripta Math. 31 (1980), 269–283.

M. Hilsum and G. Skandalis, Stabilité des C^*-algèbres de feuilletages, Ann. Inst. Fourier 33 (1983), 201–208.

M. Hilsum and G. Skandalis, Morphismes K-orienté d'espaces de feuilles et fonctorialité en théorie de Kasparov (d'après une conjecture d'A. Connes), Ann. Sci. Ecole Norm. Sup. 20 (1987), 325–390.

M. Hirsch, Foliations and non compact transformation groups, Bull. Amer. Math. Soc. 76 (1970), 1020–1023.

M. Hirsch, Stability of compact leaves of foliations, Dynamical Systems, Bahia, 1971, 135–153, Academic Press, 1973.

M. Hirsch, Foliated bundles, flat manifolds and invariant measures, Springer Lecture Notes in Math. 468 (1975), 8–9.

M. Hirsch, A stable analytic foliation with only exceptional minimal sets, Springer Lecture Notes in Math. 468 (1975), 9–10.

M. Hirsch and C. Pugh, Stable manifolds and hyperbolic sets, Global Analysis, Proc. Symp. Pure Math. Amer. Math. Soc. 14 (1970), 133–163.

M. Hirsch and C. Pugh, Smoothness of horocycle foliations, J. Diff. Geom. 10 (1975), 225–238.

M. Hirsch and W. P. Thurston, Foliated bundles, invariant measures and flat manifolds, Ann. of Math. 101 (1975), 369–390.

U. Hirsch, Some remarks on analytic foliations and analytic branched coverings, Math. Ann. 248 (1980), 139–152.

U. Hirsch and X. Wang, Foliations of planar regions and CCR $C*$-algebras with infinite composition length, Amer. J. Math. 109 (1987), 797–806.

F. Hirzebruch and H.Hopf, Felder von Flächenelementen in 4-dimensionalen Mannigfaltigkeiten, Math. Ann. 136 (1958), 156–172.

H. Holmann, Holomorphe Blätterungen komplexer Räume, Comment. Math. Helv. 47 (1972), 185–204.

H. Holmann, Analytische periodische Strömungen auf kompakten komplexen Räumen, Comment. Math. Helv. 52 (1977), 251–257.

H. Holmann, On the stability of holomorphic foliations with all leaves compact, Springer Lecture Notes in Math. 683 (1978) 217–248.

H. Holmann, Holomorphic transformation groups with compact orbits, Springer Lecture Notes in Math. 743 (1979), 419–430.

H. Holmann, On the stability of holomorphic foliations, Springer Lecture Notes in Math. 798 (1980), 192–202.

H. Holmann, Feuilletages complexes et symplectiques, Riv. Mat. Univ. Parma 10 (1984), 91–108.

H. Holmann, Analytic foliations, Geometria delle Varieta Differenziabile, Roma, 1984, Pitagora Editrice Bologna, 1985, 99–125.

R. Holubowicz and W. Mozgawa, Nonisometric transversely parallelisable foliations on four-dimensional toral bundles, Bull. Soc. Sci. Lett. Lódz 38 (1988), no 4, 8pp.

E. Hopf, Statistik der geodätischen Linien in Mannigfaltigkeiten negativer Krümmung, Ber. Verh. Sächs. Akad. Wiss. Leipzig 91 (1939), 261–304; Statistik der Lösungen geodätischer Probleme vom unstabilen Typus. II, Math. Ann. 117 (1940), 590–608.

L. Hörmander, The Frobenius-Nirenberg theorem, Ark. Mat. 5 (1964), 425–432.

J. Hounie, Minimal sets of families of vector fields on compact surfaces, J. Diff. Geom. 16 (1981), 739–744.

W. Y. Hsiang, R. S. Palais and C. L. Terng, The topology of isoparametric submanifolds, J. of Diff. Geom. 27 (1988), 423–460.

S. Hu, Equivariant homotopy and integrability, Chinese J. Math. 19 (1991), 61–74.

J. Hua Hao and H. Shima, Level surfaces of non-degenerate functions in $\mathbb{R}^{n+1}$, Geom. Dedicata 50 (1994), 193–204.

J. Hubbard and H. Masur, Quadratic differential and foliations, Acta Math. 142 (1979), 221–274.

S. Hurder, On the homotopy and cohomology of the classifying space of Riemannian foliations, Proc. Amer. Math. Soc. 81 (1981), 484–489.

S. Hurder, Dual homotopy invariants of G-foliations, Topology 20 (1981), 365–387.

S. Hurder, On the secondary classes of foliations with trivial normal bundles, Comment. Math. Helv. 56 (1981), 307–326.

S. Hurder, Independent rigid secondary classes for holomorphic foliations, Invent. Math. 66 (1982), 313–323.

S. Hurder, Exotic classes for measured foliations, Bull. Amer. Math. Soc. 7 (1982), 389–391.

S. Hurder, Global invariant for measured foliations, Trans. Amer. Math. Soc. 280 (1983), 367–391.

S. Hurder, Vanishing of secondary classes for compact foliations, J. London Math. Soc. 28 (1983), 175–183.

S. Hurder, The classifying space of smooth foliations, Ill. J. of Math. 29 (1985), 108–133.

S. Hurder, Foliation dynamics and leaf invariants, Comm. Math. Helv. 60 (1985), 319–335.

S. Hurder, The Godbillon measure for amenable foliations, J. Diff. Geom. 23 (1986), 347–365.

S. Hurder, The $\bar{\partial}$-operator, Appendix A to Global Analysis on Foliated Spaces, by C. C. Moore and C. Schochet, MSRI Publ. 9 (1988), 279–307.

S. Hurder, Ergodic theory of foliations and a theorem of Sacksteder, Springer Lecture Notes in Math. 1342 (1988), 291–328.

S. Hurder, Eta invariants and the odd index theorem for coverings, Contemp. Math. 105 (1990), 47–82.

S. Hurder, Exceptional minimal sets of $C^{1+\alpha}$-group actions on the circle, Ergodic Theory Dyn. Syst. 11 (1991), 455–467.

S. Hurder, A product theorem for $\Omega B\Gamma_G$, Topology Appl. 50 (1993), 81–86.

S. Hurder, Topology of covers and the spectral theory of geometric operators, Index theory and operator algebras, Boulder 1991, Contemp. Math. 148 (1993), 87–119.

S. Hurder, Topological rigidity of strong stable foliations for Cartan actions, Ergodic Theory and Dynamical Systems 14 (1994), 151–167.

S. Hurder, Coarse geometry of foliations, Geometric Study of Foliations, Tokyo 1993, 35–96, World Scientific, 1994.

S. Hurder, Infinitesimal rigidity for hyperbolic actions, J. Diff. Geom. 3 (1995), 515–527.

S. Hurder, Exotic index theory for foliations, to appear.

S. Hurder, Spectral theory for foliation geometric operators, to appear.

S. Hurder, Growth of leaves and secondary invariants of foliations, preprint.

S. Hurder and F. Kamber, Homotopy invariants of foliations, Topology Siegen 1979, Springer Lecture Notes in Math. 788 (1980), 49–61.

S. Hurder and A. Katok, Secondary classes and transverse measure theory of a foliation, Bull. Amer. Math. Soc. 11 (1984), 347–350.

S. Hurder and A. Katok, Ergodic theory and Weil measures for foliations, Ann. of Math. 126 (1987), 221–275.

S. Hurder and A. Katok, Differentiability, rigidity and Godbillon-Vey classes for Anosov flows, Publ. Math. IHES 72 (1990), 5–61.

S. Hurder and A. Katok, Smoothness and Godbillon-Vey classes of geodesic horocycle foliation on surfaces of variable negative curvature, to appear.

S. Hurder and D. Lehmann, Classes caractéristiques résiduelles, Differential geometry and its applications (Brno, 1989), 85–108, World Sci. Publ., 1990.

S. Hurder and D. Lehmann, Homotopy characteristic classes of foliations, Illinois J. of Math. 34 (1990), 628–6?.

S. Hurder and Y. Mitsumatsu, The intersection product of transverse invariant measures, Indiana Univ. Math. J. 40 (1991), 1169–1183.

S. Hurder and Y. Mitsumatsu, Transverse Euler class of foliations on non-atomic foliation cycles, Contemp. Math. 161 (1994), 29–39.

S. Hurder and Y. Mitsumatsu, The transverse Euler class for amenable foliations, preprint.

T. Huy Hoang, Sur le feuilletage défini par une équation de Pfaff algébrique sur $P_n\mathbb{C}$, Acta Math. Vietnam. 3 (1978), 83–97.

M. Hvidsten and Ph. Tondeur, A characterization of harmonic foliations by variations of the metric, Proc. Amer. Math. Soc. 98 (1986), 359–362.

S. Ianus, Sulla struttura fogliata di una varietà cosimpletica, Bull. Math. Soc. Sci. Math. Roumanie 20 (1977), 125–130.

V. Igosin, Decomposition theorems for double foliations that are compatible with a spray, Mat. Zametki 28 (1980), 923–934. Translation: Math. Notes 28 (1980), 916–922.

V. Igosin and Y. Shapiro, Stability of leaves of a foliation with a compatible Riemannian metric, Mat. Zametki 27 (1980), 767–778. Translation: Math. Notes 27 (1980), 367–373.

V. Igosin and Y. Shapiro, Stability theorem for leaves of Riemannian parallel foliation, Izv. Vyssh. Uchebn. Zaved. Mat. 24 (1980), 74–76. Translation: Soviet Math. 24 (1980), 87–90.

V. Igosin, Y. Shapiro and N. Zhukova, Foliations in some classes of Riemann manifolds, Izv. Vyssh. Uchebn. Zaved. Mat. 23 (1979), 93–96. Translation: Soviet Math. 23 (1979), 106–109.

G. Ikegami, Existence of regular coverings associated with leaves of codimension one foliations, Nagoya Math. J. 67 (1977), 15–34.

G. Ikegami, Vector fields tangent to foliations, Japan. J. Math. 12 (1986), 95–120.

G. Ikegami, Singular perturbations in foliations, Invent. Math. 95 (1989), 215–246.

H. Imanishi, Sur l'existence des feuilletages S^1-invariants, J. Math. Kyoto Univ. 12 (1972), 297–305.

H. Imanishi, On the theorem of Denjoy-Sacksteder for codimension one foliations without holonomy, J. Math. Kyoto Univ. 14 (1974), 607–634.

H. Imanishi, Structure of codimension one foliations which are almost without holonomy, J. Math. Kyoto Univ. 16 (1976), 93–99.

H. Imanishi, On codimension one foliations defined by closed one-forms with singularities, J. Math. Kyoto Univ. 19 (1979), 285–291.

H. Imanishi, Structure of codimension one foliations without holonomy on manifolds with abelian fundamental group, J. Math. Kyoto Univ. 19 (1979), 481–495.

H. Imanishi, Denjoy-Siegel theory of codimension one foliations, Sûgaku 32 (1980), 119–132.

H. Imanishi and K. Yagi, On Reeb components, J. Math. Kyoto Univ. 16 (1976), 313–324.

T. Inaba, On stability of proper leaves of codimension one foliations, J. Math. Soc. Japan 29 (1977), 771–778.

T. Inaba, On the structure of real analytic foliations of codimension one, J. Fac. Sci. Univ. Tokyo 26 (1979), 453–464.

T. Inaba, C^2 Reeb stability of non compact leaves of foliations, Proc. Japan Acad. 59 (1983), 158–160.

T. Inaba, Reeb stability for noncompact leaves, Topology 22 (1983), 105–118.

T. Inaba, A sufficient condition for the C^2-Reeb stability of noncompact leaves of codimension one foliations, Foliations (Tokyo, 1983), 379–394, Adv. Stud. Pure Math. 5, North-Holland, Amsterdam, 1985.

T. Inaba, Examples of exceptional minimal sets, A fête of topology, 95–100, Academic Press, Boston, 1988.

T. Inaba, Resiliant leaves in transversely affine foliations, Tôhoku Math. J. 41 (1989), 625–631.

T. Inaba, The tangentially affine structure of lagrangian foliations and the tangentially projective structure of legendrian foliations, preprint.

T. Inaba and K. Masuda, Tangentially affine foliations and leafwise affine functions on the torus, Kodai Math. J. 16 (1993), 32–43.

T. Inaba and S. Matsumoto, Some qualitative aspects of transversely projective foliations, Proc. Japan Acad. Ser. A Math. Sci. 65 (1989), 116–118.

T. Inaba and S. Matsumoto, Nonsingular expansive flows on 3-manifolds and foliations with circle prong singularities, Japan J. Math. 16 (1990), 329–340.

T. Inaba and S. Matsumoto, Resilient leaves in transversely projective foliations, J. Fac. Sci. Univ. Tokyo Sect. IA 37 (1990), 89–101.

T. Inaba, S. Matsumoto and N. Tsuchiya, Codimension one transversely affine foliations, Geometric Study of Foliations, Tokyo 1993, 263–293, World Scientific, 1994.

T. Inaba, S. Matsumoto and N. Tsuchiya, Structure of affine foliations, to appear.

T. Inaba and M. Mishchenko, On real submanifolds of Kähler manifolds foliated by complex submanifolds, Proc. Japan Acad. 70 (1994), 1–2.

T. Inaba, T. Nishimori, M. Takamura and N. Tsuchiya, Open manifolds which are nonrelizable as leaves, Kodai Math. J. 8 (1985), 112–119.

T. Inaba and N. Tsuchiya, Expansive foliations, Hokkaido Math. J. 21 (1992), 39–49.

T. Inaba and P. Walczak, Transverse Hausdorff dimension of codimension one C^2-foliations, to appear.

N. Innami, Totally flat foliations and peakless functions, Arch. Math. 41 (1983), 464–471.

J. Isenberg and V. Moncrief, The existence of constant mean curvature foliations of Gowdy 3-torus spacetime, Comm. Math. Phys. 86 (1982), 485–493.

D. Ivanenko and G. Sardanashvily, Foliation analysis of gravitation singularities, Phys. Lett. A 91 (1982), 341–344.

S. Izumiya, Smooth mapings between foliated manifolds, Bol. Soc. Brasil. Mat. 13 (1982), 3–17.

W. Jagy, Minimal hypersurfaces foliated by spheres, Michigan Math. J. 38 (1991), 255–270.

M. Jankins and W. Neuman, Lectures on Seifert manifolds, Brandeis Lecture Notes 2, 1983.

T. Januszkiewicz, Characteristic invariants of noncompact Riemannian manifolds, Topology 23 (1984), 289–301.

N. Jayne, A note on the sectional curvature of Legendre foliations, Yokohama Math. J. 41 (1994), 153–161.

N. Jayne, Legendre foliations on contact metric manifolds, Thesis, Massey Univ. (1992); Abstract in Bull. Austr. Math. Soc. 49 (1994), 173.

S. Jekel, On two theorems of A. Haefliger concerning foliations, Topology 15 (1976), 267–271.

S. Jekel, Simplicial $K(G, 1)$'s, Manuscr. Math. 21 (1977), 189–203.

S. Jekel, A note on the perfection of the fundamental group of the classifying space for codimension one real analytic foliations, Bol. Soc. Mat. Mex. 22 (1977), 58–59.

S. Jekel, Loops on the classifying space for foliations, Amer. J. Math. 102 (1980), 13–23.

S. Jekel, Simplicial decomposition of Γ-structures, Bol. Soc. Mat. Mex. 26 (1981), 13–20.

S. Jekel, Some weak equivalences for classifying spaces, Proc. Amer. Math. Soc. 90 (1984), 469–476.

D. Johnson, Kaehler submersions and holomorphic connections, J. Diff. Geom. 15 (1980), 71–79.

D. Johnson, Deformations of totally geodesic foliations, Lecture Notes in Pure and Appl. Math., Dekker, vol. 105 (1987), 167–178.

D. Johnson, Volumes of flows, Proc. Amer. Math. Soc. 104 (1988), 923–931.

D. Johnson and A. Naveira, A topological obstruction to the geodesibility of a foliation of odd dimension, Geom. Dedicata 11 (1981), 347–357.

D. Johnson and P. Smith, Regularity of volume-minimizing graphs, Indiana Univ. Math. J. 44 (1995), 45–85.

D. Johnson and P. Smith, Regularity of volume-minimizing one-dimensional foliation, Analysis and Geometry in Foliated Manifolds, Proc. VII Colloq. on Diff. Geom., Santiago de Compostela 1994, World Scientific 1995, 81–98.

D. Johnson and L. Whitt, Totally geodesic foliations on 3-manifolds, Proc. Amer. Math. Soc. 76 (1979), 355–357.

D. Johnson and L. Whitt, Totally geodesic foliations, J. Diff. Geom. 15 (1980), 225–235.

J. Jouanolou, Une preuve élémentaire d'un théorème de Thurston, Topology 17 (1978), 109–110.

J. Jouanolou, Feuilles compactes des feuilletages algébriques, Math. Ann. 241 (1979), 69–72.

G. Joubert, Construction d'un feuilletage au dessus d'un pseudogroupe, C. R. Acad. Sci. Paris 261 (1965), 3268–3271.

G. Joubert, Estructuras foliadas, Rev. Colomb. Mat. 2 (1968), 105–116.

G. Joubert and R. Moussu, El teorema de Novikov, Rev. Colomb. Mat. 2 (1968), 117–123.

G. Joubert and R. Moussu, Feuilletage sans holonomie d'une variété fermée, C. R. Acad. Sci. Paris 270 (1970), 507–509.

G. Joubert, R. Moussu, and D. Tischler, Sur les classes caractéristiques des feuilletages produits, C. R. Acad. Sci. Paris 275 (1972), 171–174.

G. Joubert and H. Rosenberg, Plongement du tore T^2 dans la sphère S^3, Cahiers Topologie Geom. Différentielle 11 (1969), 323–328.

A. Juhl, Secondary invariants and the singularity of the Ruelle zeta-function in the central critical point, Bull. Amer. Math. Soc. 32 (1995), 80–87.

A. Justino, Some properties of the quotient manifold of a Jacobi manifold by a foliation generated by infinitesimal automorphisms, C. R. Acad. Sci. Paris 298 (1984), 489–492.

A. Kabila, Sur les feuilletages logarithmiques, C. R. Acad. Sci. Paris 302 (1986), 13–15.

A. Kabila, Sur la topologie des feuilles des feuilletages logarithmiques, C. R. Acad. Sci. Paris 305 (1987), 461–464.

A. Kabila, Equidésingularisation d'un germe de feuilletages holomorphes singuliers, Geometric Study of Foliations Tokyo 1993, 295–312, World Scientific, 1994.

V. Kaimanovich, Brownian motion on foliations: entropy, invariant measures, mixing, Functional Anal. Appl. 22 (1988), 326–328.

H. Kamada, Foliations on manifolds with positive constant curvature, Tokyo J. Math. 16 (1993), 49–60.

F. Kamber, Transversal Index theory for Riemannian foliations, Foliations Tokyo 1993, to appear.

F. Kamber, E. Ruh and Ph. Tondeur, Almost transversally symmetric foliations, Proc. of the II. Int. Symp. on differential geometry, Peniscola 1985, Springer Lecture Notes in 1209 (1986), 184–189.

F. Kamber, E. Ruh and Ph. Tondeur, Comparing Riemannian foliations with transversally symmetric foliations, J. of Diff. Geom. 26 (1987), 461–475.

F. Kamber and Ph. Tondeur, Flat manifolds, Springer Lecture Notes in Math. 67 (1968).

F. Kamber and Ph. Tondeur, Invariant differential operators and the cohomology of Lie algebra sheaves, Memoirs Amer. Math. Soc. 113 (1971), 1–125.

F. Kamber and Ph. Tondeur, Cohomologie des algèbres de Weil relatives tronquées, C. R. Acad. Sci. Paris 276 (1973), 459–462.

F. Kamber and Ph. Tondeur, Algèbres de Weil semi-simpliciales, C. R. Acad. Sci. Paris 276 (1973), 1177–1179.

F. Kamber and Ph. Tondeur, Homomorphisme caractéristique d'un fibre principal feuilleté, C. R. Acad. Sci. Paris 276 (1973), 1407–1410.

F. Kamber and Ph. Tondeur, Classes caractéristiques dérivées d'un fibré principal feuilleté, C. R. Acad. Sci. Paris 276 (1973), 1449–1452.

F. Kamber and Ph. Tondeur, Characteristic invariants of foliated bundles, Manuscr. Math. 11 (1974), 51–89.

F. Kamber and Ph. Tondeur, Classes caractéristiques géneralisées des fibrés feuilletés localement homogenes, C. R. Acad. Sci. Paris 279 (1974), 847–850.

F. Kamber and Ph. Tondeur, Quelques classes caractéristiques géneralisées non-triviales de fibrés feuilletés, C. R. Acad. Sci. Paris 279 (1974), 921–924.

F. Kamber and Ph. Tondeur, Foliated bundles and characteristic classes, Springer Lecture Notes in Math. 494 (1975).

F. Kamber and Ph. Tondeur, Semisimplicial Weil algebras and characteristic classes for foliated bundles in Cech cohomology, Proc. Symp. Pure Math., Stanford, Calif., 1973, Providence, vol. 27, Part 1 (1975), 283–294.

F. Kamber and Ph. Tondeur, Non-trivial characteristic invariants of homogeneous foliated bundles, Ann. Scient. Ec. Norm. Sup. 8 (1975), 433–486.

F. Kamber and Ph. Tondeur, Classes caractéristiques et suites spectrales d'Eilenberg-Moore, C. R. Acad. Sci. Paris 283 (1976), 883–886.

F. Kamber and Ph. Tondeur, G-foliations and their characteristic classes, Bull. Amer. Math. Soc. 84 (1978), 1086–1124.

F. Kamber and Ph. Tondeur, On the linear independence of certain cohomology classes of B, Advances in Math. Suppl. Studies 5 (1979), 213–263.

F. Kamber and Ph. Tondeur, Feuilletages harmoniques, C. R. Acad. Sci. Paris 291 (1980), 409–411.

F. Kamber and Ph. Tondeur, Harmonic foliations, Proc. NSF Conference on Harmonic Maps, Tulane (1980), Springer Lecture Notes in Math. 949 (1982), 87–121.

F. Kamber and Ph. Tondeur, Infinitesimal automorphisms and second variation of the energy for harmonic foliations, Tôhoku Math. J. 34 (1982), 525–538.

F. Kamber and Ph. Tondeur, Dualité de Poincaré pour les feuilletages harmoniques, C. R. Acad. Sci. Paris 294 (1982), 357–359.

F. Kamber and Ph. Tondeur, Duality for Riemannian foliations, Proc. Symp. Pure Math., Amer. Math. Soc., vol. 40 (1983); Part 1, 609–618.

F. Kamber and Ph. Tondeur, The index of harmonic foliations on spheres, Trans. Amer. Math. Soc. 275 (1983), 257–263.

F. Kamber and Ph. Tondeur, Foliations and metrics, Proc. of a Year in Differential Geometry, University of Maryland, Birkhäuser, Progress in Mathematics vol. 32 (1983), 103–152.

F. Kamber and Ph. Tondeur, Curvature properties of harmonic foliations, Illinois J. of Math. 18 (1984), 458–471.

F. Kamber and Ph. Tondeur, Duality theorems for foliations, Astérisque 116 (1984), 108–116.

F. Kamber and Ph. Tondeur, The Bernstein problem for foliations, Proc. of the Conference on Global Differential Geometry and Global Analysis, Berlin 1984, Springer Lecture Notes in Math. 1156 (1985), 216–218.

F. Kamber and Ph. Tondeur, De Rham-Hodge theory for Riemannian foliations, Math. Annalen 277 (1987), 415–431.

F. Kamber and Ph. Tondeur, Foliations and harmonic forms, Harmonic mappings, twistors and σ-models, Adv. Series in Math. Phys., Vol. 4, World Scientific Publ. Co., Singapore, 1988, 15–25.

F. Kamber, Ph. Tondeur and G. Toth, Transversal Jacobi fields for harmonic foliations, Mich. Math. J. 34 (1987), 261–266.

J. Kamga and A. Mba, Prolongement tangent de feuilletages, Demonstr. Math. 26 (1993), 203–206.

M. Kanai, Geodesic flows of negatively curved manifolds with smooth stable and unstable foliations, Ergodic Theory and Dynam. Sys. 8 (1988), 215–240.

T. Kang, H. Kitahara and J. Pak, A formula for the radial part of the Laplace-Beltrami operator on the Riemannian foliation, Ann. Sci. Kanazawa Univ. 28 (1991), 1–8.

Y. Kanie, Cohomologies of Lie algebras of vector fields with coefficients in adjoint representations, Foliated case, Publs. Res. Inst. Math. Sci. 14 (1978), 487–501.

Y. Kanie, Some Lie algebras of vector fields on foliated manifolds and their derivation algebras, Proc. Japan Acad. 55 (1979), 409–411.

W. Kaplan, Regular curve families filling the plane I, Duke Math. J. 7 (1940), 154–185.

M. Karoubi, Classes caractéristiques de fibrés feuilletés, holomorphes on algébriques, K-Theory 8 (1994), 153–211.

A. Katok, G. Knieper and H. Weiss, Formulas for the derivative and critical points of topological entropy for Anosov and geodesic flows, Comm. Math. Phys. 138 (1991), 19–31.

B. Kaup, Ein geometrisches Endlichkeitskriterium für Untergruppen von $Aut(C,0)$ und holomorphe 1-codimensionale Blätterungen, Comment. Math. Helv. 53 (1978), 295–299.

N. Kawazumi, On the complex analytic Gelfand-Fuks cohomology of open Riemann surfaces, Ann. Inst. Fourier 43 (1993), 655–712.

N. Kawazumi, An application of the second Riemann continuation theorem to cohomology of the Lie algebra of vector fields on the complex line, Foliations Tokyo 1993, to appear.

N. Kawazumi, Moduli space and complex analytic Gelfand-Fuks cohomology of open Riemann surfaces, to appear.

M. S. Keane and M. Kellum, Topologically Riemannian foliations, preprint.

M. Kellum, Uniformly quasi-isometric foliations, Ergodic Theory Dyn. Syst. 13 (1993), 101–122.

M. Kellum, Uniform Lipschitz distortion, invariant measures and foliations, Geometric Study of Foliations, Tokyo 1993, 313–326, World Scientific, 1994.

M. Kellum, Transverse geometric structures and deformation spaces for foliated manifolds with bundle-like metrics, to appear.

M. Kellum, Orbit equivalence for uniformly quasi-isometric foliations, preprint.

M. Kellum, Transverse homogeneous structures and deformation spaces for foliated manifolds with bundle-like metrics, preprint.

S. Kerckhoff, Simplicial systems for interval exchange maps and measured foliations, Ergodic Theory and Dynam. Systems 5 (1985), 257–271.

B. Kim, On Escobales-Parker's theorem, Proc. Japan Acad. 65 (1989), 151–153.

B. Kim, N. Kim and J. Kwon, Fibred Riemannian spaces with critical Riemannian metrics, J. Korean Math. Soc. 31 (1994), 205–211.

H. Kim, Geometric and dynamical properties of riemannian foliations, Ph. D. Thesis, Univ. of Illinois, 1990.

H. Kim and Ph. Tondeur, Riemannian foliations on manifolds with non-negative curvature, Manuscripta Math. 74 (1991), 39–45.

H. Kim and G. Walschap, Riemannian foliations on compact hyperbolic manifolds, Indiana Univ. J. 41 (1992), 37–42.

I. Kim and J. Park, On transversally oriented foliations of codimension one, with trivial Godbillon-Vey classes, J. Hankuk Univ. of Foreign Studies 25 (1992), 627–636.

H. Kimura, Uniform foliation associated with the Garnier system, Sugaku Expositions 8 (1995), 103–123.

M. Kimura, Minimal hypersurfaces foliated by geodesics of 4-dimensional space forms, Tokyo J. Math. 16 (1993), 241–260.

H. Kitahara, The existence of complete bundle-like metrics, Ann. Sci. Kanazawa. Univ. 9 (1972), 37–40.

H. Kitahara, The existence of complete bundle-like metrics, II, Ann. Sci. Kanazawa Univ. 10 (1973), 51–54.

H. Kitahara, The completeness of a Clairaut foliation, Ann. Sci. Kanazawa Univ. 11 (1974), 37–40.

H. Kitahara, Remarks on square-integrable basic cohomology spaces on a foliated Riemannian manifold, Kodai Math. J. 2 (1979), 187–193.

H. Kitahara, Nonexistence of nontrivial Kählerian harmonic 1-forms on a complete foliated Riemannian manifold, Trans. Amer. Math. Soc. 262 (1980), 429–435.

H. Kitahara, On a parametrix form in a certain V-submersion, Springer Lecture Notes in Math. 792 (1980), 264–298.

H. Kitahara, Stability theorems for G-foliations associated with semisimple flat homogeneous spaces, Ann. Sci. Kanazawa Univ. 21 (1984), 7–18.

H. Kitahara and H. Pak, Construction of a complete negatively curved singular Riemannian foliation, J. Korean Math. Soc. 32 (1995), 609–614.

H. Kitahara and H. Pak, Some remarks on basic L^2-cohomology, Analysis and geometry on foliated manifolds, Santiago de Compostela 1994, World Scientific 1995, 99–111.

H. Kitahara and Y. Shinsuke, On some differential geometric characterization of a bundle-like metric, Kodai Math. J. 2 (1979), 130–138.

H. Kitahara and S. Yorozu, Godbillon-Vey invariants and its differential geometric interpretations, Ann. Sci. Kanazawa Univ. 12 (1975), 41–51.

H. Kitahara and S. Yorozu, Sur l'homomorphisme de Chern-Weil local et ses applications aux feuilletages, C. R. Acad. Sci. Paris 281 (1975), 703–706.

H. Kitahara and S. Yorozu, A formula for the normal part of the Laplace-Beltrami operator on the foliated manifold, Pacific J. Math. 69 (1977), 425–432.

H. Kitahara and S. Yorozu, On some differential geometric characterization of a bundle-like metric, Kodai Math. J. 2 (1979), 130–138.

H. Kitahara and S. Yorozu, On the Cech bicomplex associated with foliated structures, Ann. Inst. Fourier 28 (1978), 217–224; errata: ibid. 29–2 (1979) and 30–3 (1980).

R. Knill, A C^∞ flow on S^3 with Denjoy minimal set, J. Diff. Geom. 16 (1981), 271–280.

A. Kochubej, Manifolds with invariant semi-Riemannian almost product structure, Ukr. Mat. Zh. 44 (1992), 926–931.

K. Kodaira and D. Spencer, Multifoliate structures, Ann. Math. 74 (1961), 52–100.

F. Kohler, Feuilletages holomorphes singuliers sur les surfaces contenant une coquille sphérique globale, Ann. Inst. Fourier 45 (1995), 161–182.

N. Koike, Foliations on a Riemannian manifold and Ehresmann connections, Indiana Univ. Math. J. 40 (1991), 277–292.

N. Koike, Totally umbilic foliations and decomposition theorems, Saitama Math. J. 8 (1990), 1–18.

N. Koike, Totally umbilic orthogonal nets and decomposition theorems, Saitama Math. J. 10 (1992), 1–19.

N. Koike, Ehresmann connections for a foliation on a manifold with boundary, SUT J. Math. 30 (1994), 147–158.

Y. Kordyukov, A theorem on coincidence of spectra for tangentially elliptic operators on foliated manifolds, Funct. Anal. Appl. 19 (1991), 327–328.

Y. Kordyukov, L^p-theory of elliptic differential operators on manifolds of bounded geometry, Acta Appl. Math. 23 (1991), 223–260.

Y. Kordyukov, Transversally elliptic operators on G-manifolds of bounded geometry, Russ. J. of Math. Phys. 2 (1994), 175–198.

Y. Kordyukov, Functional calculus for tangentially elliptic operators on foliated manifolds, Analysis and Geometry in Foliated Manifolds, Proc. VII Colloq. on Diff. Geom., Santiago de Compostela 1994, World Scientific 1995, 113–136.

U. Koschorke, Singularities and bordism of q-plane fields and foliations, Bull. Amer. Math. Soc. 80 (1974), 762–765.

U. Koschorke, Line fields transversal to foliations, Proc. Symp. Pure Math., Stanford, Calif., 1973, Providence, Vol. 27, Part 1 (1975), 295–301.

U. Koschorke, Bordism of plane fields, Proc. of the 9th Braz. Math. Colloq. 1973, Vol. II, Inst. Mat. Pura Apl. Sao Paolo 1977, 311–320.

Y. Kosmann-Schwarzbach, Lagrangian foliations and Lax equations, Lett. Math. Phys. 9 (1985), 163–167.

J. Koszul, Homologie et cohomologie des algèbres de Lie, Bull. Soc. Math. France 78 (1950), 65–128.

J. Kubarski, Pradines-type groupoids over foliations; cohomology, connections and the Chern-Weil homomorphism, Preprint, Inst. of Math., Technical Univ. of Lodz, 1986.

J. Kubarski, About Stefan's definition of a foliation with singularities: a reduction of the axioms, Bull. Soc. Math. France 118 (1990), 391–394.

J. Kubarski, Characteristic classes of regular Lie algebroids, Suppl. al Rendiconti del Circolo Matematico di Palermo Ser. II, No. 30, 1993, 71–94.

J. Kubarski, Invariant cohomology of regular Lie algebroids, Analysis and Geometry in Foliated Manifolds, Proc. VII Colloq. on Diff. Geom., Santiago de Compostela 1994, World Scientific 1995, 137–151.

J. Kubarski, Tangential Chern-Weil homomorphism, Foliations Tokyo 1993, to appear.

J. Kubarski, Characteristic classes of flat and partially flat regular Lie algebroids over foliated manifolds, Publ. Dep. Math. Univ. Lyon, to appear.

N. Kuiper, Sur les surfaces localement affines, Colloque de géométrie différentielle, Strasbourg, 1953, 79–87.

A. Kumpera and D. Spencer, Lie Equations, vol. I: General Theory, Ann. of Math. Studies, vol. 73, Princeton, NJ: Princeton Univ. (1972).

G. Kuperberg and K. Kuperberg, Generalized counterexamples to the Seifert conjecture, preprint.

K. Kuperberg, A smooth counterexample to the Seifert conjecture, Ann. of Math. 140 (1994), 723–732.

I. Kupka, The singularities of integrable structurally stable Pfaffian forms, Proc. Nat. Acad. Sci. USA 52 (1964), 1431–1432.

I. Kupka and Ngô Van Quê, Formes différentielles fermés non singulierès, Springer Lecture Notes in Math. 484 (1975), 239–256.

J. Lacaze, Feuilletage d'une variété lorentzienne par des hypersurfaces spatiales à courbure moyenne constante, C. R. Acad. Sci. Paris 289 (1979), 771–774.

S. Laederich, $\bar{\partial}$-torsion, foliations and holomorphic vector fields, Comm. Math. Phys. 145 (1992), 447–458.

S. Laederich, Analytic torsion, flows and foliations, Hamiltonian dynamical systems (Cincinnati, OH, 1992), 181–193, IMA vol. Math. Appl., 63, Springer, New York, 1995.

F. Lalonde, Homologie de Shih d'une submersion (homologies non singulières des variétés feuilletées), Mém. Soc. Math. France 30, 1987.

C. Lamoureux, Feuilletages de codimension 1. Transversales fermées, C. R. Acad. Sci. Paris 270 (1970), 1659–1662.

C. Lamoureux, Feuilletages de codimension 1. Holonomie et homotopie, C. R. Acad. Sci. Paris 270 (1970), 1718–1721.

C. Lamoureux, Une condition pour qu'une feuille soit propre et ait une enveloppe composée de feuilles fermées, C. R. Acad. Sci. Paris 274 (1972), 31–34.

C. Lamoureux, Feuilles fermées et captage; applications, C. R. Acad. Sci. Paris 277 (1973), 579–581.

C. Lamoureux, Feuilles exceptionelles, feuilles denses, homologie et captage, C. R. Acad. Sci. Paris 277 (1973), 1041–1043.

C. Lamoureux, Sur quelques phénomènes de captage, Ann. Inst. Fourier 23 (1973), 229–243.

C. Lamoureux, The structure of foliations without holonomy on non-compact manifolds with fundamental group Z, Topology 13 (1974), 219–224.

C. Lamoureux, Quelques conditions d'existence de feuilles compactes, Ann. inst. Fourier 24 (1974), 229–240.

C. Lamoureux, Quelques propriétés globales des feuilletages par des feuilles simplement connexes non-nécessairement bornées, C. R. Acad. Sci. Paris 279 (1974), 813–816.

C. Lamoureux, Groupes d'homologie et d'homologie d'ordre supérieur des variétés compactes ou non compactes feuilletées en codimension 1, C. R. Acad. Sci. Paris 280 (1975), 411–414.

C. Lamoureux, Non-bounded leaves in codimension one foliations, Springer Lecture Notes in Math. 484 (1975), 257–272.

C. Lamoureux, Feuilles non captées et feuilles denses, Ann. Inst. Fourier 25 (1975), 285–293.

C. Lamoureux, Feuilletages des variétés compactes et non compactes, Ann. Inst. Fourier 26 (1976), 221–271.

C. Lamoureux, Holonomie et feuilles exceptionnelles, Ann. Inst. Fourier 26 (1976), 273–300.

C. Lamoureux, Quelques remarques sur les bouts des feuilles, Ann. Inst. Fourier 27 (1977), 191–196.

C. Lamoureux, Geometric properties connected with the transverse structure of codimension one foliations, Astérisque 116 (1984), 117–133.

R. Langevin, Courbure, feuilletages et singularités algébriques, Séminaire Le Dung Trang; Paris VII.

R. Langevin, Feuilletages tendus, Bulletin Soc. Math. France 107 (1979), 271–281.

R. Langevin, Tight foliations, Berlin 1979, Springer Lecture Notes in Math. 838 (1981), 181–186.

R. Langevin, Thèse d'Etat, Courbure, feuilletages et surfaces (mesures et distributions de Gauss) soutenue 1980 à l'Université de Paris XI – Orsay publications mathématique d'Orsay $N°$ 80–83.

R. Langevin, Feuilletages, énergies et cristaux liquides, Astérisque 107–108 (1983), 201–213.

R. Langevin, Energie et géométrie intégrale, Differential Geometry, Peniscola (1982), Springer Lecture Notes in Math 1045 (1984), 95–103.

R. Langevin (ed.), A list of questions about foliations, Contemp. Math. 161 (1994), 59–80.

R. Langevin, Geometry and dynamics of codimension 1 foliations, to appear.

R. Langevin and G. Levitt, Courbure totale des feuilletages des surfaces, Comment. Math. Helv. 57 (1982), 175–195.

R. Langevin and G. Levitt, Sur la courbure totale des feuilletages des surfaces à bord, Bol. Soc. Brasil. Mat. 16 (1985), 1–13.

R. Langevin, G. Levitt and H. Rosenberg, Hérissons et multihérissons (enveloppes paramétrées par leur application de Gauss), Banach Center Publ. 20 (1988), 245–253.

R. Langevin and Y. Nikolayevsky, Three viewpoints on the integral geometry of foliations, preprint.

R. Langevin and C. Possani, Courbure totale de feuilletages et enveloppes, C. R. Acad. Sci. Paris 309 (1989), 821–824.

R. Langevin and C. Possani, Total curvature of foliations, Ill. J. of Math. 37 (1993), 508–524.

R. Langevin and H. Rosenberg, On stability of compact leaves and fibrations, Topology 16 (1977), 107–111.

R. Langevin and H. Rosenberg, Integrable perturbations of fibrations and a theorem of Seifert, Springer Lecture Notes in Math. 652 (1978), 122–127.

R. Langevin and P. Walczak, Entropie d'une dynamique, C. R. Acad. Sci. Paris 312 (1991), 141–144.

R. Langevin and P. Walczak, Entropy, transverse entropy and partitions of unity, Ergodic Theory and Dynamical systmes, to appear.

R. Langevin and P. Walczak, Some invariants measuring dynamics of codimension-one foliations, Geometric Study of Foliations, Tokyo 1993, 345–358, World Scientific, 1994.

D. Lappas, On the integrable G-invariant metrics, Yokohama Math. J. 42 (1994), 87–94.

M. Lasso de la Vega, Groupoide fundamental et d'holonomie de certains feuilletages réguliers, Publ. Math. 33 (1989), 431–443.

F. Latour, Existence de 1-formes fermées non singulierès daus une classe de cohomologie de De Rham, Inst. Hautes Etudes Scient. Publ. Math. 80 (1994), 135–194.

F. Laudenbach, Formes différentielles de degré 1 fermées non singulières: classes d'homotopie de leurs noyaux, Comment. Math. Helv. 51 (1976), 447–464.

F. Laudenbach and R. Roussarie, Un exemple de feuilletage sur S^3, Topology 9 (1970), 63–70.

J. Lawrynowicz and M. Okada, Canonical diffusion and foliations involving the complex Hessian, Bull. Polish. Acad. Sci. Math. 34 (1986), 661–667.

H. Lawson, Codimension one foliations of spheres, Bull. Amer. Math. Soc. 77 (1971), 437–438.

H. Lawson, Codimension-one foliations of spheres, Ann. Math. 94 (1971), 494–503.

H. Lawson, Foliations, Bull. Amer. Math. Soc. 80 (1974), 369–418.

H. Lawson, Lectures on the quantitative theory of foliation, CBMS Regional Conf. Series, Vol. 27 (1977).

C. Lazarov, A permanence theorem for exotic classes, J. Diff. Geom. 14 (1979), 475–486.

C. Lazarov, A vanishing theorem for certain rigid classes, Mich. Math. J. 28 (1981), 89–95.

C. Lazarov, An index theorem for foliations, Ill. J. of Math. 30 (1986), 101–121.

C. Lazarov, Spectral invariants for foliations, Mich. Math. J. 33 (1986), 231–243.

C. Lazarov, A relation between index and exotic classes, Contemp. Math. 70 (1988), 125–143.

C. Lazarov, Transverse index and periodic orbit, to appear.

C. Lazarov, Characteristic classes for flat Diff(M)-bundles, preprint.

C. Lazarov, A note on the construction of elements in $H^*(\overline{B \text{ Diff}(M)})$, preprint.

C. Lazarov and J. Pasternack, Secondary characteristic classes for Riemannian foliations, J. Diff. Geom. 11 (1976), 365–385.

C. Lazarov and J. Pasternack, Residues and characteristic classes for Riemannian foliations, J. Diff. Geom. 11 (1976), 599–612.

C. Lazarov and H. Shulman, Obstructions to foliation preserving Lie group actions, Topology 18 (1979), 255–256.

C. Lazarov and H. Shulman, Obstructions to foliation preserving vector fields, J. Pure Appl. Algebra 24 (1982), 171–178.

C. Le Brun, Foliated CR manifolds, J. Differential Geom. 22 (1985), 81–96.

F. Ledrappier, Ergodic properties of the stable foliations, Springer Lecture Notes in Math. 1514 (1992), 131–145.

F. Ledrappier, Harmonic 1-forms on the stable foliation, Bol. Soc. Bras. Mat. 25 (1994), 121–138.

D. Lehmann, Quelques problèmes de classes caractéristiques secondaires, Rev. Columbiana 6 (1972), 116–124.

D. Lehmann, *J*-homotopie dans les espaces de connexions et classes exotiques de Chern-Simons, C. R. Acad. Sci. Paris 275 (1972), 835–838.

D. Lehmann, Classes caractéristiques exotiques et *J*-connexité des espaces de connexions, Ann. Inst. Fourier 24 (1974), 267–306.

D. Lehmann, Sur l'approximation de certains feuilletages nilpotents par des fibrations, C. R. Acad. Sci. Paris 286 (1978), 251–254.

D. Lehmann, Sur la généralization d'un théorème de Tischler à certains feuilletages nilpotents, Proc. K. Ned. Akad. Wet. Ser. A 82 (1979), 177–189.

D. Lehmann, Résidus des connexions à singularités et classes caractéristiques, Ann. Inst. Fourier 31 (1981), 83–98.

D. Lehmann, Structures de Maurer-Cartan et Γ_θ-structures, II, Espaces classifiants, Astérisque 116 (1984), 134–138.

D. Lehmann, Modèle minimal relatif des feuilletages, Springer Lecture Notes in Math. 1183 (1986), 250–258.

D. Lehmann, Classes caractéristiques résiduelles, Publ. IRMA-Lille 17 (1989).

D. Lehmann, Résidus des sous-variétés invariantes d'un feuilletage singulier, Ann. Inst. Fourier 41 (1991), 211–258.

D. Lehmann and G. Martinez, Classes caracteristicas exoticas. Tipo de homotopia racional *y* formas differenciales, Gac. Mat. 26 (1974), 147–157.

D. Lehmann and T. Suwa, Residues of holomorphic vector fields relative to singular invariant subvarieties, J. Diff. Geom. 42 (1995), 165–192.

J. Lehmann-Lejeune, Cohomology over the transverse bundle to a foliation, C. R. Acad. Sci. Paris 295 (1982), 495–498.

J. Lehmann-Lejeune, Cohomologies sur le fibré transverse à un feuilletage, Astérisque 116 (1984), 149–179.

M. Leon-de, Examples of stable compact symplectic foliations on compact symplectic manifolds, Portugal. Math. 46 (1989), 391–399.

J. Leslie, A remark on the group of automorphisms of a foliation having a dense leaf, J. Diff. Geom. 7 (1972), 597–601.

H. Levine and M. Shub, Stability of foliations, Trans. Amer. Math. Soc. 184 (1973), 419–437.

G. Levitt, Feuilletages des variétés de dimension 3 qui sont des fibrés en cercles, Comment. Math. Helv. 53 (1978), 572–594.

G. Levitt, Sur les mesures transverses invariantes d'un feuilletage de codimension 1, C. R. Acad. Sci. Paris 290 (1980), 1139–1140.

G. Levitt, Propriétés homologiques des feuilletages des surfaces, C. R. Acad. Sci. Paris 293 (1981), 597–600.

G. Levitt, Feuilletages de surfaces, Ann. Inst. Fourier 32 (1982), 179–217.

G. Levitt, Pantalons et feuilletages des surfaces, Topology 21 (1982), 9–33.

G. Levitt, Feuilletage des surfaces, Thèse d'Etat, Paris VII (1983).

G. Levitt, Foliation and laminations on hyperbolic surfaces, Topology 22 (1983), 119–135.

G. Levitt, 1-formes fermées singulières et groupe fondamental, Invent. Math. 88 (1987), 635–667.

G. Levitt, La décomposition dynamique et la différentiabilité des feuilletages des surfaces, Ann. Inst. Fourier 37 (1987), 85–116.

G. Levitt, Groupe fondamental de l'espace des feuilles dans les feuilletages sans holonomie, J. Diff. Geom. 31 (1990), 711–761.

G. Levitt, Feuilletages mesurés et arbres réels, Séminaire Gaston Darboux de Géom. et Topologie Différ. 1991–92, 73–80 (1993).

G. Levitt and G. Meigniez, Closed differential one-forms and R-trees, Geometric Study of Foliations, Tokyo 1993, 359–373, World Scientific, 1994.

M. Lewkowicz, On the action of diffeomorphisms on the Gelfand-Fuks cohomology, Bull. Polish Acad. Sci. Math. 32 (1984), 695–701.

Z. Li, Real projective algebraic manifolds with a foliation, Chinese Quart. J. Math. 2 (1987), 105–107.

Z. Li, Harmonic foliations on the sphere, Tsukuba J. Math. 15 (1991), 397–407.

Z. Li and Z. Sun, Complex foliations on a complex projective space, Acta Math. Sin. 37 (1994), 762–766.

C. Liang, Multifoliations on open manifolds, Math. Ann. 221 (1976), 143–146.

C. Liang, On the rank of an open manifold, Mich. Math. J. 23 (1976), 325–326.

C. Liang, On the volume preserving foliations, Math. Ann. 223 (1976), 13–17.

C. Liang, Riemannian $\mathbb{R}^k$-actions on compact manifolds, Chinese J. Math. 6 (1978), 113–120.

C. Liang, Vector fields on V-manifolds and locally free $G \times \mathbb{R}^1$-actions on manifolds, Indiana Univ. Math. J. 27 (1978), 349–352.

P. Libermann, Pfaffian systems and transverse differential geometry. Differential Geometry and Relativity: D. Reidel (1976), 107–126.

P. Libermann, Problèmes d'équivalences et géométrie symplectique, Astérisque 107–108 (1983), 43–68.

P. Libermann, Submanifolds and regular symplectic foliations, Pitman Research notes in Math. 80 (1983), 81–106.

P. Libermann, Symplectic regular foliations, Atti Accad. Sci. Torino Cl. Sci. Fis. Mat. Natur. 117 (1983), 239–246.

P. Libermann, Cartan-Darboux theorems for Pfaffian forms on foliated manifolds, Proc. of the Sixth Intern Colloq. on Differential Geometry (Santiago de Compostela, 1988), 125–144.

P. Libermann, Legendre foliations on contact manifolds, Diff. Geom. Appl. 1 (1991), 57–76.

A. Lichnerowicz, Sur l'algèbre de Lie des champs de vecteurs, Comment. Math. Helv. 51 (1976), 343–368.

A. Lichnerowicz, Feuilletage et algèbres de Lie associées, C. R. Acad. Sci. Paris 286 (1978), 1141–1145.

A. Lichnerowicz, Algèbres de Lie attachées à un feuilletage, Ann. Fac. Sci. Toulouse Math 1 (1979), 45–76.

A. Lichnerowicz, Feuilletages et déformations infinitésimales des algèbres de Lie associées, Proc. IV Int. Colloq. on Diff. Geom. Santiago de Compostela (1978), 192–204.

A. Lichnerowicz, Variétés de Poisson et feuilletages, Ann. Fac. Sci. Toulouse Math. 4 (1982), 195–262.

A. Lichnerowicz, Formes caractéristiques d'un feuilletage et classes de cohomologie de l'algèbre des vecteurs tangents à valeurs dans les formes normales, C. R. Acad. Sci. Paris 296 (1983), 67–71.

A. Lichnerowicz, Feuilletages, géométrie riemannienne et géométrie symplectique, Riv. Mat. Univ. Parma 10 (1984), 81–90.

A. Lichnerowicz, Formes caractéristiques d'un feuilletage et classes de cohomologie de l'algèbre des champs de vecteurs à valeurs dans les formes normales, Feuilletages et quantification géométrique, Lyon, 1983, Travaux en Cours 26 (1984), 9–18.

A. Lichnerowicz, Generalized foliations and local Lie algebras of Kirillov, Differential geometry (Santiago de Compostela, 1984), Pitman Res. Notes Math. Ser. 131 (1985), 198–210.

A. Lichnerowicz and Tran Van Tan, Feuilletages, géométrie riemannienne et géométrie symplectique, C. R. Acad. Sci. Paris 296 (1983), 205–210.

W. Lickorish, A foliation for 3-manifolds, Ann. of Math. 82 (1965), 414–420.

X. Li-Jost, Uniqueness of minimal surfaces in Euclidean and hyperbolic 3-space, Math. Z. 217 (1994), 275–285.

E. Lima, Common singularities of commuting vector fields on 2-manifolds, Comment. Math. Helv. 39 (1964), 97–110.

E. Lima, Commuting vector fields on S^3, Ann. of Math. 81 (1965), 70–81.

L. Lininger, Codimension 1 foliations on manifolds with even index, Springer Lecture Notes in Math. 438 (1975), 336–338.

A. Lins Neto, Construction of singular holomorphic vector fields and foliations in dimension two, J. Differential Geom. 26 (1987), 1–31.

A. Lins Neto, Complex codimension one foliations leaving a compact submanifold invariant, Dynamical systems and bifurcation theory, Pitman Research Notes in Mathematics Series, vol. 160 (1987), 295–317.

A. Lins Neto, Algebraic solutions of polynomial differential equations and foliations in dimension two, Springer Lecture Notes in Mathematics 1345 (1988).

I. Liousse, Feuilletages transversalement affines des surfaces et actions affines sur les arbres réels, Thèse, 1994.

I. Liousse, Dynamique générique des feuilletages affines des surfaces, Bull. Soc. Math. France 123 (1995), 493–516.

K. Lisiecki, Foliated groupoids, Supl. ai Rendiconti del Circulo Mat. di Palermo, Series II, no. 22, 1990, 127–149.

M. Llabres and A. Reventós, Unimodular Lie foliations, Ann. Fac. Sci. Toulouse 9 (1988), 243–255.

M. Llabres and A. Reventós, Some remarks on Lie flows, Publicaciones del Depto de Geometria y Topologia, Universidad de Santiago de Compostela, 57 (1989), 517–531.

T. Lokot, On foliations of codimension 1 with a trivial holonomy group, Usp. Mat. Nauk 27 (1972), 177–178.

T. Lokot, A topological classification of foliations of codimension 1 with trivial holonomy groups of class C^2 on the three-dimensional torus, Usp. Mat. Nauk 27 (1972), 205.

T. Lokot, Foliations of codimension 1 with trivial holonomy groups, Soviet Math. 18 (1974), 67–72.

T. Lokot, On some properties of foliations of codimension 1 with trivial holonomy groups, in: Tr. Mosk. Inzh.-Stroit. Inst. 121 (1974), 132–139.

F. Loray and R. Meziani, Classification de certains feuilletages associés à un cusp, Bol. Soc. Bras. Mat., Nova Ser. 25 (1994), 93–106.

M. Losik, Characteristic classes of structures on manifolds, Functional Anal. Appl. 21 (1987), 206–216.

M. Losik, Characteristic classes of transformation groups, Diff. Geom. and its Applications 39 (1993), 205–218.

D. Lu, Homogeneous foliations of spheres, Trans. Amer. Math. Soc. 340 (1993), 95–102.

M. Macho-Stadler, La conjecture de Baum-Connes pour un feuilletage sans holonomie de codimension un sur une variété fermée, Publicaciones del Depto de Geometria y Topologia, Universidad de Santiago de Compostela, 57 (1982), 445–457.

E. Macias, Las cohomologias diferenciable, continua y discreta de una variedad foliada, Publ. del Depto. de Geometria y Topologia Santiago de Compostela 60 (1983).

E. Macias, Une application de la suite spectrale de Grothendieck à la cohomologie des feuilletages, VII Congresso do GMEL, Coimbra (1985).

E. Macias, Continuous cohomology of linear foliations on T^2, Rend. Mat. Appl. 11 (1991), 523–528.

E. Macias, Non-closed Lie subgroups of Lie groups, Ann. Global Anal. Geom. 11 (1993), 35–40.

E. Macias, Homotopy groups in Lie foliations, Trans. Amer. Math. Soc. 344 (1994), 701–711.

E. Macias and E. Sanmartin, Minimal foliations on Lie groups, Indag. Math. New Ser. 3 (1992), 41–46.

E. Macias and E. Sanmartin, The manifold of bundle-like metrics, Foliations Tokyo 1993.

Y. Maeda, S. Rosenberg and Ph. Tondeur, The mean curvature of gauge orbits, Global Analysis in Modern Mathematics, Publish or Perish, Inc. (1993), 171–217.

Y. Maeda, S. Rosenberg and Ph. Tondeur, Minimal submanifolds in infinite dimensions, Analysis and geometry in foliated manifolds, Santiago de Compostela 1994, World Scientific 1995, 177–182.

M. Magid, Submersions from anti-de Sitter space with totally geodesic fibers, J. Diff. Geom. 16 (1981), 323–331.

A. Maier, Trajectories on the closed orientable surfaces, Math. Sb. 12 (1943), 71–84.

B. Malgrange, Frobenius avec singularités I. Codimension un. Inst. Hautes Etudes Sci. Publ. Math. 46 (1976), 163–173.

B. Malgrange, II. Le cas général, Invent. Math. 39 (1977), 67–89.

B. Malgrange, Frobenius avec singularités, Invent. Math. 39 (1977), 67–89.

D. Mall, On the topology of holomorphic foliations on Hopf manifolds, Compositio Math. 89 (1993), 243–250.

E. Malysheva and N. Zhukova, On minimal sets of Riemannian foliations, Izv. Vyssh. Uchebn. Zaved. Mat. 1986, 38–45, 83.

S. Mancini and M. Ruas, Bifurcations of generic one parameter families of functions on foliated manifolds, Math. Scand. 72 (1993), 5–19.

S. Mancini, A. Maria and R. Soares, Bifurcations of generic one parameter families of functions on foliated manifolds, Math. Scand. 72 (1993), 5–19.

I. Maniscalco and A. Portolano, The space of closed simple curves and measured foliations on noncompact orientable surfaces, Rend. Circ. Mat. Palermo 36 (1987), 401–406.

I. Maniscalco and A. Portolano, Curves as measured foliations on noncompact spaces, Rendiconti del Circolo Matematico di Palermo Series II, Vol. 42 (1993), 161–180.

P. March, M. Min-Oo and E. Ruh, Mean curvature of Riemannian foliations, preprint.

J. P. Marco, Obstructions topologiques à l'intégrabilité des flots géodésiques en classe de Bott, Bull. Sci. Math. 117 (1993), 185–209.

R. Marinosci, Pseudoconnections on foliated manifolds, Note Math. 1 (1981), 71–92.

N. Markley, The Poincaré-Bendixson theorem for the Klein bottle, Trans. Amer. Math. Soc. 135 (1969), 159–165.

J. Marsden and F. Tipler, Maximal hypersurfaces and foliations of constant mean curvatue in general relativity, Phys. Reports 66 (1980), 109–139.

J. Martinet, Classes caractéristiques des systèmes de Pfaff, Springer Lecture Notes in Math. 392 (1974), 30–36.

J. Martinet and G. Reeb, Sur une généralisation des structures feuilletées de codimension un, Dynamical Systems, Bahia, 1971, 177–184, Academic Press, 1973.

Y. Martinez, Feuilletages des surfaces et décompositions en pantalons, Bull. Soc. Math. France 112 (1984), 387–396.

P. Marty, Feuilletages Lagrangiens et métriques quasi cotangentes complètes, C. R. Acad. Sci. Paris 309 (1989), 187–190.

H. Marzougui, Sur les feuilletages des surfaces compactes et non compactes, Thèse de troisième cycle, Tunis (1993).

H. Marzougui, Structures des feuilles sur les surfaces non compactes, Proc. of the Int. Workshop on complex structures and vector fields, Pravetz, Bulgaria, Aug. 14–17, 1994, 24–35, World Scientific, 1995.

X. Masa, Sucesión espectral de cohomología asociada a variedades foliadas. Aplicaciones, Publ. del Depto. de Geometría y Topología no. 19, Santiago de Compostela (1973).

X. Masa, Non-existence d'un type de feuilletage sur S^7, in G. Reeb, Feuilletages: résultats anciens et nouveaux (Painlevé, Hector et Martinet), Montréal 1972, Presses Univ. Montreal 1974, 58–62.

X. Masa, Quelques propriétés des feuilletages de codimension 1 a connexion transverse projetable, C. R. Acad. Sci. Paris 284 (1977), 811–812.

X. Masa, Cohomology of Lie foliations, Differential Geometry, Proceedings of the Vth Colloquium on Differential Geometry, Santiago de Compostela (1984), Pitman Research Notes 131 (1986), 211–214.

X. Masa, Duality and minimality in Riemannian foliations, Comment. Math. Helvetici 67 (1992), 17–27.

X. Masa, Minimal model of nilpotent Lie foliations, Foliations Tokyo 1993, to appear.

H. Masur, Interval exchange transformations and measured foliations, Ann. of Math. 115 (1982), 169–200.

H. Masur and J. Smillie, Hausdorff dimension of sets of nonergodic measure foliations, Ann. of Math. 134 (1991), 455–543.

J. Mather, The vanishing of the homology of certain groups of diffeomorphisms, Topology 10 (1971), 297–298.

J. Mather, On Haefliger's classifying space, I, Bull. Amer. Math. Soc. 77 (1971), 1111–1115.

J. Mather, On Haefliger's classifying space, II, Harvard University.

J. Mather, Integrability in codimension 1, Comment. Math. Helv. 48 (1973), 195–233.

J. Mather, Simplicity of certain groups of diffeomorphisms, Bull. Amer. Math. Soc. 80 (1974), 271–273

J. Mather, Commutators of diffeomorphisms, Comment. Math. Helv. 49 (1974), 512–518.

J. Mather, Commutators of diffeomorphisms, II, Comment. Math. Helv. 50 (1975), 33–40.

J. Mather, Foliations and local homology of groups of diffeomorphisms, Proc. Int. Congr. Math., Vancouver 1974, vol. 2,S. 1 (1975), 35–37.

J. Mather, Loops and foliations, Manifolds, Tokyo 1973, U. of Tokyo Press 1975, 175–180.

J. Mather, On the homology of Haefliger's classifying spaces, C.I.M.E. Varenna 1976, Liguori 1979, 73–116.

J. Mather, Foliations of surfaces, I, An ideal boundary, Ann. Inst. Fourier 32 (1982), 235–261.

K. Matsuda, An analogy of the theorem of Hector and Duming, Hokkaido Math. J. 21 (1992), 151–158.

S. Matsumoto, Foliations of Seifert fibered spaces over S^2, Foliations (Tokyo, 1983), 325–339, Adv. Stud. Pure Math. 5, North-Holland, Amsterdam, 1985.

S. Matsumoto, Some remarks on foliated S^1-bundles, Invent. Math. 90 (1987), 343–358.

S. Matsumoto, Measure of exceptional minimal sets of codimension one foliations, A Fête of Topology, Academic Press, Boston, 1988, 81–94.

S. Matsumoto, Codimension one foliations on solvable manifolds, Comment. Math. Helv. 68 (1993), 633–652.

S. Matsumoto, Codimension one Anosov flows, Lecture Note Series, Seoul. 27, Seoul National Univ. v (1995), 95 pp.

S. Matsumoto, Codimension one foliations and Anosov flows on solvable manifolds, to appear.

S. Matsumoto and A. Sato, Flows and transverse foliations, Stability theory and related topics in dynamical systems, World Scientific Publishing, 1989, 113–123.

S. Matsumoto and A. Sato, Perturbation of the Hopf flow and transverse foliations, Nagoya Math. J. 122 (1991), 75–82.

S. Matsumoto and N. Tsuchiya, The Lie affine foliations on 4-manifolds, Invent. Math. 109 (1992), 1–16.

T. Matsuoka and S. Morita, On characteristic classes of Kähler foliations, Osaka J. Math. 16 (1979), 539–550.

J. Mattei, Holonomie et intégrales premières, Ann. Scien. Ec. Norm. Sup. 4° S., 12 (1980), 469–523.

J. Mattei, Modules de feuilletages holomorphes singuliers. I. Equisingularité, Invent. Math. 103 (1991), 297–325.

J. Mattei and M. Nicolau, Equisingular unfoldings of foliations by curves, Astérisque 222 (1994), 285–302.

H. Matuszczyk, On foliations of differential spaces, Ann. Polon. Math. 48 (1988), 269–273.

H. Matuszczyk, On generalized foliations with locally conected leaves, Demonstr. Math. 22 (1989), 967–971.

L. Maxim-Raileanu, On the prolongations of the foliated Banach manifolds and of the foliate fibre bundles, An. Stiint. Univ. "Al. I. Cuza" Iasi 21 (1975), 73–84.

L. Maxim-Raileanu, The equivariant stability of the differentiable mappings between G-equivariant foliated manifolds, An. Stiint. Univ. "Al. I. Cuza" Iasi 31 (1985), 239–245.

L. Maxim-Raileanu, Smooth mappings between foliated manifolds with boundary, An. Stiint. Univ. "Al. I. Cuza" Iasi 37 (1991), 53–64.

A. Mba and J. Wouafo Kamga, Prolongement tangent de feuilletages, Demonstratio Math. 26 (1993), 203–206.

J. McCarthy and A. Papadopoulos, Dynamics on Thurston's sphere of projective measured foliations, Comment. Math. Helv. 65 (1989), 133–166.

D. McDuff, Foliations and monoids of embeddings, Geometric Topology, Athens 1977, Academic Press 1979, 429–444.

D. McDuff, The homology of some groups of diffeomorphisms, Comment. Math. Helv. 55 (1980), 97–129.

D. McDuff, On groups of volume preserving diffeomorphisms and foliations with transverse volume form, Proc. London Math. Soc. 43 (1981), 295–320.

A. Medeiro, Structural stability of integrable differential forms, Springer Lecture Notes in Math. 597 (1977), 395–428.

A. Medina, Invariants d'un feuilletage et champs de vecteurs feuilletés, Bol. Soc. Mat. Mexicana 18 (1973), 94–99.

A. Medina, Nombres caractéristiques du fibré transverse à un feuilletage et champs de vecteurs feuilletés, C. R. Acad. Sci. Paris 276 (1973), 863–865.

A. Medina, Quelques remarques sur les feuilletages affines, C. R. Acad. Sci. Paris 282 (1976), 1159–1162.

A. Medina, Quelques remarques sur les feuilletages affines, Cah. Topol. Geom. Different., Ch. Ehresmann 17 (1976), 59–68.

A. Medina and Perea, Connection on the transverse bundle to a foliation, Rev. Colombiana Mat. 7 (1973), 85–92.

A. Medina and Perea, Une note sur les invariants caractéristiques de certaines Γ_q-structures, Rev. Colombiana Mat. 7 (1973), 109–112.

X. Mei, Note on the residues of the singularities of a Riemannian foliation, Proc. Amer. Math. Soc. 89 (1983), 359–366.

X. Mei, Killing vector fields and characteristic forms, Japan J. Math. 21 (1995), 471–486.

G. Meigniez, Actions de groupes sur la droite réelle et feuilletages de codimension 1, Thèse, 1988.

G. Meigniez, Bouts d'un groupe opérant sur la droite. II. Théorie algébrique. Ann. Inst. Fourier 40 (1990), 271–312.

G. Meigniez, Bouts d'un groupe opérant sur la droite. I. Applications à la topologie des feuilletages, Tôhoku Math. J. 43 (1991), 473–500.

G. Meigniez, On completeness of transversely projective foliations, Topology 31 (1992), 421–432.

G. Meigniez, Submersions et fibrations localement triviales, C. R. Acad. Sci. Paris 321 (1995), 1363–1365.

G. Meigniez, Trying to characterize holonomy groups of Lie foliations, to appear.

I. Melnikova, A test for non-compactness of the foliation of a Morse form, Russ. Math. Surveys 50 (1995), 444–445.

M. Mesmoudi, Métriques plates singulières et feuilletages mesurés sur les surfaces, Thèse Strasbourg, 1994.

J. Meyer, e-foliations of codimension two, J. Diff. Geom. 12 (1977), 583–594.

A. Miernowski and W. Mozgawa, A note on the theorem of Pasternak, Serdica Bulgaricae mathematicae publicationes 13 (1987), 197.

A. Miernowski and W. Mozgawa, On Molino lifting of Riemannian vector fields, Acta Math. Hungar. 55 (1990), 185–191.

A. Mihai, Sur les feuilletages des surfaces reglées, Rev. Roumaine Math. Pures Appl. 23 (1978), 211–214.

I. Mihut, Chern classes of foliated manifolds, Inst. Politehn. Traian Vuia Timisoara. Lucrar. Sem. Mat. Fiz. (1983), 29–32.

I. Mihut, Differential forms on foliated fibre bundles, Proc. of the National Conf. on Geometry and Topology (Tirgoviste, 1986), 149–152, Univ. Bucuresti, Bucharest, 1988.

I. Mihut, Some remarks on analysis on foliated manifolds, Proc. of the 22nd conference on differential geometry and topology, Bucharest 1991; Bucharest: Polytechnic Institute of Bucharest (1991), 207–210.

W. Mikulski, Natural liftings of foliations to the tangent bundle, Math. Bohem. 117 (1992), 409–412.

W. Mikulski, Natural transformations of foliations into foliations on the cotangent bundle, Proc. Winter School in Geometry and Topology Srni 1992, Rend. Circ. Mat. Palermo (2) Suppl. 32 (1993), 61–67.

K. Millet, Compact foliations, Springer Lecture Notes in Math. 484 (1975), 277–287.

K. Millet, Transversality for surfaces in foliated 3-manifolds, Springer Lecture Notes 1144 (1985), 115–133.

K. Millet, Generic properties of proper foliations, Fund. Math. 128 (1987), 131–138.

J. Milnor, On the existence of a connection with curvature zero, Comment. Math. Helv. 32 (1958), 215–223.

J. Milnor, Foliations and foliated vector bundles, M. I. T. Notes, 1970.

H. Minakawa, Examples of exceptional homomorphisms which have nontrivial Euler numbers, Topology 30 (1991), 429–438.

H. Minakawa, Piecewise linear homeomorphisms of a circle and foliations, Tôhoku Math. J. 43 (1991), 69–74.

H. Minakawa, Transversely piecewise linear foliations by planes and cylinders; PL version of a theorem of E. Ghys, Hokkaido Math. J. 20 (1991), 531–538.

M. Miniconi, Familles de feuilletages analytiques, C. R. Acad. Sci. Paris 278 (1974), 273–276.

M. Min-Oo and E. Ruh, Comparison theorems for compact symmetric space, Ann. Scient. Ec. Norm Sup., 4^esérie, 12 (1979), 335–353.

M. Min-oo, E. Ruh and Ph. Tondeur, Vanishing theorems for the basic cohomology of Riemannian foliations, J. für die reine und angewandte Mathematik 415 (1991), 167–174.

M. Min-Oo, E. Ruh and Ph. Tondeur, A comparison theorem for almost Lie foliations, Ann. Global Anal. and Geom. 9 (1991), 61–66.

M. Min-Oo, E. Ruh and Ph. Tondeur, Transversal curvature and tautness for Riemannian foliations, Proc. of the Conference on Global Analysis and Global Differential Geometry, Berlin 1990, Springer Lecture Notes in Math. 1481 (1991), 145–146.

N. Mishachev, Elimination of the standard singularities of flags of foliations, Syktyvkar Univ. (1978).

N. Mishachev, The construction of flags of foliations, Usp. Mat. Nauk. 34 (1979), 237–238. Translation: Russ. Math. Surveys 34 (1979), 233–234.

N. Mishachev, Classification of flags of foliations, Uspekhi Mat. Nauk 35 (1980), 217–218. Translation: Russ. Math. Surveys 35 (1980), 253–254.

Y. Mitsumatsu, A relation between the topological invariance of the Godbillon-Vey invariant and the differentiability of Anosov foliations, Foliations (Tokyo, 1983), 159–167, Adv. Stud. Pure Math. 5, North-Holland, Amsterdam, 1985.

Y. Mitsumatsu, Self-intersections and transverse Euler numbers of foliation cycles, Thesis, Univ. of Tokyo, 1985.

Y. Mitsumatsu, On the self-intersections of foliation cycles, Trans. Amer. Math. Soc. 334 (1992), 851–860.

Y. Mitsumatsu, $SL(2, \mathbb{R})$-actions on surfaces, Geometric Study of Foliations, Tokyo 1993, 375–389, World Scientific, 1994.

Y. Mitsumatsu, Anosov flows and non-stein symplectic manifolds, Ann. Inst. Fourier 45 (1995), 1407–1421.

Y. Mitsumatsu, Tightness and tautness of projectively Anosov flows, to appear.

S. Miyoshi, On the placement problem of Reeb components, Comment. Math. Helv. 57 (1982), 260–281.

S. Miyoshi, Foliated round surgery of codimension one foliated manifolds, Topology 21 (1982), 245–261.

S. Miyoshi, Existence of Sullivan's vanishing cycles in codimension-one foliations, Topology and Computer Science (Atami, 1986), Kinokuniya, Tokyo (1987), 395–406.

S. Miyoshi, On Sullivan's vanishing cycles in codimension-one foliations, Tokyo J. Math. 11 (1988), 387–404.

T. Mizutani, Remarks on codimension one foliations of spheres, J. Math. Soc. Japan. 24 (1972), 732–735.

T. Mizutani, Foliated cobordisms of S^3 and examples of foliated 4-manifolds, Topology 13 (1974), 353–362.

T. Mizutani, Foliations and foliated cobordisms of spheres in codimension one, J. Math. Soc. Japan. 27 (1975), 264–280.

T. Mizutani, Foliated cobordism of PA foliations, Foliations (Tokyo, 1983), 121–134, Adv. Stud. Pure Math. 5, North-Holland, Amsterdam, 1985.

T. Mizutani, The Godbillon-Vey cocycles of Diff $\mathbb{R}^n$, A fête in topology, Academic Press, 1988, 49–62.

T. Mizutani, A formula for characteristic classes of foliated $\mathbb{R}^n$-products, Saitama Math. J. 6 (1988), 33–43.

T. Mizutani, Characteristic classes of foliated bundles, Saitama Math. J. 10 (1992), 25–31.

T. Mizutani, Characteristic classes of foliations and the group cocycles of Diff F, Geometric Study of Foliations, Tokyo 1993, 391–409, World Scientific, 1994.

T. Mizutani, S. Morita and T. Tsuboi, The Godbillon-Vey class of codimension one foliations which are almost without holonomy, Ann. of Math. 113 (1981), 515–527.

T. Mizutani, S. Morita and T. Tsuboi, On the cobordism classes of codimension one foliations which are almost without holonomy, Topology 22 (1983), 325–343.

T. Mizutani and I. Tamura, Foliations of even dimensional manifolds, Manifolds, Tokyo 1973, U. of Tokyo Press 1975, 189–194.

T. Mizutani and I. Tamura, Null-cobordant codimension one foliation on S^{4n-1}, J. Fac. Sci. Univ. Tokyo 24 (1977), 93–96.

T. Mizutani and T. Tsuboi, Foliations without holonomy and foliated bundles, Sci. Rep. Saitama Univ. 9 (1979), 45–55.

I. Moerdijk, Classifying toposes and foliations, Ann. Inst. Fourier 41 (1991), 189–209.

I. Moerdijk, Foliations, groupoids and Grothendieck étendues, Rev. Acad. Cienc. Zaragoza (2) 48 (1993), 5–33.

N. Mok, Foliation techniques and vanishing theorems, Contemp. Math. 49 (1986), 79–118.

P. Molino, Connexions et G-structures sur les variétés feuilletées, Bull. Sci. Math. 92 (1968), 59–63.

P. Molino, G-structures plates et classes caractéristiques, C. R. Acad. Sci. Paris 269 (1969), 917–919.

P. Molino, Classe d'Atiyah d'un feuilletage et connexions transverses projetables, C. R. Acad. Sci. Paris 272 (1971), 779–781.

P. Molino, Classes caractéristiques et obstructions d'Atiyah pour les fibrés principaux feuilletés, C. R. Acad. Sci. Paris 272 (1971), 1376–1378.

P. Molino, Feuilletages et classes caractéristiques, Geometria Differenziale, Roma, 1971 Symp. Math. 10 (1972), 59–68.

P. Molino, Propriétés cohomologiques et propriétés topologiques des feuilletages à connexions transverses projetables, Topology 12 (1973), 317–325.

P. Molino, La classe d'Atiyah d'un feuilletage comme cocycle de déformation infinitésimale, C. R. Acad. Sci. Paris 278 (1974), 719–721.

P. Molino, Γ_q-structures partielles et classes de Bott-Haefliger, C. R. Acad. Sci. Paris 281 (1975), 203–206.

P. Molino, Sur la géométrie transverse des feuilletages, Ann. Inst. Fourier 25 (1975), 279–284.

P. Molino, Feuilletages transversement parallélisables et feuilletages de Lie, Applications, C. R. Acad. Sci. Paris 282 (1976), 99–101.

P. Molino, Etude des feuilletages transversalement complets et applications, Ann. Sci. Ecole Norm. Sup. 70 (1977), 289–307.

P. Molino, Feuilletages Riemanniens sur les variétés compactes; champs de Killing transverses, C. R. Acad. Sci. Paris 289 (1979), 421–423.

P. Molino, Invariants structuraux des feuilletages, Bull. Sci. Math. 105 (1981), 337–347.

P. Molino, Géometrie globale des feuilletages riemanniens, Proc. Kon. Nederland Akad. Ser. A, 1, 85 (1982), 45–76.

P. Molino, Feuilletages riemanniens, Secrétariat des Mathématiques, Université des Sciences et Technique du Languedoc (1982–1983).

P. Molino, Flots riemanniens et flots isométriques, Séminaire de géometrie différentielle (1982–1983), Montpellier.

P. Molino, Désingularisation des feuilletages riemanniens, Amer. J. Math. 106 (1984), 1091–1106.

P. Molino, Géométrie de polarisations, feuilletages et quantification géométrique, Lyon, 1983, Travaux en Cours 6 (1984), 37–53.

P. Molino, Espace des feuilles des feuilletages riemanniens, Astérisque 116 (1984), 180–189.

P. Molino, Feuilletages riemanniens réguliers et singuliers, Géom. Diff. (Paris, 1986), 173–201, Travaux en Cours, 33, Hermann, Paris, 1988.

P. Molino, Riemannian foliations, Progress in Mathematics Vol. 73, Birkhäuser Boston Inc. (1988).

P. Molino, Dualité symplectique, feuilletages et géométrie du moment, Publicacions Matematiques 33 (1989), 533–541.

P. Molino, La géométrie différentielle des feuilletages dans l'oeuvre de B. Reinhart, Ann. Global Anal. Geom. 9 (1991), 5–7.

P. Molino, Orbit-like foliations, Geometric Study of Foliations, Tokyo 1993, 97–119, World Scientific, 1994.

P. Molino, Lagrangian holonomy, Analysis and Geometry in Foliated Manifolds, Proc. VII Colloq. on Diff. Geom., Santiago de Compostela 1994, World Scientific 1995, 183–194.

P. Molino, Feuilletages presque-isométriques et structures de Poisson-Riemann, preprint.

P. Molino and M. Pierrot, Théorèmes de slice et holonomie des feuilletages, Ann. Inst. Fourier 379 (1987), 207–223.

P. Molino and V. Sergiescu, Deux remarques sur les flots riemanniens, Manuscripta Math. 51 (1985), 145–161.

P. Molino and F. Turiel, Une observation sur les actions de $\mathbb{R}^m$ sur les variétés compactes de caractéristique nulle, Comment. Math. Helv. 61 (1986), 370–375.

P. Molino and F. Turiel, Dimension des orbites d'une action de $\mathbb{R}^p$ sur une variété compacte, Comment. Math. Helv. 63 (1988), 253–258.

G. Monna, Feuilletages de k-contact sur les variétés compactes de dimension 3, Publ. Sec. Mat. Univ. Autonoma Barcelona 28 (1984), 81–87.

G. Monna, Feuilletage de contact défini par une forme de contact régulière, Riv. Mat. Univ. Parma 11 (1985), 197–202.

A. Montesinos, On certain classes of almost product structures, Mich. Math. J. 30 (1983).

A. Montesinos, Conformal curvature for the normal bundle of a conformal foliation, Ann. Inst. Fourier 32 (1982), 261–274.

R. Montgomery, Generic distributions and Lie algebras of vector fields, J. Diff. Equations 103 (1993), 387–393.

C. Moore and C. Schochet, Global analysis on foliated spaces, MSRI Publications 9, 1988.

C. Moore, C. Schochet and R. Zimmer, L^2-harmonic forms on non-compact manifolds, in C. Moore and C. Schochet, Global analysis on foliated spaces, Springer (1988), 308–315.

F. Morel, Cohomologie singulière et cohomologie des Q-variétés, C. R. Acad. Sci. Paris 281 (1975), 309–312.

F. Morel, Type d'homotopie rationelle des Q-variétés et formes simpliciales basiques, C. R. Acad. Sci. Paris 285 (1977), 11–14.

A. Morgan, Holonomy and metric properties of foliations in higher codimension, Proc. Amer. Math. Soc. 58 (1976), 255–261.

A. Morimoto, Prolongations of geometric structures, Lecture Notes Math. Inst. Nagoya Univ. 1969.

A. Morimoto, Almost complex foliations and its applications to contact geometry, Natur. Sci. Rep. Ochanomizu Univ. 43 (1992), 11–23.

T. Morimoto, Geometric structures on filtered manifolds, Hokkaido Math. J. 22 (1993), 263–347.

S. Morita, A remark on the continuous variations of secondary characteristic classes for foliations, J. Math. Soc. Japan. 29 (1977), 253–260.

S. Morita, Cartan connections and characteristic classes of foliations, Proc. Japan. Acad. 53 (1977), 211–214.

S. Morita, On characteristic classes of Riemannian foliations, Osaka J. Math. 16 (1979), 161–172.

S. Morita, On characteristic classes of conformal and projective foliations, J. Math. Soc. Japan 31 (1979), 693–718.

S. Morita, On the splitting problem of the boundary functionals for the characteristic classes of foliations, Topology 20 (1981), 411–420.

S. Morita, Nontriviality of the Gelfand-Fuks characteristic classes for flat S^1-bundles, Osaka J. Math. 21 (1984), 545–563.

S. Morita, Discontinuous invariants of foliations, Foliations (Tokyo, 1983), 169–193, Adv. Stud. Pure Math. 5, North-Holland, Amsterdam, 1985.

S. Morita and T. Tsuboi, The Godbillon-Vey class of codimension one foliations without holonomy, Topology 19 (1980), 43–49.

Y. Moriyama, Remarks on manifolds which admit locally free nilpotent Lie group actions, Hokkaido Math. J. 17 (1988), 235–240.

Y. Moriyama, Polycyclic groups of diffeomorphisms on the half-line, to appear.

H. Moriyoshi, Secondary characteristic numbers and locally free S^1-actions, Contemp. Math. 105 (1990), 119–144.

H. Moriyoshi, On cyclic cocycles associated with general Godbillon-Vey classes, Geometric Study of Foliations, Tokyo 1993, 411–423, World Scientific, 1994.

H. Moriyoshi and T. Natsume, The Godbillon-Vey cocycle and longitudinal Dirac operators, to appear.

J. Morvan, Connection and foliation in symplectic geometry, C. R. Acad. Sci. Paris 296 (1983), 765–768.

J. Moser, Minimal solutions of variational problems on a torus, Ann. Inst. H. Poincaré Anal. Non Linéaire 3 (1986), 229–272.

J. Moser, A stability theorem for minimal foliations on a torus, Ergodic Theory Dynamical Systems 8 (1988), Charles Conley Memorial Issue, 251–281.

J. Moser, Tiling the projective foliation space of a punctured surface, Trans. Amer. Math. Soc. 306 (1988), 1–70.

J. Moser, Minimal foliations on a torus, Springer Lecture Notes in Math. 1365 (1989), 62–99.

I. Moskowitz, Nonvanishing local cohomology classes, Trans. Amer. Math. Soc. 286 (1984), 831–837.

I. Moskowitz, A note on the Bott vanishing theorem, Proc. Amer. Math. Soc. 94 (1985), 529–530.

M. Mostow, Continuous cohomology of spaces with two topologies, Mem. Amer. Math. Soc. 175 (1976).

M. Mostow, Variations, characteristic classes, and the obstruction to mapping smooth to continuous cohomology, Trans. Amer. Math. Soc. 240 (1978), 163–182.

R. Moussu, Sur un théorème de Novikov, Rev. Colomb. Mat. 3 (1969), 51–81.

R. Moussu, Feuilletage sans holonomie d'une variété fermée, C. R. Acad. Sci. Paris 270 (1970), 1308–1311.

R. Moussu, Feuilletage de codimension 1 transverse au bord, C. R. Acad. Sci. Paris 271 (1970), 15–18.

R. Moussu, Feuilletages presque sans holonomie, C. R. Acad. Sci. Paris 272 (1971), 114–117.

R. Moussu, Sur les feuilletages de codimension 1, Thèse, Orsay, 1971.

R. Moussu, Sur les problèmes de classification de feuilletages, Rev. Colomb. Mat. 6 (1972), 59–68.

R. Moussu, Sur les classes exotiques des feuilletages, Springer Lecture Notes in Math. 392 (1974), 37–42.

R. Moussu, Holonomie évanescente des équations différentielles dégénerées transverses, Singularities and dynamical systems, Iráklion, 1983; North-Holland Math. Studies 103 (1985), 161–173.

R. Moussu, Les conjectures de R. Thom sur les singularités de feuilletages holomorphes, Equations diff. daus le champ complexe, Colloque Franco-Japonais 1985, Vol. III (1988), 105–113.

R. Moussu and F. Pelletier, Sur le théorème de Poincaré-Bendixson, Ann. Inst. Fourier 24 (1974), 131–148.

R. Moussu and R. Roussarie, Une condition suffisante pour qu'un feuilletage soit sans holonomie, C. R. Acad. Sci. Paris 271 (1970), 240–243.

R. Moussu and R. Roussarie, Relations de conjugaison et de cobordisme entre certains feuilletages, Publ. Math. Inst. Hautes Etudes Sci. 43 (1973), 143–168.

W. Mozgawa, Feuilletages de Killing, Collect. Math. 36 (1985), 285–290.

W. Mozgawa, Riemannian vector fields and Pontrjagin numbers, Ann. Univ. Mariae Curie-Sklodowska Sect. A 39 (1985), 113–115.

M. Mukherjee and R. Tucker, Relativistic elastodynamics of self-gravitating foliations, Classical Quantum Gravity 5 (1988), 849–860.

M. Muller, Quelques propriétés des feuilletages polynomiaux du plan, Bol. Soc. Mat. Mex. 21 (1976), 6–14.

M. Muller, An analytic foliation of the plane without weak first integrals of class C^2, Bol. Soc. Mex. 21 (1976), 1–5.

M. Muller, Sur les composantes de Novikov des feuilletages, Topology 19 (1980), 199–201.

T. Müller, Beispiel einer periodischen instabilen holomorphen Strömung, Enseign. Math. 25 (1979), 309–312.

H. Münzner, Isoparametrische Hyperflächen in Sphären, Math. Ann. 251 (1980), 57–71.

H. Münzner, Isoparametrische Hyperflächen in Sphären, II, Über die Zerlegung der Sphäre in Ballbündel, Math. Ann. 256 (1981), 215–232.

S. Nagai, Reduction in codimension of mixed foliate CR-submanifolds of a quaternion manifold, Saitama Math. J. 12 (1994), 1–6.

S. Nagai, Superintegrable foliations ane web structure, Geometry and analysis in dynamical systems (Kyoto, 1993), 126–139, Adv. Ser. Dyn. Systems 14, World Sci. Publ., 1994

S. Nagamine, A remark on minimal foliations of Lie groups, Tsukuba J. Math. 9 (1985), 317–320.

S. Nagamine, Totally umbilical Riemannian foliations on Lie groups, Math. J. Toyama Univ. 18 (1995), 195–198.

S. Nagano and K. Yagi, The affine structures on the real two-torus, I, Osaka J. Math. 11 (1974), 181–210.

R. Naimi, Constructing essential laminations in some 3-manifolds, Caltech thesis, 1992.

R. Naimi, Foliations transverse to fibers of Seifert manifolds, Comment. Math. Helv. 69 (1994), 155–162.

H. Nakagawa and R. Takagi, Harmonic foliations on a compact Riemannian manifold of non-negative constant curvature, Tôhoku Math. J. 40 (1988), 465–471.

I. Nakai, A rigidity theorem for transverse dynamics of real analytic foliations of codimension one, Astérisque 222 (1994), 327–343.

H. Nakayama, Codimension one foliations of S^3 with only one compact leaf, J. Fac. Sci. Univ. Tokyo 37 (1990), 189–199.

H. Nakayama, On cutting pseudo-foliations along incompressible surfaces, J. Math. Soc. Japan 43 (1991), 483–499.

H. Nakayama, Transversely affine foliations of some surface bundles over S^1 of pseudo-Anosov type, Ann. Inst. Fourier 41 (1991), 755–778.

H. Nakayama, Some remarks on non-Hausdorff sets for flows, Geometric Study of Foliations, Tokyo 1993, 425–429, World Scientific, 1994.

F. Narita, The integrability tensors of Riemannian submersions and submanifolds, Res. Rep. Akita Nat. College Tech. 22 (1987), 66–74.

F. Narita, Riemannian submersions and isometric reflections with respect to submanifolds, Math. J. Toyama Univ. 15 (1992), 83–94.

F. Narita, Riemannian submersion with isometric reflections relative to the fibers, Kodai Math. J. 16 (1993), 416–427.

F. Narita, Riemannian submersions of locally conformal Kaehler manifolds, to appear.

A. Narmanov, A stability theorem for non compact leaves of a foliation of codimension one, Vestnik Leningrad. Univ. 1983, 100–102.

A. Narmanov, Limit sets of leaves of a foliation of codimension 1, Vestnik Leningrad. Univ. Mat. Mekh. Astronom. 3 (1983), 21–25.

A. Narmanov, Controllability sets of control systems that are fibers of a foliation of codimension one, Differentsial'nye Uravneniya 19 (1983), 1627–1630.

A. Narmanov, The holonomy group of codimension one foliations, Izv. Akad. Nauk Uz SSR 1989, no. 6, 35–37, 77.

H. Natsume and T. Natsume, A remark on the DeRham map for foliated manifolds, Kodai Math. J. 3 (1980), 364–373.

H. Natsume and T. Natsume, On a theorem of Bott and Haefliger, Sci. Rep. Saitama Univ. Ser. A 9 (1980), 81–95.

T. Natsume, The C^*-algebra of codimension one foliations without holonomy, Math. Scand. 56 (1985), 96–104.

T. Natsume, Topological K-theory for codimension one foliations without holonomy, Foliations (Tokyo, 1983), 15–27, Adv. Stud. Pure Math. 5, North-Holland, Amsterdam, 1985.

T. Natsume, The C^1-invariance of the Godbillon-Vey map in analytical K-theory, Canad. J. Math. 39 (1987), 1210–1222.

T. Natsume, Another look into the foliation index theorem, Foliations Tokyo 1993, to appear.

A. Naveira, Variedades foliadas con metrica casi-fibrada, Collect. Math. 21 (1970), 5–61.

A. Naveira, A classification of Riemannian almost product manifolds, Rend. Math. 3 (1983), 577–592.

A. Naveira and A. Rocamora, A geometric obstruction to the existence of two totally umbilic complementary foliations in compact manifolds, Springer Lecture Notes in Mathematics 1139 (1985), 263–279.

M. Nicolau and A. Reventós, Compact Hausdorff foliations, I, Int. Symp. on differential geometry, Peniscola 1982, Springer Lecture Notes in Math. 1045 (1984), 147–153.

M. Nicolau and A. Reventós, On some geometrical properties of Seifert bundles, Israel J. Math. 47 (1984), 323–334.

S. Nishikawa, Residues and secondary characteristic classes for projective foliations, Proc. Japan. Acad. 54 (1978), 79–82.

S. Nishikawa, Residues and characteristic classes for projective foliations, Japan J. Math. 7 (1981), 45–108.

S. Nishikawa, M. Ramachandran and Ph. Tondeur, Heat conduction for Riemannian foliations, Bull. Amer. Math. Soc. 21 (1989), 265–267.

S. Nishikawa, M. Ramachandran and Ph. Tondeur, The heat equation for Riemannian foliations, Trans. Amer. Math. Soc. 319 (1990), 619–630.

S. Nishikawa and H. Sato, On characteristic classes of Riemannian, conformal and projective foliations, J. Math. Soc. Japan. 28 (1976), 224–241.

S. Nishikawa and M. Takeuchi, Γ-foliations and semisimple flat homogeneous spaces, Tôhoku Math. J. 30 (1978), 307–335.

S. Nishikawa and Ph. Tondeur, Transversal infinitesimal automorphisms for harmonic Kähler foliations, Tôhoku Math. J. 40 (1988), 599–611.

S. Nishikawa and Ph. Tondeur, Transversal infinitesimal automorphisms of harmonic foliations on complete manifolds, Annals of Global Analysis and Geometry 7 (1989), 47–57.

S. Nishikawa, Ph. Tondeur and L. Vanhecke, Spectral geometry for Riemannian foliations, Annals of Global Analysis and Geometry 10 (1992), 291–304.

S. Nishikawa and S. Yorozu, Transversal infinitesimal automorphisms for compact Riemannian foliations, to appear.

T. Nishimori, Isolated ends of open leaves of codimension-one foliations, Q. J. Math. 26 (1975), 159–167.

T. Nishimori, Compact leaves with abelian holonomy, Tôhoku Math. J. 27 (1975), 259–272.

T. Nishimori, Behaviour of leaves of codimension one foliations, Tôhoku Math. J. 29 (1977), 255–273.

T. Nishimori, Ends of leaves of codimension one foliations, Tôhoku Math. J. 31 (1979), 1–22.

T. Nishimori, Octahedral webs on closed manifolds, Tohoku Math. J. 32 (1980), 399–400.

T. Nishimori, SRH-decompositions of codimension one foliations and the Godbillon-Vey class, Tôhoku Math. J. 32 (1980), 9–34.

T. Nishimori, Some remarks on octahedral webs, Japan J. Math. 7 (1981), 168–179.

T. Nishimori, Existence problem of transverse foliations for some foliated 3-manifolds, Tôhoku Math. J. 34 (1982), 179–238.

T. Nishimori, Foliations transverse to the turbulized foliations of punctured torus bundles over a circle, Hokkaido Math. J. 13 (1984), 1–25.

T. Nishimori, Average Euler characteristic of leaves of codimension-one foliations, Foliations (Tokyo, 1983), 395–415, Adv. Stud. Pure Math. 5, North-Holland, Amsterdam, 1985.

T. Nishimori, Average signatures of PA-leaves of codimension-one foliations, A fête of topology, 101–126, Academic Press, Boston, 1988.

T. Nishimori, A qualitative theory of similarity pseudogroups and an analogy of Sacksteder's theorem, Hokkaido Math. J. 21 (1992), 141–150.

T. Nishimori, A note on the classification of nonsingular flows with transverse similarity structures, Hokkaido Math. J. 21 (1992), 318–393.

J. Noakes, Foliations by manifolds with boundaries, J. Diff. Geom. 16 (1981), 129–136.

F. Noebel, On integral manifolds for vector distributions, Math. Ann. 294 (1992), 1–17.

F. Noelker, Isometric immersion of warped products, Diff. Geom. Appl., to appear.

K. Nomizu, Some results in E. Cartan's theory of isoparametric families of hypersurfaces, Bull. Amer. Math. Soc. 79 (1973), 1184–1188.

V. Norden, Description canonique de champs de vecteurs sur une surface, Ann. Inst. Fourier 32 (1982), 151–166.

S. Novikov, Foliations of codimension 1 on manifolds, Soviet Math. Dokl. 5 (1964), 540–544.

S. Novikov, Foliations of codimension 1, Soviet Math. Dokl. 5 (1964), 1023–1025.

S. Novikov, Smooth foliations on three manifolds, Russian Math. Surveys 19-6 (1964), 79–81.

S. Novikov, Topology of foliations, Trudy Moskov. Mat. Obsc. 14 (1965), 248–278; AMS Translation, Trans. Moscow Math. Soc. 14 (1965), 268–304.

S. Novikov, The Hamiltonian formalism and a multivalued analog of Morse theory, Uspekhi Math. Nauk. (Russian Math. Surveys) 37, 5 (1982), 3–49.

S. Novikov, Critical points and level surfaces of multivalued functions, Proceedings of the Steklov Institute of Mathematics, 1986, AMS, Issue 1, 223–232.

S. Novikov, Quasiperiodic structures in topology, Proceedings of the Conference "Topological methods in Modern Mathematics," Milnor 60th birthday, 223–233, Publish or Perish, 1993.

S. Novikov, The semiclassical electron in a magnetic field and lattice; Some problems of low dimensional "periodic" topology, Geom. and Funct. Anal. 5 (1995), 434–444.

U. Oertel and A. Papadopoulos, Feuilletages affines et structures hyperboliques brisées sur les surfaces, Institut de Recherche Math. avancée, Strasbourg, 1992.

G. Oh and J. Pak, A note on L^2-transverse conformal fields on complete foliated Riemannian manifolds, Math. J. Toyama Univ.13 (1990), 51–62.

G. Oh and J. Pak, Transverse conformal fields on foliated Riemannian manifolds, Math. J. Toyama Univ. 13 (1990), 63–75.

H. Ohsato, Characteristic classes of S-foliated vector bundles and Gelfand-Fuks cohomology, J. Fac. Sci. Univ. Tokyo 26 (1979), 279–301.

K. Ohshika, Incompressibility of measured laminations in 3-manifolds, Tokyo J. Math. 12 (1989), 145–157.

K. Ohshika, Minimal measured laminations in geometric 3-manifolds, Pacific J. Math. 144 (1990), 327–344.

M. Oka, Singular foliations on cross-sections of expansive flows on 3-manifolds, Osaka J. Math. 27 (1990), 863–883.

J. Omegar, Persistencia de folheaçoes definidas por formas logarítmicas, thesis IMPA, 1990.

B. O'Neill, The fundamental equations of a submersion, Mich. Math. J. 13 (1966), 459–469.

L. Ornea and G. Romani, The fundamental equations of conformal submersions, Beiträge Algebra Geom. 34 (1993), 233–243.

G. Oshikiri, The surgery of codimension one foliations, Tôhoku Math. J. 31 (1979), 63–70.

G. Oshikiri, Foliated cobordism of suspended foliations, Tohoku Math. J. 32 (1980), 375–392.

G. Oshikiri, A remark on minimal foliations, Tôhoku Math. J. 33 (1981), 133–137.

G. Oshikiri, Jacobi fields and the stability of leaves of codimension one minimal foliations, Tôhoku Math. J. 34 (1982), 417–424.

G. Oshikiri, Totally geodesic foliations and Killing fields, Tôhoku Math. J. 35 (1983), 387–392.

G. Oshikiri, Totally geodesic foliations and Killing fields, II, Tôhoko Math. J. 38 (1986), 351–356.

G. Oshikiri, Some remarks on minimal foliations, Tôhoku Math. J. 39 (1987), 223–229.

G. Oshikiri, Mean curvature functions of codimension-one foliations, Comment. Math. Helv. 65 (1990), 79–84.

G. Oshikiri, Mean curvature functions of codimension-one foliations, II, Comment. Math. Helv. 66 (1991), 512–520.

G. Oshikiri, Tautness of codimension-one foliations and isoperimetric constants, Interdisciplinary Information Sciences 1 (1995), 47–50.

G. Osipenko, Integrability of invariant plane fields, separation and partial hyperbolicity, Differential Equations 19 (1983), 1251–1255.

M. O'uchi, Coverings of foliations and associated C^*-algebras, Math. Scand. 58 (1986), 69–76.

K. Ove, Anomalous foliations of Einstein space-times, J. Math. Phys. 31 (1990), 2688–2693.

H. Pak, On the transversal conformal curvature tensor on Hermitian foliations, Bull. Korean Math. Soc. 28 (1991), 231–241.

H. Pak, One dimensional metric foliations in Einstein spaces, Ill. J. of Math. 36 (1992), 594–599.

H. Pak, λ-automorphisms of a Riemannian foliation, Ann. of Global Anal. and Geom. 13 (1995), 281–288.

H. Pak and J. Pak, Stability theorem for a holomorphic Riemannian foliation, Math. Rep. Toyama Univ. 12 (1989), 181–187.

H. Pak and J. Pak, Normal holonomy group of a Riemannian foliation, Bull. Korean Math. Soc. 30 (1993). 17–23.

H. Pak and J. Pak, Notes on dense leaves of Riemannian foliations, Math. J. Toyama Univ. 18 (1995), 199–208.

H. Pak and H. Yoo, L^2-transverse harmonic fields on complete foliated Riemannian manifolds, Kyungpook Math. J. 31 (1991), 253–262.

J. Pak and H. Yoo, Transverse harmonic fields on Riemannian manifolds, Bull. Korean Math. Soc. 29 (1992), 73–80.

J. Pak, Y. Shin and H. Yoo, L^2-transverse fields preserving the transverse Ricci field of a foliation, J. Korean Math. Soc. 32 (1995), 51–60.

J. Pak and S. Yorozu, Transverse fields on foliated Riemannian manifolds, J. Korean Math. Soc. 25 (1988), 83–92.

J. Pak and S. Yorozu, The Laplace-Beltrami operator on a Riemannian manifold with a Clairaut foliation, Ann. Sci. Kanazawa Univ. 26 (1989), 13–15.

R. Palais, A global formulation of the Lie Theory of Transformation Groups, Mem. Amer. Math. Soc. 22 (1957).

R. Palais and C. L. Terng, Critical point theory and submanifold geometry, Springer Lecture Notes in Math 1353 (1988).

J. Palis, Rigidity of the centralizers of diffeomorphisms and structural stability of suspended foliations, Springer Lecture Notes in Math. 652 (1978), 114–121.

J. Palis, C. Pugh and R. Robinson, Non differentiability of invariant foliations, Dynamical systems, Warwick 1974, Springer Lecture Notes in Math. 468 (1975), 234–240.

C. Palmeira, Variétés ouvertes feuilletées par plans, C. R. Acad. Sci. Paris 283 (1976), 237–239.

C. Palmeira, Open manifolds foliated by planes, Ann. of Math. 107 (1978), 109–131.

C. Palmeira, Feuilletages par cylindres fermés de $\mathbb{R}^3$, C. R. Acad. Sci. Paris 290 (1980), 419–421.

C. Palmeira, Erratum, Feuilletages par cylindres fermés de $\mathbb{R}^3$, C. R. Acad. Sci. Paris 290 (1980), 929.

C. Palmeira, Foliations by closed cylinders in 3-dimensional manifolds, Bol. Soc. Brasil. Mat. 13 (1982), 55–78.

C. Palmeira, Line fields defined by eigenspaces of derivatives of maps from the plane to itself, Proc. of the sixth Intern. Colloq. on Differential Geometry (Santiago de Compostela, 1988), 177–205.

C. Palmeira and S. Schecter, Feuilletages de $\mathbb{R}^3$ définis par des équations de pfaff polynomiales homogènes, Ann. Inst. Fourier 32 (1982), 241–250.

R. Palovsky, Geometry of the leaf, In: Differential geometry and its applications. Int. Conf., Brno, 27 Aug.–2 Sept. 1989, World Scientific 1990, 389–391.

M. Pang, The structure of Legendre foliations, Trans. Amer. Math. Soc. 320 (1990), 417–455.

P. Pang, Minimal models and Riemannian foliations, Ph.D. thesis, University of Illinois at Urbana-Champaign, 1988.

P. Pang, Basic dual homotopy invariants of Riemannian foliations, Trans. Amer. Math. Soc. 322 (1990), 189–199.

P. Pang, On the signature of generalised Seifert fibrations, Bull. Austral. Math. Soc. 46 (1992), 55–58.

P. Pansu, Le flot géodesique des variétés riemanniennes à courbure négative, Séminaire Bourbaki, Vol. 1990/91, Astérisque 201–203 (1991), exp. 738, 269–298 (1992).

P. Pansu and R. Zimmer, Rigidity of locally homogeneous metrics of negative curvature on the leaves of a foliation, Israel J. Math. 68 (1989), 56–62.

A. Papadopoulos, Geometric intersection functions and Hamiltonian flows on the space of measured foliations on a surface, Pacific J. Math. 124 (1986), 375–402.

A. Papadopoulos and R. Penner, A characterization of pseudo-Anosov foliations, Pacific J. Math. 130 (1987), 359–377.

A. Papadopoulos and R. Penner, Enumerating pseudo-Anosov foliations, Pacific J. Math. 142 (1990), 159–173.

N. Papaghiuc, On the geometry of leaves on a semi-invariant $\xi^\perp$-submanifold in a Kenmotsu manifold, An. Stiint. Univ. Al. I. Cuza Iaşi Sect. I a Mat. 38 (1992), 111–119.

C. Parenti, Operatori pseudo-differenziali anasitropi su varieta fogliettate, Rend. Sem. Mat. Univ. Padova 52 (1974), 275–298.

E. Park and K. Richardson, The basic Laplacian of a Riemannian foliation, to appear.

J. Park, The Laplace-Beltrami operator and Riemannian submersion with minimal and not totally geodesic fibers, Bull. Korean Math. Soc. 27 (1990), 39–47.

J. Park and S. Yorozu, Transverse fields preserving the transverse Ricci field of a foliation, J. Korean Math. Soc. 27 (1990), 167–175.

J. Park and S. Yorozu, Transversal conformal fields of foliations, Nihonkai Math. J. 4 (1993), 73–85.

K. Park, On ergodic foliations, Ergodic Theory and Dynamical Systems 8 (1988), 437–457.

Ph. Parker, Geometry of leaves and the heat equation, Teubner Texte zur Mathematik 57 (1983), 247–251.

J. Pasternack, Topological obstructions to integrability and Riemannian geometry of foliations, Thesis, Princeton University, Princeton (1970).

J. Pasternack, Foliations and compact group actions, Comment. Math. Helv. 46 (1971), 467–477.

J. Pasternack, Classifying spaces for Riemannian foliations, Proc. Symp. Pure. Math. AMS 1973, vol. 27 (1975), Part 1, 303–310.

C. Patrizio and P. Wong, Stability of the Monge-Ampère foliation, Math. Ann. 263 (1983), 13–29.

E. Paul, Etude topologique des formes logarithmiques fermées, Invent. Math. 95 (1989), 395–420.

M. Pauly, Les actions quasi-libres de S^1 en dimension 4, Travaux Math., Fasc. VII, 1995, Sém. des Math. de Luxembourg, 81–99.

A. Pazhitnov, Morse theory of closed forms, Alg. Top. Poznán 1989, 98–110, Eds. Jackowski, Oliver and Pawalowski, Springer Lecture Notes 1474, 1991.

H. Pederson and A. Swann, Riemannian submersions, four-manifolds and Einstein-Weyl geometry, Proc. London Math. Soc. 66 (1993), 381–399.

W. Pelletier, The secondary characteristic classes of solvable foliations, Proc. Amer. Math. Soc. 88 (1983), 651–659.

J. Perchik, Cohomology of Hamiltonian and related formal vector field Lie algebras, Topology 15 (1976), 395–404.

J. Pereira da Silva, Feuilletages contravariants à feuilles orientables, J. Math. Pures Appl. 68 (1989), 349–354.

G. Perelman, Proof of the soul conjecture of Cheeger and Gromoll, J. Differential Geometry 40 (1994), 209–212.

R. Pérez-Marco and J. C. Yoccoz, Germes de feuilletages holomorphes à holonomie prescrite, Astérisque 222 (1994), 345–371.

G. Peric, Index theorem for foliated manifolds with bundary and cyclic cocycles, preprint, 1990.

G. Peric, Eta invariants of Dirac operators on foliated manifolds, Trans. Amer. Math. Soc. 334 (1992), 761–782.

N. Petrov, A generalization of the notion of foliation, Differential Equations 16 (1980), 649–652.

A. Phillips, Submersions of open manifolds, Topology 6 (1967), 171–206.

A. Phillips, Foliations of open manifolds, I, Comment. Math. Helv. 43 (1968), 204–211.

A. Phillips, Foliations of open manifolds, II, Comment. Math. Helv. 44 (1969), 367–370.

A. Phillips, Smooth maps transverse to a foliation, Bull. Amer. Math. Soc. 76 (1970), 792–797.

A. Phillips, Smooth maps of constant rank, Bull. Amer. Math. Soc. 80 (1974), 513–517.

A. Phillips and D. Stone, The Euler cycle of a foliation, J. Diff. Geom. 15 (1980), 39–50.

A. Phillips and D. Sullivan, Geometry of leaves, Topology 20 (1981), 209–218.

J. Phillips, The holonomic imperative and the homotopy groupoid of a foliated manifold, Rocky Mountain J. Math. 17 (1987), 151–165.

A. Piatkowski, A stability theorem for foliations with singularities, Dissertationes Math. (Rozprawy Mat.) 267 (1988), 52 pp.

A. Piatkowski, Reeb stability for nonsingular foliations derived from that for singular ones, Acta Univ. Lodz. Folia Math. 4 (1991), 109–115.

A. Piatkowski, On the topological foliation defined by a system of relations, Bull. Soc. Sci. Lett. Lodz Ser. Rech. Deform. 15 (1993), no. 1–10, 67–72.

A. Piatkowski and K. Spallek, Foliated differential spaces. Stability and quotient structure. Rend. Mat. Appl. VII Ser. 13 (1993), 673–700.

P. Piccinni, A Weitzenböck formula for the second fundamental form of a Riemannian foliation, Atti. Accad. Naz. Lincei. Rend. Cl. Sci. Fis. Mat. Natur. 77 (1984), 102–110.

M. Pierrot, Orbites des champs feuilletés pour un feuilletage riemannien sur une variété compacte, C.R.Acad. Sci. Paris 301 (1985), 443–445.

G. Pitiş, On the cohomology of foliated manifolds, Bul. Univ. Brasov. Ser. C 22 (1980), 101–106.

G. Pitiş, A class of chain complexes of a foliated manifold, An. Stiint. Univ. Al. I. Cuza Iaci Sect. I a Mat. 27 (1981), 63–66.

G. Pitiş, A cohomology theory on the category of foliated manifolds, Studia Univ. Babes-Bolyai Math. 29 (1984), 33–38.

G. Pitiş, Feuilletages et sous-variétés d'une classe de variété de contact, C. R. Acad. Sci. Paris 310 (1990), 197–202.

G. Pitiş, Integration on Riemannian foliated manifolds, Bul. Univ. Braşiv Ser. C32 (1990), 63–73.

H. Pittie, Characteristic classes of foliations, Pitman Research Notes in Math. 10 (1976).

H. Pittie, The secondary characteristic classes of parabolic foliations, Comment. Math. Helv. 54 (1979), 601–614.

M. Piu, Sur les flots riemanniens des espaces de D'Atri de dimension 3, Rend. Sem. Mat. Univ. Politec. Torino 46 (1988), 171–187.

D. Pixton, Nonsmoothable, unstable group actions, Trans. Amer. Math. Soc. 229 (1977), 619–627.

J. Plante, Diffeomorphisms with invariant line bundles, Invent. Math. 13 (1971), 325–334.

J. Plante, Anosov flows, Amer. J. Math. 94 (1972), 729–754.

J. Plante, Asymptotic properties of foliations, Comment. Math. Helv. 47 (1972), 449–456.

J. Plante, A generalization of the Poincaré-Bendixson theorem for foliation of codimension one, Topology 12 (1973), 177–181.

J. Plante, On the existence of exceptional minimal sets in foliations of codimension one, J. Diff. Eq. 15 (1974), 178–194.

J. Plante, Foliations transverse to fibers of a bundle, Proc. Amer. Math. Soc. 24 (1974), 631–635.

J. Plante, Foliations with measure preserving holonomy, Springer Lecture Notes in Math. 468 (1975), 6–7.

J. Plante, Foliations with measure preserving holonomy, Ann. of Math. 102 (1975), 327–361.

J. Plante, Measure preserving pseudogroups and a theorem of Sacksteder, Ann. Inst. Fourier 25 (1975), 237–249.

J. Plante, Foliations of 3-manifolds with solvable fundamental group, Invent. Math. 51 (1979), 219–230.

J. Plante, Locally free affine group actions, Trans. Amer. Math. Soc. 259 (1980), 449–456.

J. Plante, Anosov flows, transversely affine foliations and a conjecture of Verjovsky, J. London Math. Soc. 23 (1981), 359–362.

J. Plante, Diffeomorphisms without periodic points, Proc. Amer. Math. Soc. 88 (1983), 716–718.

J. Plante, Solvable groups acting on the line, Trans. Amer. Math. Soc. 278 (1983), 401–414.

J. Plante, Stability of codimension one foliations by compact leaves, Topology 22 (1983), 173–177.

J. Plante, Subgroups of continuous groups acting differentiably on the half-line, Ann. Inst. Fourier 34 (1984), 47–56.

J. Plante, Polycyclic groups and transversely affine foliations, J. Diff. Geom. 35 (1992), 521–534.

J. Plante and W. Thurston, Anosov flows and the fundamental group, Topology 11 (1972), 147–150.

J. Plante and W. Thurston, Polynomial growth in holonomy groups of foliations, Comment. Math. Helv. 51 (1976), 567–584.

F. Pluvinage, Espaces des feuilles de certaines structures feuilletées planes, Colloq. Math. 13 (1967), 89–102.

V. Podolski and A. Sirokov, Some types of Riemannian foliations, Gravitacija i Teor. Otnositelnosti 10–11 (1975), 232–236.

V. Poénaru, Travaux de Thurston sur les difféomorphismes des surfaces, Séminaire Orsay 1979, Astérisque 66–67.

Z. Pogoda, Γ-foliations and Weil prolongations, Proc. Winter School Geometry and Topology, Srni 1992, Rend. Circ. Mat. Palermo (2) Suppl. No. 32 (1993), 69–79.

Z. Pogoda, Horizontal lifts and foliations, Proc. of the Winter School on Geometry and Physics (Srni, 1988), Rend. Circ. Mat. Palermo, Suppl. 1989, no. 21, 279–289.

R. Ponge and H. Reckziegel, Twisted products in pseudo-riemannian geometry, Geometriae Dedicata 48 (1993), 15–25.

P. Popescu, Sur une classe de groupoides riemanniens, preprint.

A. Portolano and I. Maniscalco, Curves as measured foliation on noncompact surfaces, Rend. Circ. Mat. Palermo, II. ser. 42 (1993), 161–180.

C. Possani, A theorem of the total curvature of two orthogonal foliations of a hyperbolic surface, An. Acad. Brasil. Cienc. 60 (1988), 253–254.

C. Possani, Curvatura total de folheacoes, Thesis, Instituto de Matematica e Estastistica Universidade de Sao Paulo, 1989.

G. Pourcin, Deformations of coherent foliations on a compact normal space, Ann. Inst. Fourier 37 (1987), 33–48.

G. Pourcin, Deformations of singular holomorphic foliations on reduced compact C-analytic spaces, Springer Lecture Notes in Math. 1345 (1988), 246–255.

J. Pradines, Remarque sur le théorème d'annulation de Bott-Martinet, C. R. Acad. Sci. Paris 282 (1976), 527–529.

J. Pradines, Feuilletages dout les groupes d'holonomie sont de Coxeter, C. R. Acad. Sci. Paris 286 (1978), 255–258.

J. Pradines, Un feuilletage sans holonomie transversale, dont le quotient n'est pas une Q-variété, C. R. Acad. Sci. Paris 288 (1979), 245–248.

J. Pradines, Echelles et faisceaux sur les quotients de feuilletages, Cahiers Topologie Géom. Différentielle 22 (1981), 73–83.

J. Pradines, Foliations: holonomy and local graphs, C. R. Acad. Sci. Paris 298 (1984), 297–300.

J. Pradines, How to define the graph of a singular foliation, Cahiers Topol. Géom. Différentielle Catégoriques 26 (1985), 339–380.

J. Pradines, Graph and holonomy of singular foliations, Differential geometry (Santiago de Compostela, 1984), Pitman Res. Notes Math. Ser 131 (1985), 215–219.

J. Pradines, Variétés d'orbites, Publ. Dép. Math. Nouvelle Sér. A, 2, 65–70, Univ. Claude-Bernard, Lyon, 1987.

J. Pradines, Morphisms between spaces of leaves viewed as fractions, Cahiers Topologie Geom. Différentielle Catégoriques 30 (1989), 229–246.

J. Pradines and A. Alta'ai, Caractérisation universelle du groupe de Haefliger-van Est d'un espace de feuilles ou d'orbites, et théorème de van Kampen, C. R. Acad. Sci. Paris 309 (1989), 503–506.

J. Pradines and J. Wuafo Kanga, Relations d'équivalence transversalement différentiables, C. R. Acad. Sci. Paris 283 (1976), 25–28.

M. Puta, The Godbillon-Vey invariant of L_p-structures, Rend. Mat. 3 (1983), 107–112.

M. Puta, Differential forms on a complex foliated manifold and geometric quantization, Rend Sem. Mat. Torino 42 (1985), 59–71.

M. Puta, Some remarks on the cohomology of a real foliated manifold, Rend. Math. 5 (1985), 189–201.

M. Puta, Some remarks on the relative cohomology of a foliated manifold, An. Univ. Timisoara Ser. Stiint. Mat. 23 (1985), 56–60.

M. Puta, A remark on the basic cohomology of de Rham currents, Proc. of the National Conference on Geometry and Topology (Tirgoviste, 1986), 235–238, Univ. Bucuresti, Bucharest, 1988.

Ngô van Quê, Feuilletage à singularités de variétés de dimension 3 (théorème de J. Wood), J. Diff. Geom. 6 (1972), 473–478.

Ngô van Quê, Formes singulières génériques, Proc. 9th Braz. Math. Coll. 1973, Vol. I, Inst. Mat. Pura Apl. Sao Paulo (1977), 217–226.

R. Roussarie and Ngô Van Quê, Sur l'isotopie des formes fermées en dimension 3, Invent. Math. 64 (1981), 69–87.

R. Quiroga, Hermitian metric rigidity on compact foliated manifolds, Geom. Dedicata 57 (1995), 305–315.

G. Raby, Invariance des classes de Godbillon-Vey par C^1-difféomorphismes, Ann. Inst. Fourier 38 (1988), 205–213.

L. Raileanu, Smooth mappings between foliated manifolds with boundary, An. Stiint. Univ. Al. I. Cuza Iaci 37 (1991), 53–64.

A. Ramsay, The Mackey-Glimm dichotomy for foliations and other Polish groupoids, J. Funct. Anal. 94 (1990), 358–374.

D. Randall and P. Schweitzer, On foliations, concordance spaces, and the Smale conjectures, Contemp. Math. 161 (1994), 235–258.

A. Ranjan, Structural equations and integral formula for foliated manifolds, Geometriae Dedicata 20 (1986), 85–91.

O. Rasmussen, Foliations with bundle-like metric, Prepr. Ser. Mat. Inst. Aarhus Univ. No. 26, 33 (1971–1972).

O. Rasmussen, Locally free R^{n-1} actions on M^n, Prepr. Ser. Mat. Inst. Aarhus Univ. No. 32, 13 (1971–1972).

O. Rasmussen, Reeb foliations, Prepr. Ser. Mat. Inst. Aarhus Univ. No. 58, 8 (1973).

O. Rasmussen, The horocyclic foliation, Prepr. Ser. Mat. Inst. Aarhus Univ. No. 3, 21 (1975–1976).

O. Rasmussen, Exotic characteristic classes for holomorphic foliations, Invent. Math. 46 (1978), 153–171.

O. Rasmussen, Continuous variation of foliations in codimension two, Topology 19 (1980), 335–349.

O. Rasmussen, Foliations with integrable transverse G-structures, J. Diff. Geom. 16 (1981), 699–710.

A. Ras-Sabidó, Killing fields preserving totally geodesic, codimension-one foliations, Kodai Math. J. 14 (1991), 477–484.

B. Raymond, Ensembles de Cantor et feuilletages, Thèse Doct. Sci. Univ. Paris, 1976.

F. Raymond, Local triviality for Hurewicz fiberings of manifolds, Topology 3 (1965), 43–57.

C. Rea, Levi-flat submanifolds and holomorphic extension of foliations, Ann. Suola Norm. Super. Pisa., Sci. Fis. Mat. 26 (1972) 665–681.

C. Rea, Varietá pseudo-platte, Symp. Math. 1st. Naz. Alta Mat. Conv., Nov. 1971–Maggio 1972, London-New York, vol. 11 (1973), 347–354.

H. Reckziegel, A fiber bundle theorem, Manuscripta Math. 76 (1992), 105–110.

H. Reckziegel, On the decomposition theorem of Blumenthal and Hebda, Geometry and Topology of Submanifolds, IV, Eds. F. Dillen and L. Verstraelen, World Scientific 1992, 174–182.

H. Reckziegel, Twisted products in Pseudo-Riemannian geometry, Geometriae Dedicata 48 (1993), 15–25.

G. Reeb, Sur les variétés intégrales des champs d'éléments de contact complètement intégrables, C. R. Acad. Sci. Math. 220 (1945), 236–237.

G. Reeb, Sur les points singuliers d'une forme de Pfaff complètement intégrable ou d'une fonction numérique, C. R. Acad. Sci. Paris 222 (1946), 847–849.

G. Reeb, Variétés feuilletées, feuilles voisines, C. R. Acad. Sci. Paris 224 (1947), 1613–1614.

G. Reeb, Remarques sur les variétés feuilletées contenant une feuille compacte à groupe de Poincaré fini, C. R. Acad. Sci. Paris 226 (1948), 1337–1339.

G. Reeb, Sur les singularités d'une forme de Pfaff analytique complètement integrable, C. R. Acad. Sci. Paris 227 (1948), 1201–1203.

G. Reeb, Stabilité des feuilles compactes à groupe de Poincaré fini, C. R. Acad. Sci. Paris 228 (1949), 47–48.

G. Reeb, Sur la courbure moyenne des variétés intégrales d'une equation de Pfaff $\omega = 0$, C.R. Acad. Sci. Paris 231 (1950), 101–102.

G. Reeb, Sur certaines propriétés topologiques des variétés feuilletées, Actualité Sci. Indust. 1183, Hermann, Paris (1952).

G. Reeb, Les espaces localement numériques non séparés et leurs applications à un problème classique, Colloque de Topologie, Strasbourg, 1955.

G. Reeb, Sur la théorie générale des systèmes dynamiques, Ann. Inst. Fourier 6 (1955), 89–115.

G. Reeb, Estructuras folheadas, Notas de Matematica, Rio de Janeiro, 12, 1958.

G. Reeb, Remarques sur les structures feuilletés, Bull. Soc. Math. France 179 (1959), 445–450.

G. Reeb, Sur les feuilletages analytiques, Séminaire Bourbaki, exp. 192 (1959).

G. Reeb, Sur les structures feuilletées de codimension un et sur un théorème de M. A. Denjoy, Ann. Inst. Fourier 11 (1961), 185–200.

G. Reeb, Sur un Théorème de Seifert sur les trajectoires fermées de certains champs de vecteurs, Nonlinear Differential Equations and Nonlinear Mechanics, Colorado Springs (1961), Academic Press 85 (1963), 16–21.

G. Reeb, Formes de Pfaff polynômes complètement intégrables, Structures feuilletées, Grenoble (1963), Ann. Inst. Fourier 14 (1964), 37–42.

G. Reeb, Feuilletages: résultats anciens et nouveaux (Painlevé, Hector et Martinet), Montréal 1972, Presses Univ. Montréal 1974.

G. Reeb, Structures feuilletées, Springer Lecture Notes in Math. 652 (1978), 104–113.

G. Reeb, Souvenirs, Gazette des Mathématiciens 46 (1990), 6–7.

G. Reeb and P. Schweitzer, Un théorème de Thurston établi au moyen de l'analyse non standard, Springer Lecture Notes in Math. 652 (1978), 138; addendum by W. Schachermayer, Une modification standard de la démonstration non standard de Reeb et Schweitzer, 139–140.

H. Reiffen, The variety of a moduli of foliations on a complex space, Enseign. Math. 33 (1987), 191–197.

H. Reiffen, Leaf spaces and integrability, Springer Lecture Notes in Math. 1345 (1988), 271–293.

B. Reinhart, Harmonic integrals on almost product manifolds, Trans. Amer. Math. Soc. 88 (1958), 243–276.

B. Reinhart, Foliated manifolds with bundle-like metrics, Ann. of Math. 69 (1959), 119–132.

B. Reinhart, Harmonic integrals on foliated manifolds, Amer. J. Math. 81 (1959), 529–536.

B. Reinhart, Line elements on the torus, Amer. J. Math. 81 (1959), 617–631.

B. Reinhart, Closed metric foliations, Mich. Math. J. 8 (1961), 7–9.

B. Reinhart, Structures transverse to a vector field, Int. Symp. on nonlinear differential equations and nonlinear mechanics, Academic Press, New York (1963), 442–444.

B. Reinhart, Cobordism and foliations, Ann. Inst. Fourier 14 (1964), 49–52.

B. Reinhart, Characteristic numbers of foliated manifolds, Topology 6 (1967), 467–472.

B. Reinhart, Algebraic invariants of foliations, Symposium on Differential Equations and Dynamical Systems 1968–69, University of Warwick, Berlin, Heidelberg, Springer-Verlag, New York (1971), 119–120.

B. Reinhart, Automorphisms and integrability of plane fields, J. Diff. Geom. 6 (1971), 263–266.

B. Reinhart, A metric formula for the Godbillon-Vey invariant for foliations, Proc. Amer. Math. Soc. 38 (1973), 427–430.

B. Reinhart, Indices for foliations of the two dimensional torus, Dynamical Systems, Bahia, 1971, 421–424, Academic Press, 1973.

B. Reinhart, Maximal foliations of extended Schwarzschild space, J. Math. Phys. 14 (1973), 719.

B. Reinhart, Holonomy invariants for framed foliations, Colloque de Géométrie Différentielle, Santiago de Compostela, 1972, Springer Lecture Notes in Math. 392 (1974), 47–52.

B. Reinhart, Foliation invariants and leaves, Holiday Symposium on Foliations and the Gelfand-Fuks cohomology, Las Cruces, 11 pages; New Mexico State University (1975).

B. Reinhart, The second fundamental form of a plane field, J. Diff. Geom. 12 (1977), 619–627.

B. Reinhart, Foliations and second fundamental form, Fourth colloquium on differential geometry, Santiago de Compostela, Spain (1978), 246–253.

B. Reinhart, Differential geometry of foliations, Ergeb. Math. 99 (1983), Springer-Verlag, New York.

B. Reinhart and J. Wood, A metric formula for the Godbillon-Vey invariant for foliations, Proc. Amer. Math. Soc. 38 (1973), 427–430.

C. Remsing, Introducere in Teoria Geometrica a Foliatiilor, Monografii Matematice, Universitatae dim Timisoara, 1984.

C. Remsing, On a generalized version of the Godbillon-Vey invariant, Proc. of the Seminar on Mathematics and Physics (Timişoara, 1988), 61–65, Inst. Politehnic "Traian Veia", Timişoara, 1988.

C. Remsing, h-feuilletages de codimension 1, Proc. of the Seminar on Mathematics and Physics (Timişoara, 1988), 66–70, Inst. Politehnic "Traian Veia", Timişoara, 1988.

C. Remsing, Tangentially symplectic foliations, Thesis, Rhodes University (Namibia), 1994.

C. Remsing, The tangential geometry of a foliation, to appear.

J. Renault, C^*-algebras of groupoids and foliations, AMS Proc. Symp. Pure Math. 38 (1982), Part 1, 339–350.

V. Resnetikov, On the cohomology of the Lie algebra of vector fields on a manifold with non trivial coefficients, Soviet Math. Dokl. 14 (1973), 234–240.

A. Reznikov, Complete geodesic foliations of Lie groups (Russian), Differentsia'naya Geom. Mnogoobraz. Figur. 16 (1985), 67–70, 124.

R. Roberts, Constructing taut foliations, Cornell thesis, 1992.

A. Rocamora, Some geometric consequences of the Weitzenböck formula on Riemannian almost-product manifolds. Harmonic distributions, Ill. J. Math. 32 (1988), 654–671.

A. Rocamora, Weak-harmonic distributions with respect to the characteristic connection, Rev. Roumaine Math. Pures et Appl. 35 (1990), 563–576.

C. Roche, Densities for certain leaves of real analytic foliations, Astérisque 22 (1994), 373–?.

J. Roe, Finite propagation speed and Connes' foliation algebra, Math. Proc. Cambridge Phil. Soc. 102 (1987), 459–466.

J. Roe, Elliptic operators, topology and asymptotic methods, Pitman Research Notes in Mathematics Series 179, Longman Scientific and Technical, 1988.

J. Roe, Remark on a paper of E. Ghys: "Gauss-Bonnet theorem for 2-dimensional foliations", J. Funct. Anal. 89 (1990), 150–153.

J. Roe, From foliations to coarse geometry and back, Analysis and Geometry in Foliated Manifolds, Proc. VII Colloq. on Diff. Geom., Santiago de Compostela 1994, World Scientific 1995, 195–205.

C. Roger, Sur les classes caractéristiques des feuilletages donnés par des isométries, C. R. Acad. Sci. Paris 276 (1973), 1185–1188.

C. Roger, Etude des Γ-structures de codimension 1 sur la sphère S^2, Ann. Inst. Fourier 23 (1973), 213–227.

C. Roger, Méthodes homotopiques et cohomologiques en théorie des feuilletages, Thèse Doct. Sc. Math., Univ. Paris (1976).

C. Roger, Homology of affine groups and classifying spaces of piecewise linear foliations, Funkts. Anal. Ego Prilozhen 13 (1979), 47–52. Translation: Funct. Anal. Appl. 13 (1979), 273–278.

C. Roger, Sur la cohomologie de l'espace classifiant des feuilletages symplectiques, C. R. Acad. Sci. Paris, 290 (1980), 617–619.

C. Roger, Foliations with a symplectic or contact transverse structure, Symplectic Geometry, Toulouse (1981), Pitman Research Notes in Math. 80 (1983), 243–250.

C. Roger, Cohomologie (p,q) des feuilletages et applications, Astérisque 116 (1984), 195–213.

H. Rosenberg, The rank of $S^2 \times S^1$, Amer. J. Math. 87 (1965), 11–24.

H. Rosenberg, Actions of $\mathbb{R}^n$ on manifolds, Comment. Math. Helv. 41 (1966), 170–178.

H. Rosenberg, Foliations by planes, Topology 7 (1968), 131–138.

H. Rosenberg, Singularities of $\mathbb{R}^2$-actions, Topology 7 (1968), 143–145.

H. Rosenberg, Feuilletages sur des sphères (d'après H. Blaine Lawson), Springer Lecture Notes in Math. 383 (1974), 294–306.

H. Rosenberg, Un contre exemple à la conjecture de Seifert (d'après P. Schweitzer), Sém. Bourbaki, exp 434, Springer Lecture Notes in Math. 383 (1974), 294–306.

H. Rosenberg, Labyrinths in the disc and surfaces, Ann. of Math. 117 (1983), 1–33.

H. Rosenberg and R. Roussarie, Reeb foliations, Ann. Math. 91 (1970), 1–24.

H. Rosenberg and R. Roussarie, Topological equivalence of Reeb foliations, Topology 9 (1970), 231–242.

H. Rosenberg and R. Roussarie, Les feuilles exceptionelles ne sont pas exceptionelles, Comment. Math. Helv. 45 (1970), 517–523.

H. Rosenberg and R. Roussarie, Some remarks on stability of foliations, J. Diff. Geom. 10 (1975), 207–219.

H. Rosenberg and W. Thurston, Some remarks on foliations, Dynamical systems, Bahia 1971, Academic Press 1973, 463–478.

J. Rosenberg, K-theory of group C^*-algebras, foliation C^*-algebras and crossed products, Contemp. Math. 70 (1988), 251–301.

R. Roussarie, Sur les feuilletages des variétés de dimension trois, Thèse Doct. Sci. Math., Fac. Sci. Orsay Univ., Paris, 70 (1969).

R. Roussarie, Sur les feuilletages des variétés de dimension trois, Ann. Inst. Fourier 21 (1971), 13–82.

R. Roussarie, Plongement dans les variétés feuilletées et classification de feuilletages sans holonomie, Inst. Hautes Etudes Sc. Publ. Math. 43 (1973), 101–104.

R. Roussarie, Phénomènes de stabilité et d'instabilité dans les feuilletages, Springer Lecture Notes in Math. 392 (1974), 53–60.

R. Roussarie, Constructions de feuilletages (d'après W. Thurston), Springer Lecture Notes in Math. 677 (1978), 138–154.

V. Rovenskii, Totally geodesic foliations, Sibirsk. Mat. Zh. 23 (1982), 217–219, 224.

V. Rovenskii, Geodesic foliations on the three-dimensional sphere, Theor. and applied problems in differential equations, 115–119, Karagand. Gos. Univ. Karaganda, 1986.

V. Rovenskii, Totally geodesic foliations with an orthogonal distribution that is close to integrable, Sibirsk. Mat. Zh. 32 (1991), 217–219 and 224.

V. Rovenskii, Totally geodesic foliations, close to the riemannian ones, Ukr. Geom. Sb. 35 (1992), 114–118.

V. Rovenskii, Metric decomposition of foliations with nonnegative curvatures, Dokl. Acad. of Science Russia 334 (1994), 699–701.

V. Rovenskii, Classes of submersions with compact fibers, Siber. Math. J. 35 (1994), 1027–1035.

V. Rovenskii, Metric decomposition of foliations with nonnegative curvature, Dokl. Akad. Nauk 334 (1994), 699–701. Translation: Russ. Acad. Sci. Dokl. Math. 49 (1994), 202–205.

V. Rovenskii, Foliations with nonnegative sectional curvature in mixed directions, Siber. Math. J., to appear.

V. Rovenskii and V. Topogonov, Great sphere foliations and manifolds with curvature bounded above, preprint.

V. Rubanov, Transversal and projectable invariant connections, Vestsi Akad. Navuk BSSR Ser. Fiz. Mat. Navuk (1983), 47–51.

V. Rubanov, Invariant foliations and transversal connections, Dokl. Akad. Nauk BSSR 28 (1984), 696–697.

D. Ruelle, Integral representation of measures associated with a foliation, Inst. Hautes Etudes Sci. Publ. Math. 48 (1978), 127–132.

D. Ruelle, Invariant measures for a diffeomorphism which expands the leaves of a foliation, Inst. Hautes Etudes Sci. Publ. Math. 48 (1978), 133–135.

D. Ruelle and D. Sullivan, Currents, flows and diffeomorphisms, Topology 14 (1975), 319–327.

E. Ruh and Ph. Tondeur, Almost Lie foliations and the heat equation method, Proc. of the VI. Int. Coll. on Differential Geometry, Santiago de Compostela (Spain) 1988, Cursos y Congresos 61, Univ. Santiago de Compostela, 1989, 239–246.

E. Ruh and J. Vilms, The tension field of the Gauss map, Trans. Amer. Math. Soc. 149 (1970), 569–573.

Ph. Rukimbira, The dimension of leaf closures of K-contact flows, Annals of Global Anal. and Geom. 12 (1994), 103–108.

Ph. Rukimbira, Vertical sectional curvature and K-contactness, J. Geom. 53 (1995), 163–166.

H. Rummler, Quelques notions simples en géométrie riemannienne et leurs applications aux feuilletages compacts, Comment. Math. Helv. 54 (1979), 224–239.

H. Rummler, Kompakte Blätterungen durch Minimalflächen, Habilitationsschrift Universität Freiburg i.Ve. (1979).

H. Rummler, Differential forms, Weitzenböck formulae and foliations, Publicaciones del Depto de Geometria y Topologia, Universidad de Santiago de Compostela, 57 (1982), 543–554.

T. Rybicki, On the Lie algebra of a transversally complete foliation, Publ. Soc. Mat. Univ. Autònoma Barcelona 31 (1987), 5–16.

T. Rybicki, Lie algebras of vector fields and codimension one foliations, Publicacions Matematiques 34 (1990), 311–321.

T. Rybicki, On Lie algebras of vector fields related to Riemannian foliations, Ann. Pol. Math. 58 (1993), 111–122.

R. Sacksteder, On the existence of exceptional leaves in foliations of codimension one, Ann. Inst. Fourier 14 (1964), 221–226.

R. Sacksteder, Some properties of foliations, Structures feuilletées, Grenoble (1963), aNN. iNST. fOURIER 14 (1964), 31–35.

R. Sacksteder, Foliations and pseudo-groups, Amer. J. Math. 87 (1965), 79–102.

R. Sacksteder, Degeneracy of orbits of actions of $\mathbb{R}^n$ on a manifold, Comment. Math. Helv. 41 (1966), 1–9.

R. Sacksteder, A remark on Thurston's stability theorem, Ann. Inst. Fourier 25 (1975), 219–220.

R. Sacksteder, Foliations and separation of variables, Astérisque 116 (1984), 214–222.

R. Sacksteder and A. Schwartz, Limit sets of foliations, Ann. Inst. Fourier 15 (1965), 201–214.

A. Saeki, On foliations of complex spaces, Proc. Japan Acad., 68, ser. A (1992), 261–265.

A. Saeki, On foliations of complex spaces. II, Proc. Japan Acad., 69, ser. A (1993), 5–9.

A. Saeki, On some foliations on ruled surfaces, Proc. Japan Acad. 70, ser. A (1994), 17–21.

A. Saeki, Some foliations on ruled surfaces II, J. Math. Sci. Univ. Tokyo 2 (1995), 291–301.

E. Salem, Feuilletages riemanniens et pseudogroupes d'isométries, Thèse, Geneva (1987).

E. Salem, Riemannian foliations and pseudogroups of isometries, in P. Molino, Riemannian foliations, Birkhäuser (1988), 265–296.

E. Salem, Une généralisation du théorème de Myers-Steenrod aux pseudogroupes d'isométries locales, Ann. Inst. Fourier 38 (1988), 185–200.

E. Salhi, On local minimal sets, C. R. Acad. Sci. Paris 295 (1982), 691–694.

E. Salhi, Problèmes de structure dans les feuilletages de codimension un en classe C^0, Thèse d'Etat, Strasbourg (1984).

E. Salhi, Sur un théorème de structure des feuilletages de codimension 1, C. R. Acad. Sci. Paris 300 (1985), 635–638.

E. Salhi, Niveau des feuilles, C. R. Acad. Sci. Paris 301 (1985), 219–222.

E. Sallum, Vector fields tangent to a Reeb foliation on S^3, J. Diff. Eq. 34 (1979), 204–211.

J. Sampson, Foliations from quadratic and hermitian differential forms, Arch. Rat. Mech. and Anal. 70 (1979), 91–99.

M. Samuélidès, Tout feuilletage à croissance polynomiale est hyperfini, J. Funct. Anal. 34 (1979), 363–369.

A. Sanini, Minimal and totally geodesic foliations, Atti. Accad. Sci. Torino Cl. Sci. Fis. Mat. Natur. 116 (1982), 117–126.

A. Sanini and F. Tricerri, Prolungamenti di enti geometrici definiti su una varietà fogliettate, Rend. Sem. Mat. Univ. Politec. Torino 34 (1976), 211–221.

A. Sanini and F. Tricerri, Connessioni e varietà fogliettate, Cooperative Libraria Universitaria Torinese-Editrice, 1977.

M. Sanmartin, La variedad de las metricas casi-fibradas para una foliacion Riemaniana, Publ. del Dpto. de Geometria y Topologia, Univ. de Santiago de Compostela, 83 (1995), 197 pp.

G. Santhanam, A class of totally geodesic foliations of Lie groups, Proc. Indian Acad. Sci. Math. Sci. 100 (1990), 57–64.

N. dos Santos, Differentiable conjugation of actions of $\mathbb{R}^p$ on torus T^n, Geometric Study of Foliations Tokyo 1993, 181–191, World Scientific, 1994.

N. dos Santos, Foliated cohomology and characteristic classes, Contemp. Math. 161 (1994), 41–57.

N. dos Santos and M. do Socorro Pereira, On the cohomology of foliated bundles, to appear.

M. Saralegui, The Euler class for flows of isometries, Pitman Research Notes in Math. 131 (1985), 220–227.

M. Saralegui, A Gysin sequence for semifree actions of S^3, Proc. Amer. Math. Soc. 118 (1993), 1335–1345.

M. Saralegui, Gysin sequences, Analysis and Geometry in Foliated Manifolds, Proc. VII Colloq. on Diff. Geom., Santiago de Compostela 1994, World Scientific 1995, 207–222.

G. Sardanashvily, Space-time foliations, Acta Physica Hungarica 57 (1985), 31–40.

G. Sardanashvily, Caustics of space-time foliations, Soviet Phys. J. 31 (1988), 716–719.

G. Sardanashvily and V. Yanchevskii, Space-time foliations in the theory of gravitation, Izv. Vyss. Uchebn. Zaved. Fiz. (1982), 20–23.

G. Sardanashvily and V. Yanchevskii, Caustics of space-time foliations in general relativity, Acta Phys. Polon. B 17 (1986), 1017–1027.

K. Sarkaria, A finiteness theorem for foliated manifolds, J. Math. Soc. Japan. 30 (1978), 687–696.

K. Sarkaria, Non degenerescence of some spectral sequences, Ann. Inst. Fourier 34 (1984), 39–46.

K. Sasano, Foliations transverse to nonsingular Morse-Smale flows, A fête of topology, 127–160, Academic Press, Boston, 1988.

I. Satake, On a generalization of the notion of a manifold, Proc. NAS 42 (1956), 359–363.

A. Sato, Every 3-manifold admits a transverse pair of codimension one foliations which cannot be raised to a total foliation, Foliations (Tokyo, 1983), 395–415, Adv. Stud. Pure Math. 5, North-Holland, Amsterdam, 1985.

A. Sato, Stably solitary foliations, Hokkaido Math. J. 14 (1985), 143–147.

A. Sato, Flows and transverse foliations, Stability theory and related topics in dynamical systems (Nagoyo, 1988), 113–123, World Sci. Adv. Ser. Dyn. Syst., 6, World Sci. Publishing, 1989.

A. Sato and I. Tamura, On transverse foliations, Inst. Hautes Etudes Sci. Publ. Math. 54 (1981), 205–235.

J. Sauvageot, Semi-groupe de la chaleur transverse sur la C^*-algèbre d'un feuilletage Riemannien, C. R. Acad. Sci. Paris 310 (1990), 531–536.

S. Schecter and M. Singer, Planar polynomial foliations, Proc. Amer. Math. Soc. 79 (1980), 649–656.

S. Schecter and M. Singer, Addendum to Planar polynomial foliations, Proc. Amer. Math. Soc. 83 (1981), 220.

T. Schmitt, Submersions, foliations, and RC-structures on smooth supermanifolds, Teubner-Texte Math. 96 (1987), 305–322.

A. Schwartz, A generalization of the Poincaré-Bendixson theorem to closed two-dimensional manifolds, Amer. J. Math. 85 (1963), 453–458.

S. Schwartzman, Asymptotic cycles, Ann. of Math. 66 (1957), 270–284.

G. Schwarz, On the De Rham cohomology of the leaf space of a foliation, Topology 13 (1974), 185–187.

P. Schweitzer, Counterexamples to the Seifert conjecture and opening leaves of foliations, Ann. of Math. 100 (1974), 386–400.

P. Schweitzer, Compact leaves of foliations, Springer Lecture Notes in Math. 468 (1975), 4–6.

P. Schweitzer, Compact leaves of codimension one foliations, Springer Lecture Notes in Math. 484 (1975), 273–276.

P. Schweitzer, Codimension one plane fields and foliations, Proc. Symp. Pure Math., Stanford, Calif., 1973, Providence, vol. 27, Part 1 (1975), 311–312.

P. Schweitzer, Compact leaves of foliations, Proc. Int. Congr. Math., Vancouver, 1974, vol. 1, S. 1 (1975), 543–546.

P. Schweitzer (Ed), Some problems in foliation theory and related areas, Springer Lecture Notes in Math. 652 (1978), 240–252.

P. Schweitzer, Stability of compact leaves with trivial linear holonomy, Topology 27 (1988), 37–56.

P. Schweitzer, Existence of codimension one foliation with minimal leaves, Ann. Global Anal. Geom. 9 (1991), 77–81.

P. Schweitzer (ed.), Differential topology, foliations, and group actions, Proc. of the Workshop on Topology held at the Pontificia Universidade Católica, Rio de Janeiro, January 6–17, 1992, Contemp. Math. 161 (1994).

P. Schweitzer, Codimension one foliations without compact leaves, Comment. Math. Helv. 70 (1995), 171–209.

P. Schweitzer, Surfaces not quasi-isometric to leaves of foliations of compact 3-manifolds, Analysis and Geometry in Foliated Manifolds, Proc. VII Colloq. on Diff. Geom., Santiago de Compostela 1994, World Scientific 1995, 223–238.

P. Schweitzer, Riemannian manifolds not quasi-isometric to leaves in codimension one, to appear.

P. Schweitzer and P. Whitman, Pontryagin polynomial residues of isolated foliation singularities, Springer Lecture Notes in Math. 652 (1978), 95–103.

P. Scofield, Symplectic and complex foliations, Ph.D. thesis, Univ. of Illinois at Urbana-Champaign, 1990.

P. Scofield, Some deformations of the Hopf foliation are also Kähler, Proc. Amer. Math. Soc. 119 (1993), 251–253.

M. Sebastiani, Sur les nombres caractéristiques des feuilles d'un certain type de variétés feuilletées, C. R. Acad. Sci. Paris 260 (1965), 1055–1058.

M. Sebastiani, Sur les feuilletages à groupe d'holonomie fini et feuilles compactes, Ann. Inst. Fourier 18 (1968), 331–336.

M. Sebastiani, Sur la rigidité des feuilletages analytiques, C. R. Acad. Sci. Paris 318 (1994), 481–482.

A. Sec, Sur certaines équations de Pfaff complètement intégrables dans le champ complexe (propriétés du feuilletage associé), Springer Lecture Notes in Math. 484 (1975), 224–233.

G. Segal, Classifying spaces related to foliations, Topology 17 (1978), 367–382.

H. Seifert, Topologie dreidimensionaler gefaserter Raüme, Acta Math. 60 (1933), 147–238.

H. Seifert, Closed integral curves in 3-space and isotopic deformations, Proc. Amer. Math. Soc. 1 (1950), 287–302.

A. Seitoh, Remarks on stability for semiproper exceptional leaves, Tokyo J. Math. 6 (1983), 95–108.

B. Seke, Sur les structures transversalement affines des feuilletages de codimension un, Ann. Inst. Fourier 30 (1980), 1–29.

B. Seke, Structures transverses affines trivialisables, Publ. IRMA Strasbourg 188 P-108 (1982).

M. Sekizawa, Completeness of the k-th nullity foliations, J. Differential Geom. 11 (1976), 461–465.

W. Senn, Über Mosers regularisiertes Variationsproblem für minimale Blätterungen des n-dimensionalen Torus, Z. Angew. Math. Phys. 42 (1991), 527–546.

F. Sergeraert, Feuilletages et difféomorphismes infiniment tangent à l'identité, Invent. Math. 39 (1977), 253–275.

F. Sergeraert, $B\Gamma$ (d'après Mather et Thurston), Springer Lecture Notes in Math. 710 (1979), 300–315.

F. Sergeraert, La classe de cobordisme des feuilletages de Reeb de S^3 est nulle, C. I. M. E. Varenna 1976, Liguori 1979, 141–150.

V. Sergiescu, Cohomologie basique et dualité des feuilletages Riemanniens, Ann. Inst. Fourier 35 (1985), 137–158.

V. Sergiescu, Sur la suite spectrale d'un feuilletage riemannien, Proc. of the XIXth Nat. Congress of the Mexican Math. Soc., Vol. 2 (Guadalajara, 1986), 33–39; Comun., 4, Soc. Mat. Mixicana, Mexico City, 1987.

V. Sergiescu, Basic cohomology and tautness of Riemannian foliations, Appendix B in P. Molino, Riemannian foliations, Birkhäuser (1988), 235–248.

C. Series, Foliations of polynomial growth are hyperfinite, Israel J. Math. 34 (1979), 245–258.

C. Series, The Poincaré flow of a foliation, Amer. J. Math. 102 (1980), 93–128.

S. Sertöz, On Bott's vanishing theorems and applications to singular foliations, Doga Math. 11 (1987), 62–67.

S. Sertöz, Residues of singular holomorphic foliations, Compositio Math. 70 (1989), 227–243.

S. Sertöz, Generic singularities of holomorphic foliations, to appear.

N. Shananin, Foliations of manifolds and weighting of derivatives, Amer. Math. Soc. Transl. (2) vol. 167 (1995), 207–215.

A. Sheu, Singular foliation C^*-algebras, Proc. Amer. Math. Soc. 104 (1988), 1197–1203.

K. Shibata, On Haefliger's model for the Gelfand-Fuks cohomology, Japan J. Math. 7 (1981), 379–415.

Y. Shikata, On a homology theory associated to foliations, Nagoya Math. J. 38 (1970), 53–61.

Y. Shikata, On the cohomology of bigraded forms associated with foliated structures, Bull. Soc. Math. Grèce 15 (1974), 68–76.

H. Shulman, Characteristic classes and foliations, Thesis, University of California, Berkeley (1972).

H. Shulman, Secondary obstruction to foliations, Topology 13 (1974), 177–183.

H. Shulman, The double complex of Γ_k, AMS Proc. Symp. Pure Math., Stanford, Calif., 1973, vol. 27, Part 1 (1975), 313–314.

H. Shulman, Covering dimension and characteristic classes for foliations, Proc. Symp. Pure Math., Stanford 1976, Providence, Vol. 32, Part 2 (1978), 189–190.

H. Shulman and J. Stasheff, De Rham theory for B, Springer Lecture Notes in Math. 652 (1978), 62–74.

H. Shulman and D. Tischler, Leaf invariants for foliations and the van Est isomorphism, J. Diff. Geom. 11 (1976), 535–546.

C. Siegel, Note on differential equations on the torus, Ann. of Math. 46 (1945), 423–428.

J. Sikorav, Formes différentielles fermées non singulières sur le tore de dimension n, C. R. Acad. Sci. Paris 292 (1980), 829–832.

J. Sikorav, Classes d'homotopies des formes fermées non singulières, C. R. Acad. Sci. Paris 294 (1982), 413–416.

J. Sikorav, Formes différentielles fermées sur le n-tore, Comment. Math. Helv. 57 (1982), 79–106.

E. Silberstein, Multifoliations on $M^n \times S^1$ where M^n is a stably parallelizable manifold, Proc. London Math. Soc. 35 (1977), 463–482.

A. Silva, Atiyah sequences and complete closed pseudogroups preserving a local parallelism, Springer Lecture Notes in Math. 1345 (1988), 302–316.

D. Simen, $\mathbb{R}^k$-actions on manifolds, Amer. J. Math. 104 (1982), 1–7.

M. Simonnet, Feuilletages et sous-fibrés intégrables, Esquis. Math. 28 (1977), 1–65.

C. Simpson, Lefschetz theorems for the integral leaves of a holomorphic one-form, Compositio Math. 87 (1993), 99–113.

K. Sithanantham, On the cohomology of the Lie algebra of formal vectorfields preserving a flag, Ill. J. Math. 28 (1984), 487–494.

V. Slukhaev, Nonholonomic polytope foliations and nonholonomic polyhedra, Geom. Sb. 31 (1993), 79–83.

L. Slutskin, Computations on the transverse measured foliations associated with a pseudo-Anosov automorphism, J. Diff. Geom. 39 (1994), 359–378.

S. Smale, Differentiable dynamical systems, Bull. Amer. Math. Soc. 73 (1967), 747–817.

J. Smillie, Flat manifolds with nonzero Euler Characteristic, Comment. Math. Helv. 52 (1977), 453–455.

J. Smith, Commuting vector fields on open manifolds, Bull. Amer. Math. Soc. 75 (1969), 1013–1016.

J. Smith, Extending regular foliations, Ann. Inst. Fourier 19 (1969), 155–168.

M. Soares, A note on algebraic solutions of foliations in dimension two, Pitman Res. Notes Math. Ser. 285 (1993), 250–254.

M. Soares, On algebraic sets invariant by one-dimensional foliations on $CP(3)$, Ann. Inst. Fourier 43 (1993), 143–162.

M. Soares, Invariant algebraic sets of foliations, Symp. on Singularity Theory, ICTP, Aug. 19–Sept. 6, 1991, World Scientific (1995), 800–816.

M. do Socorro Pereira, Obstructions á la représentation d'un feuilletage comme intersection transverse de deux feuilletages riemanniens, Bull. Soc. Math. Belg. 43 (1991), 199–210.

V. Solodov, On mappings of the circle into foliations, Geometrich. Metody v Zadachakh Analiza i Algebry, Yaroslavl' (1978), 100–107.

V. Solodov, A geometric proof of a theorem of Mather, Uspekhi Mat. Nauk. 34 (1979), 243–244. Translation: Russ. Math. Surveys 34 (1979), 241–242.

V. Solodov, Foliations on manifolds whose fundamental group has a nontrivial center, Uspekhi Mat. Nauk 36-6 (1981), 229–230. Translation: Russ. Math. Surveys 36-6 (1981), 201–202.

V. Solodov, Foliations on manifolds with a special fundamental group, Uspekhi Mat. Nauk 36-3 (1981), 225–226. Translation: Russ. Math. Surveys 36-3 (1981), 264–265.

V. Solodov, Components of topological foliations, Mat. Sb. 119 (1982), 340–354. Translation: Math. USSR Sbornik 47 (1984), 329–343.

V. Solodov, Homeomorphisms of the line and a foliation, Izv. Akad. Nauk 46 (1982). Translation: Math. USSR Izv. 21 (1983), 341–354.

V. Solodov, Homeomorphisms of the circle and foliations, Izv. Akad. Nauk. SSSR Ser. Mat. 48 (1984), 599–631. Translation: Math. USSR Izv. 24 (1985), 553–566.

V. Solodov, Pregroups and foliations, Algebraic problems in analysis and topology (Russian), 147–152, 160, Novoe Global. Anal., Voronezh. Gos. Univ. Voronezh, 1990.

V. Solodov, On the universal cover of Anosov flows, to appear.

B. Solomon, On foliations of $\mathbb{R}^{n+1}$ by minimal hypersurfaces, Comment. Math. Helv. 61 (1986), 67–83.

V. Solovov, Riemannian foliations with constant transversal curvature, Sibirsk. Mat. Zh. 28 (1987), 160–166. Translation: Siberion Math. J. 28 (1987).

V. Solovov, Pontrjagin classes of completely geodesic subfoliations, Siberian Math. J. 29 (1988), 510–512.

F. Sommer, Komplex-analytische Blaetterung reeller Mannigfaltigkeiten in $\mathbb{C}^n$, Math. Ann. 136 (1958), 111–133.

F. Sommer, Komplex-analytische Blaetterung reeller Hyperflächen im $\mathbb{C}^n$, Math. Ann. 137 (1959), 392–411.

J. Sondow, The Godbillon-Vey invariant of a product foliation is zero, Dynamical Systems, Bahia 1971, Academic Press 1973, 545–547.

J. Sondow, When is a manifold a leaf of some foliation?, Bull. Amer. Math. Soc. 81 (1975), 622–624.

K. Spallek, Foliations on singularities, Complex analysis and applications (Varna, 1985), 643–657, Bulgar. Acad. Sci., Sofia, 1986.

K. Spallek, Continuation of foliations and integration of arbitrary distributions, Rev. Ronmaine Math. Pures Appl. 36 (1991), 271–286.

K. Spallek, Almost complex structures on spaces, associated singularities and foliations, Proc. of the Int. Workshop on Almost Complex Structures, Sofia, Aug. 20–25, 1992, 1–19, World Scientific, 1994.

M. Stadler, Sur la conjecture de Baum-Connes, to appear

D. Stasheff, Continuous cohomology of groups and classifying spaces, Bull. Amer. Math. Soc. 84 (1978), 513–530.

P. Stefan, Accessible sets, orbits, and foliations with singularities, Proc. London Math. Soc. 29 (1974), 699–713.

P. Stefan, Integrability of Systems of Vector Fields, J. London Math. Soc. 21 (1980), 544–556.

S. Stepanov, A class of Riemannian almost-product structures, Izv. Vyssh. Uchebn. Zaved. Mat. 1989, 40–46.

S. Stepanov, A geometric obstruction to the existence of completely umbilical distribution on a compact manifold, Webs and quasigroups, 135–137, Kalinin. Gos. Univ., Kalinin, 1990.

S. Stepanov, The Bochner technique in the theory of Riemannian almost product structures, Math. Notes 48 (1990), 778–781.

S. Stepanov, A minimal hyperdistribution on a compact manifold, Differentsial'naya Geom. Mnogoobraz. Figur 1991, no. 22, 101–104.

S. Stepanov, Weyl submersions, Russ. Math. 36 (1992), 87–89.

S. Stepanov, An integral formula for a compact manifold with a Riemannian almost product structure, Russ. Math. 38 (1994), No. 7, 66–70.

S. Stepanov, An integral formula for a Riemannian almost-product manifold, Tensor 55 (1994), 209–214.

S. Sternberg, Local C^∞ transformations of the real line, Duke Math. J. 24 (1957), 97–102.

D. Stowe, The stationary set of a group action, Proc. Amer. Math. Soc. 79 (1980), 139–146.

D. Stowe, Stable orbits of differentiable group actions, Trans. Amer. Math. Soc. 277 (1983), 665–684.

P. Stredder, Morse foliations, Thesis, Warwick, 1976.

J. Strelcyn, Flots sur le tore et nombres de rotations, Bull. Soc. Math. France 100 (1972), 195–208.

G. Stuck, On the characteristic classes of actions of lattices in higer rank Lie groups, Trans. Amer. Math. Soc. 324 (1991), 181–200.

G. Stuck, Un analogue feuilleté du théorème de Cartan-Hadamard, C. R. Acad. Sci. Paris 313 (1991), 519–522.

D. Sullivan, La classe d'Euler réelle d'un fibré vectoriel à groupe structural $SL(n, \mathbb{Z})$ est nulle, C. R. Acad. Sci. Paris 281 (1975), 17–18.

D. Sullivan, A new flow, Bull. Amer. Math. Soc. 82 (1976), 331–332.

D. Sullivan, A counterexample to the periodic orbit conjecture, Inst. Hautes Etudes Sci. Publ. Math. 46 (1976), 5–14.

D. Sullivan, A generalization of Milnor's inequality concerning affine foliations and affine manifolds, Comment. Math. Helv. 51 (1976), 183–189.

D. Sullivan, Cycles for the dynamical study of foliated manifolds and complex manifolds, Invent. Math. 36 (1976), 225–255.

D. Sullivan, Infinitesimal computations in topology, Inst. Hautes Etudes. Sci. Publ. Math. 47 (1977), 269–331.

D. Sullivan, A foliation of geodesics is characterized by having no tangent homologies, J. Pure Appl. Algebra 13 (1978), 101–104.

D. Sullivan, A homological characterization of foliations consisting of minimal surfaces, Comment. Math. Helv. 54 (1979), 218–223.

D. Sullivan, Bounds, quadratic differentials, and renormalization conjectures, Mathematics into the 21st Century, vol. 2, Amer. Math. Soc. Centennial Publications, Providence, 1991.

D. Sullivan and R. Williams, Homology of attractors, Topology 15 (1976), 259–262.

T. Sunada, Magnetic flows on a Riemann surface, Proc. of KAIST Math. Workshop 8 (1993), 93–108.

D. Sundararaman, Compact Hausdorff transversally holomorphic foliations, Springer Lecture Notes 950 (1982), 360–374.

D. Sundararaman, On holomorphic vector fields (maps) with singularities (fixed points) of Poincaré type, Coll. on dyn. systems, Guanajuato, 1983; Aportaciones Mat., Soc. Mat. Mexicana 1985, 127–151.

V. Surygin, Families of totally geodesic hypersurfaces in Weyl spaces, Trudy Geom. Sem. Kazan Univ. 10 (1978), 140–147.

H. Sussmann, Orbits of vector fields and integrability of distributions, Trans. Amer. Math. Soc. 180 (1973), 171–188.

H. Sussmann, A generalization of the closed subgroup theorem to quotients of arbitrary manifolds, J. Differential Geom. 10 (1975), 151–166.

T. Suwa, A theorem of versality for unfoldings of complex analytic foliation singularities, Invent. Math. 65 (1981/82), 29–48.

T. Suwa, Unfoldings of meromorphic functions, Math Ann. 262 (1983), 215–224.

T. Suwa, Kupka-Reeb phenomena and universal unfoldings of certain foliation singularities, Osaka J. Math. 20 (1983), 373–382.

T. Suwa, Unfoldings of foliations with multiform first integrals, Ann. Inst. Fourier 33 (1983), 99–112.

T. Suwa, Unfoldings of complex analytic foliations with singularities, Japan J. Math. 9 (1983), 181–206.

T. Suwa, Singularities of complex analytic foliations, Proc. Symp. Pure Math. 40, Part 2, 1983, 551–559.

T. Suwa, Residues of complex analytic foliation singularities, J. Math. Soc. Japan 36 (1984), 37–45.

T. Suwa, Determinacy of analytic foliation germs, Foliations (Tokyo, 1983), 427–460, Adv. Study Pure Math. 5, North-Holland, Amsterdam, 1985.

T. Suwa, The versatility theorem for RL-morphisms of foliation unfoldings, Complex analytic singularities, 599–631, Adv. Stud. Pure Math., 8, North-Holland, Amsterdam, 1987.

T. Suwa, $\mathcal{D}$-modules associated to complex analytic singular foliations, J. Fac. Sci. Univ. Tokyo 37 (1990), 297–320.

T. Suwa, Unfoldings of codimension one complex analytic foliation singularities, Symp. on Singularity Theory, ICTP, Aug. 19–Sept. 6, 1991, World Scientific (1995), 800–816.

H. Suzuki, Characteristic classes of foliated principal GL_r-bundle, Hokkaido Math. J. 4 (1975), 159–168.

H. Suzuki, A property of a characteristic class of an orbit foliation, London Math. Soc. Lect. Notes Series 26 (1977), 190–203.

H. Suzuki, Construction of transverse projectable connections in some foliated bundles, Publ. Res. Inst. Math. Sci. 17 (1981), 215–233.

H. Suzuki, Foliation preserving Lie group actions and characteristic classes, Proc. Amer. Math. Soc. 85 (1982), 633–637.

H. Suzuki, Lift foliations in flat principal bundles and modular functions, Foliations (Tokyo, 1983), 29–36, Adv. Stud. Pure Math. 5, North-Holland, Amsterdam, 1985.

H. Suzuki, An interpretation of the Weil operator $\chi(y_1)$, Differential Geometry (Santiago de Compostela, 1984), Pitman Res. Notes Math. Ser. 131 (1985), 228–244.

H. Suzuki, Modular cohomology class from the viewpoint of characteristic class, Geometric methods in operator algebras (Kyoto, 1983), Pitman Res. Notes Math. Ser. 123 (1986), 375–386.

H. Suzuki, Differentiable singular cohomology for foliations, A fête of topology, 63–80, Academic Press, Boston, 1988.

H. Suzuki, Holonomy groupoids of generalized foliations, II. Transverse measures and modular classes. Math. Sci. Res. Inst. Publ. 20 (1991), 267–279.

M. Suzuki, Sur les intégrales premières de certains feuilletages analytiques complexes, Springer Lecture Notes in Math. 670 (1978), 53–79.

J. Szenthe, Orthogonally transversal submanifolds and the generalization of the Weyl group, Period. Math. Hungar. 15 (1984), 281–299.

S. Tabachnikov, Characteristic classes of homogeneous foliations, Uspekhi Mat. Nauk 39 (1984), 189–190. Translation: Russian Math. Surveys 39 (1984), 203–204.

S. Tabachnikov, Characteristic classes of Grassman bundles, Funktsional. Anal. i Prilozen 19 (1985), 83–84.

S. Tabachnikov, Characteristic classes of parabolic foliations of series B, C, and D and degrees of isotropic Grassmannians, Funkt. Anal. Appl. 20 (1986), 158–160.

S. Tabachnikov, Characteristic classes of Lagrange foliations, Functional Anal. Appl. 23 (1989), 162–163.

B. Tabak, A geometric characterization of harmonic diffeomorphisms between surfaces, Math. Ann. 270 (1985), 147–157.

R. Takagi and S. Yorozu, Minimal foliations on Lie groups, Tôhoku Math. J. 36 (1984), 541–554.

R. Takagi and S. Yorozu, Notes on the Laplace-Beltrami operator on a foliated Riemannian manifold with a bundle-like metric, Nihonkai Math. J. 1 (1990), 89–106.

H. Takai, C^*-algebras of Anosov foliations, Lecture Notes in Math. 1132 (1985), 509–516.

H. Takai, KK-theory for the C^*-algebras of Anosov foliations, Pitman Res. Notes Math. Ser. 123 (1986), 387–399.

H. Takai, A counterexample of Baum-Connes conjectures for foliated manifolds, The study of dynamical systems (Kyoto, 1989), 149–154, World Sci. Adv. Ser. Dyn. Syst. 7, World Sci. Publ., 1989.

M. Takeuchi, On foliations with the structure group of automorphisms of a geometric structure, J. Math. Soc. Japan 32 (1980), 119–152.

I. Tamura, Spinnable structures of differentiable manifolds, Proc. Japan. Acad. 48 (1972), 293–296.

I. Tamura, Every odd dimensional sphere has a foliation of codimension one, Comment. Math. Helv. 47 (1972), 164–170.

I. Tamura, Foliations of total spaces of sphere bundles over spheres, J. Math. Soc. Japan. 24 (1972), 698–700.

I. Tamura, Foliations and spinnable structures of manifolds, Ann. Inst. Fourier 23 (1973), 197–214.

I. Tamura, Specially spinnable manifolds, Manifolds, Tokyo 1973, U. of Tokyo Press 1975, 181–187.

I. Tamura, The Topology of Foliations, Iwanami Shoten Publ., Tokyo, 1976; AMS Translations of Math. Monographs 97 (1992).

I. Tamura (Ed), Foliations, Proc. Symp. Univ. of Tokyo, 1983.

I. Tamura, Dynamical systems on foliations and existence problems of transverse foliations, Foliations (Tokyo, 1983), 229–293, Adv. Stud. Pure Math. 5, North-Holland, Amsterdam, 1985.

I. Tamura, Topology of foliations: an introduction, 1976, AMS Translations of Mathematical Monographs 97 (1992).

S. Tan, Nullity and generalized characteristic classes of foliations, Trans. Amer. Math. Soc. 243 (1978), 75–88.

C. Tanasi, Sur les feuilletages mesurés arationnels, Cahiers Topol. Géom. Diff. Catégories 25 (1984), 303–310.

C. Tanasi, A rational measured foliation, II, Rend. Circ. Mat. Palermo 34 (1985), 300–309.

C. Tanasi, A Hopf-type theorem for measured foliations of two-dimensional orbifolds, Rend. Sem. Mat. Univ. Politec. Torino 45 (1987), 83–98.

C. Tanasi, The Euler-Poincaré characteristic of two-dimensional orbifolds, Rend. Sem. Mat. Univ. Politec. Torino 45 (1987), 133–155.

T. Tanemura and S. Yorozu, Differential geometry of foliated manifolds with harmonic foliations and bundle-like metrics, Tech. Rep. Series in Kanazawa Univ. Japan, 1988.

T. Tanemura and S. Yorozu, Green's theorem on a foliated Riemannian manifold and its applications, Acta Math. Hung. 56 (1990), 239–245.

S. Tanno, A theorem on totally geodesic foliations and its applications, Tensor 24 (1972), 116–122.

S. Tanno, Totally geodesic foliations with complete leaves, Hokkaido Math. J. 1 (1972), 7–11.

D. Tanré, Groupes feuilletés, Esquis Math. 20 (1973), 1–35.

J. Tapia, Etale topos associated with the quotient of a foliation, C. R. Acad. Sci. Paris 303 (1986), 753–755.

J. Tapia, Sur la pente du feuilletage de Kronecker et la cohomologie étale de l'espace des feuilles, C. R. Acad. Sci. Paris 305 (1987), 427–429.

A. Teleman, Asupra clasei caracteristice Godbillon-Vey, Stud. Cerc. Mat. 37 (1985), 472–475.

C. Terng, Natural vector bundles and natural differential operators, Amer. J. Math. 100 (1978), 775–828.

C. Terng, Isoparametric submanifolds and their Coxeter groups, J. Diff. Geom. 21 (1985), 79–107.

C. Terng, A convexity theorem for isoparametric submanifolds, Invent. Math. 85 (1986), 487–492.

C. Terng, Recent progress in submanifold geometry, Proc. Symp. Pure Math. Amer. Math. Soc. 54, Part I (1993), 439–484.

R. Thom, Généralisation de la théorie de Morse aux variétés feuilletées, Ann. Inst. Fourier 14 (1964), 173–190.

R. Thom, On singularities of foliations, Manifolds Tokyo 1973, 171–173, Univ. of Tokyo, 1975.

R. Thom, Limit sets of leaves of foliations, Sügaku 30 (1978), 132–136.

R. Thom, La marche au chaos vu par un topologue, 1988.

R. Thom, Sur les bouts d'une feuille d'une feuilletage au voisinage d'un point singulier isolé, Springer Lecture Notes in Math. 1345 (1988), 317–321.

C. Thomas, A classifying space for the contact pseudogroup, Mathematika 25 (1978), 191–201.

E. Thomas, Vector fields on manifolds, Bull. Amer. Math. Soc. 75 (1969), 643–683.

E. Thomas, Secondary obstructions to integrability, Dynamical Systems, Bahia 1971, Academic Press 1973, 655–661.

G. Thorbergsson, Isoparametic foliations and their buildings, Ann. of Math. 133 (1991), 429–446.

W. Thurston, Non-cobordant foliations of S^3, Bull. Amer. Math. Soc. 78 (1972), 511–514.

W. Thurston, Foliations of 3-manifolds which are circle bundles, Thesis, University of California, Berkeley (1972).

W. Thurston, Foliations and groups of diffeomorphisms, Bull. Amer. Math. Soc. 80 (1974), 304–307.

W. Thurston, The theory of foliations of codimension greater than one, Comment. Math. Helv. 49 (1974), 214–231.

W. Thurston, A generalization of the Reeb stability theorem, Topology 13 (1974), 347–352.

W. Thurston, Foliations and groups of diffeomorphisms, Bull. Amer. Math. Soc. 80 (1974), 304–307.

W. Thurston, The theory of foliations of codimension greater than one, AMS Proc. Symp. Pure Math., Stanford, Calif., 1973, vol. 27, Part 1 (1975), 321.

W. Thurston, A local construction of foliations for three-manifolds, AMS Proc. Symp. Pure Math., Stanford, Calif., 1973, vol. 27, Part 1 (1975), 315–319.

W. Thurston, On the construction and classification of foliations, Proc. Int. Congr. Math., Vancouver, vol. 1, S. 1 (1975), 547–549.

W. Thurston, Existence of codimension one foliations, Ann. of Math. 104 (1976), 249–268.

W. Thurston, A norm for the homology of 3-manifolds, Memoirs of the American Math. Soc. 339 (1986), 99–130.

W. Thurston, Hyperbolic structures on 3-manifolds, II: Surface groups and 3-manifolds which fiber over the circle, preprint, Princeton Univ. 1986.

W. Thurston, On the geometry and dynamics of diffeomorphisms of surfaces, I, Bull. Amer. Math. Soc. 19 (1988), 417–431.

C. Tian, Foliation on a surface of constant curvature and some nonlinear evolution equations, Chinese Ann. Math. Ser. B 9 (1988), 118–122.

M. Tibar, The graph of a Reeb foliation, Stud. Cerc. Mat. 36 (1984), 262–266.

W. Ting, On nontrivial characteristic classes of contact foliations, Proc. Amer. Math. Soc. 75 (1979), 131–138.

D. Tischler, Totally parallelizable 3-manifolds, Topological dynamics, Ft. Collins (1967), Benjamin (1968), 471–492.

D. Tischler, On fibering certain foliated manifolds over S^2, Topology 9 (1970), 153–154.

D. Tischler, Locally free actions of $\mathbb{R}^{n-1}$ on M^n without compact orbits, Topology 13 (1974), 215–217.

D. Tischler, Manifolds M of rank $n-1$, Proc. Amer. Math. Soc. 94 (1985), 158–160.

D. Tischler and R. Tischler, Topological conjugacy of locally free $\mathbb{R}^{n-1}$ actions on n-manifolds, Ann. Inst. Fourier 24 (1974), 213–227; erratum: ibid. 26 (1976), 301.

Ph. Tondeur, The mean curvature of Riemannian foliations, Feuilletages riemanniens, quantification géométrique et mécanique, Travaux en Cours 26, 1988, Herman, Paris, 41–52.

Ph. Tondeur, Foliations on Riemannian manifolds, Universitext, Springer Verlag, New York, 1988.

Ph. Tondeur, Riemannian foliations and tautness, AMS Symp. Pure Math., 54 (1993), Part 3, 667–672.

Ph. Tondeur, Riemannian foliations and the heat equation, Geometry and its applications, ed. by T. Nagano et al., World Scientific, Singapore, 1993.

Ph. Tondeur, Geometry of Riemannian foliations, Seminar on Mathematical Sciences 20, Keio University, Yokohama, 1994.

Ph. Tondeur, A characterization of Riemannian flows, preprint.

Ph. Tondeur and G. Toth, On transversal infinitesimal automorphisms for harmonic foliations, Geom. Dedicata 24 (1987), 229–236.

Ph. Tondeur and L. Vanhecke, Reflections in submanifolds, Geom. Dedicata 28 (1988), 77–85.

Ph. Tondeur and L. Vanhecke, Isometric reflections with respect to submanifolds, Simon Stevin 63 (1989), 107–116.

Ph. Tondeur and L. Vanhecke, Isometric reflections with respect to submanifolds and the Ricci operator of geodesic spheres, Monatshefte für Mathematik 108 (1989), 211–217.

Ph. Tondeur and L. Vanhecke, Transversally symmetric Riemannian foliations, Tôhoku Math. J. 42 (1990), 307–317.

Ph. Tondeur and L. Vanhecke, Characterizing special Riemannian foliations, Simon Stevin, 67 (1993), 227–234.

Ph. Tondeur and L. Vanhecke, A characterization of Riemannian foliations and totally umbilical submanifolds, Bull. Austral. Math. Soc. 48 (1993), 101–108.

Ph. Tondeur and L. Vanhecke, Jacobi fields, Riccati equation and Riemannian foliations, Ill. J. Math., to appear.

A. Torpe, K-theory for the leaf space of foliations by Reeb components, J. Funct. Anal. 61 (1985), 15–71.

F. Torres Lopera, Grassmann structures and transversally Grassmann foliations, Proc. of the ninth Spanish-Portuguese Conf. on Math., Salamanca, 1982; Acta Salmanticensia Ciencias 46 (1982), 549–552.

P. Trauber, The continuous cohomology of the Lie algebra of vector fields on a smooth manifold, Thesis, Princeton University (1973).

A. Treibergs, Entire spacelike hypersurfaces of constant mean curvature in Minkowski space, Invent. Math. 66 (1982), 39–56.

F. Tricerri, Sulla geometria differenziale delle varietà multi fogliettate, Boll.break U. M. I. 16 (1978), 76–84.

F. Tricerri, Sulle G-structture fortemente integrabili, Rend. Sem. Mat. Univ. Politec. Torino 37 (1979), 81–91.

G. Tsagas, Some properties of closed 1-forms on a special riemannian manifold, Proc. Amer. Math. Soc. 81 (1981), 104–106.

G. Tsagas, On the foliation of a Riemannian regular s-manifold, Tensor 36 (1982), 150–154.

T. Tshikuna-Matamba, Submersions métriques presque de contact dont l'espace total est une variété de Kenmotsu, An. Stiint. Univ. Al. I. Cuza Iasi, Ser. Nouă, Mat. 37 (1991), 197–206.

T. Tsuboi, Foliations with trivial $\mathcal{F}$-subgroups, Topology 18 (1979), 223–233.

T. Tsuboi, On 2-cycles of BDiff(S^1) which are represented by foliated S^1-bundles over T^2, Ann. Inst. Fourier 31-2 (1981), 1–59.

T. Tsuboi, Cobordism of foliations of codimension one, Algebraic Topology, Aarhus 1981, Sem. Notes Aarhus 1982, 107–115.

T. Tsuboi, Homology of diffeomorphism groups, and foliated structures (Japanese), Sūgaku 36 (1984), 320–343.

T. Tsuboi, Γ_1-structures avec une seule feuille, Astérisque 116 (1984), 222–234.

T. Tsuboi, Foliated cobordism classes of certain foliated S^1-bundles over surfaces, Topology 23 (1984), 233–244.

T. Tsuboi, On the homology of classifying spaces for foliated products, Foliations (Tokyo, 1983), 37–120, Adv. Stud. Pure Math. 5, North-Holland, Amsterdam, 1985.

T. Tsuboi, On the homomorphism $H^*(B\mathbb{R}^d) \simeq H^*(B\mathrm{Diff}^\infty(\mathbb{R}^d))$, A fête of topology (1988), Academic Press, 33–47.

T. Tsuboi, On the foliated products of class C^1, Ann. of Math. 130 (1989), 227–271.

T. Tsuboi, On the Hurder-Katok extension of the Godbillon-Vey invariant, J. Fac. Sci. Univ. Tokyo, Sect. IA 37 (1990), 255–262.

T. Tsuboi, Area functionals and Godbillon-Vey cocycles, Ann. Inst. Fourier 42 (1992), 421–447.

T. Tsuboi, Rationality of piecewise linear foliations and homology of the group of piecewise linear homeomorphisms, Enseign. Math. 38 (1992), 329–344.

T. Tsuboi, The Godbillon-Vey invariant and the foliated cobordism group, Proc. Japan Acad. Ser. A Math. Sci 68 (1992), 85–90.

T. Tsuboi, Hyperbolic compact leaves are not C^1-stable, Geometric Study of Foliations, Tokyo 1993, 437–455, World Scientific, 1994.

T. Tsuboi, Characterization of Godbillon-Vey classes, Sugaku 8 (1995), 165–182.

N. Tsuchiya, Lower semi-continuity of growth of leaves, J. Fac. Sci. Univ. Tokyo, Sec. I.A. Math. 26 (1979), 465–471.

N. Tsuchiya, Growth and depth of leaves, J. Fac. Sci. Univ. Tokyo, Sect. I.A. Math. 26 (1979), 473–500.

N. Tsuchiya, Leaves of finite depth, Japan, J. Math. 6 (1980), 343–364.

N. Tsuchiya, Leaves with nonexact polynomial growth, Tôhoku Math. J. 32 (1980), 71–77.

N. Tsuchiya, The Nishimori decompositions of codimension-one foliations and the Godbillon-Vey classes, Tôhoku Math. J. 34 (1982), 343–365.

N. Tsuchiya, On decompositions and approximations of foliated manifolds, Foliations (Tokyo, 1983), 135–158, Adv. Stud. Pure Math. 5, North-Holland, Amsterdam, 1985.

N. Tsuchiya, On elementary transversely affine foliations. I, Kodai Math. J. 13 (1990), 289–297.

N. Tsuchiya, On elementary transversely affine foliations. II, Kodai Math. J. 13 (1990), 402–408.

N. Tsuchiya, On elementary transversely affine foliations, III, to appear.

T. Tsujishita, Continuous cohomology of the Lie algebra of vector fields, Mem. Amer. Math. Soc. 253 (1981).

T. Tsujishita, Characteristic classes of families of foliations, Foliations (Tokyo, 1983), 195–210, Adv. Stud. Pure Math. 5, North-Holland, Amsterdam, 1985.

F. Turiel, Transport de champs de vecteurs et feuilletages, C. R. Acad. Sci. Paris 280 (1975), 1021–1023.

F. Turiel, Couple de feuilletages de translation, de dimension 1 et de codimension 1 respectivement, sur une variété affine, Sém. Gaston Darboux de Géom. et Topologie Différentielle 1992–1993, Univ. de Montpellier 1994, 27–38.

K. Ueno, On foliations associated with differential equations of conformal type, Publ. Res. Inst. Math. Sci. Kyoto Univ. 22 (1986), 177–207.

R. Uomini, A proof of the compact leaf conjecture for foliated bundles, Proc. Amer. Math. Soc. 59 (1976), 381–382.

H. Uzuki, Characteristic classes of foliated principal $G\ell_r$-bundles, Hokkaido Math. J. 4 (1975), 159–168.

I. Vaisman, Sur la cohomologie des variétés riemanniennes feuilletées, C. R. Acad. Sci. Paris 268 (1969), 720–723.

I. Vaisman, Sur la cohomologie des variétés analytiques complexes feuilletées, C. R. Acad. Sci. Paris 273 (1971), 1067–1070.

I. Vaisman, Variétés riemanniennes feuilletées, Czechoslovak Math. J. 21 (1971), 46–75.

I. Vaisman, Sur l'existence des opérateurs différentiels feuilletés à symbole donné, C. R. Acad. Sci. Paris 276 (1973), 1165–1168.

I. Vaisman, Cohomology and differential forms, Marcel Dekker Inc., New York, 1973.

I. Vaisman, From the geometry of Hermitian foliated manifolds, Bull. Math. Soc. Sci. Math. RSR 17 (1973), 71–100.

I. Vaisman, Remarks about differential operators on foliated manifolds, An. Sti. Univ. Iasi 20 (1974), 327–350.

I. Vaisman, On the differential geometry of the transverse bundle of a foliation, Rev. Roum. Math. Pures Appl. 20 (1975), 89–101.

I. Vaisman, On the analytic distributions and foliations of a Kähler manifold, Proc. Amer. Math. Soc. 59 (1976), 221–228.

I. Vaisman, A class of complex analytic foliated manifolds with rigid structure, J. Diff. Geom. 12 (1977), 119–131.

I. Vaisman, On some spaces which are covered by a product space, Ann. Inst. Fourier 27 (1977), 107–134.

I. Vaisman, Conformal foliations, Kodai Math. J. 2 (1979), 26–37.

I. Vaisman, The Bott obstruction to the existence of nice polarizations, Mh. Math. 92 (1981), 231–238.

I. Vaisman, Basics of Lagrangian foliations, Publicaciones del Depto de Geometria y Topologia, Universidad de Santiago de Compostela 57 (1982), 559–575.

I. Vaisman, A note on projective foliations, Publ. Sec. Mat. Univ. Autònoma Barcelona 27 (1983), 109–128.

I. Vaisman, Obstructions to the existence of transverse volume elements of foliations, Dekker Lecture Notes in Pure and Appl. Math. 90 (1984), 525–534.

I. Vaisman, Lagrangian foliations and characteristic classes, Differential Geometry (Santiago de Compostela, 1984), Pitman Res. Notes Math. Ser. 131 (1985), 245–256.

I. Vaisman, d_f-cohomology fo Lagrangian foliations, Monatsh. Math. 106 (1988), 221–244.

I. Vaisman, Foliated cohomology of Lagrangian foliations, Differential Geometry and its applications (Dubrovnik, 1988), 377–386, Univ. Novi Sad, Novi Sad, 1989.

I. Vaisman, Foliate partial holomorphic structures on principal bundles, Monatsh. Math. 111 (1991), 307–321.

I. Vaisman, Secondary characteristic classes: a survey, Colloq. Math. Soc. János Bolyai 56 (1992), 671–789.

O. Valdivia Gutiérrez, On stability of symplectic foliations, Dynamical systems and partial differential equations (Caracas, 1984), 161–174, Univ. Simon Bolivar, Caracas, 1986.

T. Van Duc, Connexions induites et classes caractéristiques, Atti Accad. Naz. Lincei Rend. Cl. Sci. Fis. Mat. Natur. 63 (1977), 513–517.

T. Van Duc, Fibrés vectoriels feuilletés, Boll. Unione Mat. Ital. 15 (1978), 52–60.

T. Van Duc, Fibrés vectoriels feuilletés, Kodai Math. J. 1 (1978), 205–212.

T. Van Duc, Connexions induites et classes caractéristiques, Acta. Math. Vietnam. 3 (1979), 23–27.

T. Van Duc, Structure presque-transverse, J. Diff. Geom. 14 (1979), 215–219.

T. Van Duc, Feuilletage et distributions définies par une connexion, Rev. Roumaine Math. Pures Appl. 25 (1980), 1019–1025.

T. Van Duc, Un théorème d'équivalence pour les G-structures feuilletées, Comment. Math. Univ. Carolinae 24 (1983), 281–286.

T. Van Duc, Feuilletages transversalements tangents, Annali di Mat. 136 (1984), 227–239.

T. Van Duc, Forme canonique sur le dual du fibré transverse, Ann. Fac. Sci. Toulouse Math. 7 (1985), 169–177.

T. Van Duc, Algèbre de Lie attachée à la forme canonique sur le fibré des repères transverses, J. Math. Pures Appl. 66 (1987), 265–271.

T. Van Duc, Structure presque-transverse intégrable, Bull. Austral. Math. Soc. 37 (1988), 173–177.

T. Van Duc, Une caractérisation du dual du fibré transverse, Bull. Sci. Math. 114 (1990), 351–360.

T. Van Duc, Une caractérisation du fibré transverse, Collect. Math. 41 (1990), 35–44.

T. Van Duc, Relèvements dans le fibré transverse d'un feuilletage d'une variété riemannienne, Rend. Circ. Mat. Palermo 39 (1990), 235–248.

T. Van Duc, Relèvements dans le fibré conormal d'un feuilletage d'une variété riemannienne, Publ. Inst. Math. (Beograd) 49 (63) (1991), 155–162.

L. Vanhecke, Geometry in normal and tubular neighborhoods, Lecture Notes, Workshop on Differential Geometry and Topology, Cala Gonone (Sardinia), 1988.

J. Vanzura, A note on product structures and the Gelfand-Fuks cohomology, Colloq. Math. Soc. Trans. Bolyai 31 (1982), 755–768.

W. Veech, Quasimiminal invariants for foliations of orientable closed surfaces, Proc. Indian Acad. Sci. Math. Sci. 99 (1989), 27–48.

P. Ver Eecke, On a theorem of Charles Ehresmann concerning foliations, Cahiers Topologie Géom. Diff. 22 (1981), 453–455.

P. Ver Eecke, Introduction à la théorie des variétés feuilletées, Esq. Math. 31 (1982).

P. Ver Eecke, Sur le groupe fondamental d'un feuilletage, Cahiers Topol. Géom. Diff. Catégoriques 25 (1984), 381–428.

P. Ver Eecke, Sur le classifiant du groupoïde d'holonomie d'un feuilletage, C. R. Acad. Sci. Paris 300 (1985), 639–642.

P. Ver Eecke, Sur le groupe foundamental d'un feuilletage, C. R. Acad. Sci. Paris 300 (1985), 55–58.

P. Ver Eecke, Sur le classifiant du groupoïde d'holonomie et sur le groupoïde fondamental d'un feuilletage, Nederl. Akad. Wetensch. Indag. Math. 48 (1986), 179–200.

P. Ver Eecke, Sur le groupe fondamental d'un feuilletage obtenu par suspension, C. R. Acad. Sci. Paris 304 (1987), 547–550.

M. Vergne, Sur l'indice des opérateurs transversalement elliptiques, C. R. Acad. Sci. Paris 310 (1990), 329–332.

A. Verjovsky, A note on foliations on spheres, Bol. Soc. Mat. Mexicana 19 (1974), 36–37.

A. Verjovsky, Codimension one Anosov flows, Bol. Soc. Mat. Mex. 19 (1977), 49–77.

A. Verjovsky, A uniformization theorem for holomorphic foliations, Contemp. Math. 58 (1987), 233–253.

A. Verjovsky and F. Vila Freyer, The Witten-Jones invariant for flows on a 3-dimensional manifold, Commun. Math. Phys. 163 (1994), 73–88.

A. Verona, A de Rham type theorem for orbit spaces, Proc. Amer. Math. Soc. 104 (1988), 300–302.

L. Verstraelen, Foliations of quasi-umbilical submanifolds, Simon Stevin 51 (1977/78), 65–69.

J. Vey, Quelques constructions relatives aux Γ-structures, C. R. Acad. Sci. Paris 276 (1973), 1151–1153.

E. Vidal, Sobre algunos problemas en relacion con la medida en espacios foliados, In: I Coloquio Internacional de Geometria Diferencial, pp. 63–77, Santiago Universidad, (1964).

E. Vidal, Sur les feuilletages réguliers et les problèmes qui s'y rapportent, Acta Cientifica Compostelana III (2) 69078 (1966).

E. Vidal, Mesures définies sur les espaces des feuilles d'un feuilletage, Rend. Circ. Mat. Palermo 15 (1966), 247–256.

E. Vidal, Sobre los sistemas diferenciales completamente integrables regulares, VI Reunión Mat. espanoles, Sevilla (1967), 73–75.

E. Vidal, On regular foliations, Ann. Inst. Fourier 17 (1967), 129–133.

E. Vidal, Sur les variétés à structure de presque produit complexe avec métrique presque feuilletée, C. R. Acad. Sci. Paris 273 (1971), 1152–1155.

E. Vidal, Sur les connexions basiques transversales et projetables dans les variétés presque-produit complexes, C. R. Acad. Sci. Paris 277 (1973), 461–463.

E. Vidal, Metricas casi-foliadas en variedades con estructura casi-producto compleja, Actas de las primeras jornalas mat. luso-espanolas (1973), 279–284.

E. Vidal and E. Vidal Costa, Special connections and almost foliated metrics, J. Diff. Geom. 8 (1973), 297–304.

V. Viflyantsev, The Frobenius theorem for distributions with singularities, Uspekhi Mat. Nauk 32 (1977), 177–178.

V. Viflyantsev, Integrable distributions with singularities, Mat. Zametki 26 (1979), 921–930. Translations: Math. Notes 26 (1979), 965–970.

V. Viflyantsev, Local structure of a completely integrable differential system with singularities, Mat. Zametki 29 (1981), 685–690. Translations: Math. Notes 29 (1981), 349–351.

V. Viflyantsev, Conditions for the integrability of distributions with singularities defined by differential forms, Mat. Zametki 35 (1984), 589–597. Translations: Math. Notes 35 (1984), 311–315.

S. Vishik, On characteristic classes and singularities of foliations, Funkts. Anal. Ego Prilozhen. 6 (1972), 71–72. Translations: Funct. Anal. Appl. 6 (1972), 313–314.

S. Vishik, Singularities of analytic foliations and characteristic classes, Funkts. Anal. Ego Prilozhen 7 (1973), 1–15. Translations: Funct. Anal. Appl. 7 (1973), 1–12.

J. Viviente, Sur la géométrie transverse des feuilletages, Comun. I.C.M. Varsovia 1982.

J. Viviente, Una excursion por la teoria de foliaciones, Academia de Ciencias Exactas, Fisicas, Quimicas y Naturales de Zaragoza, 1984.

J. Viviente, Examples of nonbasic transversal invariants, Proc. of the tenth Spanish-Portuguese conference on mathematics (Murcia, 1985), Univ. Murcia (1985), 48–50.

J. Viviente, Estructura cociente e invariantes básicos, Contribuciones Matematicas en honor del profesor D. F. Botella Raduan, Editorial Univ. Complutense Madrid, Madrid 1986, 237–249.

E. Vogt, Stable foliations of 4-manifolds by closed surfaces. I. Local structure and free actions of finite cyclic and dihedral groups on surfaces, Invent. Math. 22 (1973), 321–348.

E. Vogt, Foliations of codimension two with all leaves compact, Manuscr. Math. 18 (1976), 187–212.

E. Vogt, A periodic flow with infinite Epstein hierarchy, Manuscr. Math. 22 (1977), 403–412.

E. Vogt, Foliations of codimension 2 with all leaves compact on closed 3-, 4-, and 5-manifolds, Math. Z. 157 (1977), 201–223.

E. Vogt, The first cohomology group of leaves and local stability of compact foliations, Manuscripta Math. 37 (1982), 229–267.

E. Vogt, Comparison between the Kodaira-Spencer and Heitsch invariant associated to deformations of foliations, Resultate Math. 8 (1985), 88–91.

E. Vogt, Examples of circle foliations on open 3-manifolds, Differential Geometry (Santiago de Compostela, 1984), Pitman Res. Notes Math. Ser. 131 (1985), 257–275.

E. Vogt, A foliation of $\mathbb{R}^3$ and other punctured 3-manifolds by circles, Publ. Math. Inst. Hautes Etud. Sci. 69 (1989), 215–232.

E. Vogt, Bad sets of compact foliations of codimension two, Low-dimensional Topology, Knoxville 1992, Int. Press Company, Boston.

E. Vogt, Existence of foliations of Euclidean spaces with all leaves compact, Math. Ann. 296 (1993), 159–178.

E. Vogt, Negative Euler characteristic as an obstruction to the existence of periodic flows on open 3-manifold, Geometric Study of Foliations, Tokyo 1993, 457–474, World Scientific, 1994.

E. Vogt, Real analytic circle foliations on 3-manifolds, to appear.

A. Wadsley, Geodesic foliations by circles, J. Diff. Geom. 10 (1975), 541–549.

E. Wagneur, A generalization of Novikov's theorem to foliations with isolated generic singularities, Topology and Its Appl., Proc. Conf. Mem. Univ. Newfoundland, St. John's, Canada, 1973, New York (1975), 189–198.

E. Wagneur, Réduction des points singuliers des feuilletages à singularités non-dégénérées de M^3, Can. Math. Bull. 19 (1976), 221–230.

E. Wagneur, Formes de Pfaff à singularités non-dégénérées, Ann. Inst. Fourier 28 (1978), 165–176.

P. Walczak, On minimal Riemannian foliations, Inst. Math. Pol. Acad. Sci., Warsaw 253 (1981).

P. Walczak, On foliations with leaves satisfying some geometrical condition, Dissertationes Math. (Rozprawy Mat.) 226 (1983), 47 pp.

P. Walczak, Mean curvature functions for codimension one foliations with all leaves compact, Czech. Math. J. 34 (1984), 146–155.

P. Walczak, Dynamics of the geodesic flow of a foliation, Ergod. Th. and Dyn. Syst. 8 (1988), 637–650.

P. Walczak, Mean curvature functions for foliated bundles, Topics in differential geometry, Vol I, II (Debrecen, 1984), 1309–1317, Colloq. Math. Soc. Janos Bolyai, 46, North-Holland, Amsterdam, 1988.

P. Walczak, An estimate for the second fundamental tensor of a foliation, Proceedings VI th Int. Coll. On Diff. Geom. in Santiago (Spain), Cursos y Congress, 61, Univ. Santiago de Compostela, 1989, 247–252.

P. Walczak, On the geodesic flow of a compact manifold of negative constant curvature, Proc. of the Winter School on Geometry and Physics (Srni, 1988), Rend. Circ. Mat. Palermo, Suppl. 21 (1989), 349–354.

P. Walczak, An integral formula for a Riemannian manifold with two orthogonal complementary distributions, Colloq. Math. 58 (1990), 243–252.

P. Walczak, On quasi-Riemannian foliations, Ann. Global Anal. Geom. 9 (1991), 83–95; Erratum, ibid., 325.

P. Walczak, A finiteness theorem for Riemannian submersions, Ann. Polon. Math. 57 (1992), 283–290; erratum, ibid. 58 (1993), 319.

P. Walczak, Foliations invariant under the mean curvature flow, Ill. J. of Math. 37 (1993), 609–623.

P. Walczak, Jacobi operator for leaf geodesics, Colloq. Math. 65 (1993), 213–226.

P. Walczak, Existence of smooth invariant measures for geodesic flows of foliations of Riemannian manifolds, Proc. Amer. Math. Soc. 120 (1994), 903–906.

P. Walczak, Transverse dynamics of flows tangent to foliations, Foliations Tokyo 1993, to appear.

P. Walczak, Foliations which admit the most mean curvature functions, to appear.

P. Walczak, Hausdorff dimension of Markov invariant sets, to appear.

P. Walczak, Commuting vector fields and curvature estimates, preprint.

P. Walczak, Losing Hausdorff dimension while generating pseudogroups, preprint.

W. Waliszewski, Complex premanifolds and foliations, Deformations of mathematical structures (Lódz/Lublin, 1985/87), 65–78, Kluwer Acad. Publ., Dordrecht, 1989.

W. Waliszewski, On foliations in Sikorski differential spaces with Brouwerian leaves, Ann. Polon. Math. 54 (1991), 179–182.

W. Waliszewski, On foliations with coregular factor mapping, Ann. Polon. Math. 52 (1991), 281–286.

A. Walker, Connexions for parallel distributions in the large I, Quart. J. Math. Oxford 6 (1955), 301–308.

A. Walker, Connexions for paralles distributions in the large II, Quart. J. Math. Oxford 9 (1958), 221–231.

R. Walker, Morse and generic contact between foliations, Trans. Amer. Math. Soc. 254 (1979), 265–281.

R. Walker, Transversal to laminations, Pacific J. Math. 99 (1982), 483–490.

G. Wallet, Nullité de l'invariant de Godbillon-Vey d'un tore, C. R. Acad. Sci. 283 (1976), 821–823.

G. Wallet, Godbillon-Vey invariant and commuting diffeomorphisms, C.I.M.E. Varenna (1976), Liguori (1979), 151–159.

G. Wallet, Holonomy and vanishing cycle, Ann. Inst. Fourier 31-4 (1981), 181–186.

G. Walschap, Metric foliations and curvature, J. of Geom. Anal. 2 (1992), 373–381.

G. Walschap, Foliations of symmetric spaces, Amer. J. Math. 115 (1993), 1189–1195.

G. Walschap, Some rigidity aspects of Riemannian fibrations, Proc. Symp. Pure Math. 54 (1993), Part 3, 679–683.

G. Walschap, Soul-preserving submersions, Mich. Math. J. 41 (1994), 609–617.

G. Walschap, On the metric structure of nonnegatively curved manifolds, Geometry and topology of submanifolds, VII, Belgium 1994, World Scientific 1995, 268–270.

G. Walschap, On measure-invariant flows in constant curvature, preprint.

G. Walschap, A nonexistence theorem for Riemannian foliations, preprint.

G. Walschap, Umbilic foliations and curvature, preprint.

S. Walter Wei and S. Wong, Bernstein conjecture in hyperbolic geometry, Seminar on minimal submanifolds, Ann. of Math. Studies 103 (1983), 339–358.

P. Wang, Decomposition theorems of Riemannian manifolds, Trans. Amer. Math. Soc. 184 (1973), 327–341.

X. Wang, On the C^*-algebras of foliations in the plane, Springer Lecture Notes in Math. 1257 (1987), 165 pp.

X. Wang, On the relation between C^*-algebras of foliations and those of their coverings, Proc. Amer. Math. Soc. 102 (1988), 355–360.

B. Watson, Almost Hermitian submersions, J. Diff. Geom. 11 (1976), 147–165.

B. Watson, The differential geometry of two types of almost contact metric submersions, The mathematical heritage of C. F. Gauss, Collect. Pap. Mem. C. F. Gauss (1991), 827–861.

G. Whiston, Cobordism through one-codimensional foliations, J. Diff. Geom. 11 (1976), 475–478.

A. Whitley, A remark on the K-theory of $B\Gamma_q$, Math. Proc. Cambridge Philos. Soc. 78 (1975), 309–314.

T. Willmore, Parallel distributions on manifolds, Proc. London Math. Soc. 6 (1956), 191–204.

T. Willmore, Connexions for systems of parallel distributions, Quart. J. Math. Oxford 7 (1956), 269–276.

T. Willmore, Systems of parallel distributions, J. London Math. Soc. 32 (1957), 153–156.

T. Willmore, Connections associated with foliated structures, Structures feuilletées, Grenoble (1963), Ann. Inst. Fourier 14 (1964), 43–48.

T. Willmore, Connexions and foliated structures, Sitzungsber, Berlin Math. Ges., S. 1, S. A. (1969–1971), 14–15.

F. Wilson, On the minimal sets of non singular vector fields, Ann. of Math. 84 (1966), 529–536.

F. Wilson, Vector fields tangent to foliations, II, Handlebody foliations, J. Diff. Eq. 27 (1978), 46–63.

H. Winkelnkemper, Locally free actions and Stiefel-Whitney numbers, Differential Geometry, Stanford 1973, AMS Proc. Symp. Pure Math. 27, Part 1 (1975), 323–330.

H. Winkelnkemper, The graph of a foliation, Ann. Global Anal. Geom. 1 (1983), 51–75.

H. Winkelnkemper, The number of ends of the universal leaf of a Riemannian foliation, Diff. Geom. Maryland 1981/82, Birkhäuser Progr. Math. 32 (1983), 247–254.

H. Winkelnkemper, Estimating $\|d\varphi^t\|$ for unit vector fields whose orbits are geodesics, J. Diff. Geom. 31 (1990), 847–857.

H. Winkelnkemper, Infinitesimal obstructions to weakly mixing, Ann. Global Anal. Geom. 10 (1992), 209–218.

D. Witte, Rigidity of horospherical foliations, Ergodic Theory Dynamical Systems 9 (1989), 191–205.

D. Witte, Topological equivalence of foliations of homogeneous spaces, Trans. Amer. Math. Soc. 317 (1990), 143–166.

R. Wolak, On ∇-G-foliations, Rend. Circ. Mat. Palermo 1984, Suppl. No. 6, 329–341.

R. Wolak, Normal bundles of foliations of order r, Demonstratio Math. 18 (1985), 977–994.

R. Wolak, On G-foliations, Ann. Polon. Math. 46 (1985), 371–377.

R. Wolak, On transverse structures of foliations, Rend. Circ. Mat. Palermo 1985, Suppl. No. 9, 227–243.

R. Wolak, Open almost-multifoliated manifolds, Topology, theory and applications (Eger, 1983), 679–686, Colloq. Math. Soc. János Bolyai, 41, North-Holland, Amsterdam, 1985.

R. Wolak, Some remarks on ∇-G-foliations, Pitman Research Notes 131 (1985), 276–289.

R. Wolak, Infinitesimal automorphisms of distributions, Ann. Polon. Math. 47 (1986), 1–16.

R. Wolak, Characteristic classes of almost-flag structures, Geom. Dedicata 24 (1987), 207–220.

R. Wolak, Characteristic classes of multifoliations, Colloq. Math. 52 (1987), 77–91.

R. Wolak, Maximal ideals in $Loc(M, F)$, Collect. Math. 38 (1987), 157–160.

R. Wolak, Foliations admitting transverse systems of differential equations, Compositio Math. 67 (1988), 89–101.

R. Wolak, Transverse completeness of foliated systems of ordinary differential equations, Proc. of the Sixth Intern. Colloq. on Differential Geometry (Santiago de Compostela, 1988), 253–262.

R. Wolak, Foliated G-structures and Riemannian foliations, Manuscripta Math. 66 (1989), 45–59.

R. Wolak, Foliated and associated geometric structures on foliated manifolds, Ann. Fac. Sci. Toulouse 10 (1989), 337–360.

R. Wolak, Leaves of foliations with a transverse structure of finite type, Publ. Mat. 33 (1989), 153–162.

R. Wolak, Le graphe d'un feuilletage admettant un système transverse d'equations différentielles, Math. Z. 201 (1989), 177–182.

R. Wolak, Maximal subalgebras in the algebra of foliated vector fields of a Riemannian foliation, Comm. Math. Helv. 64 (1989), 536–541.

R. Wolak, The structure tensor of a transverse G-structure on a foliated manifold, Boll. Un. Mat. Ital. A 4 (1990), 1–15.

R. Wolak, Transversely affine foliations compared with affine manifolds, Quart. J. Math. Oxford 41 (1990), 369–384.

R. Wolak, Closures of leaves in transversely affine foliations, Canad. Math. Bull. 34 (1991), 553–558.

R. Wolak, Compact leaves in minimal Riemannian foliations, Indag. Math. 5 (1994), 375–379.

R. Wolak, Pierrot's theorem for singular Riemannian foliations, Publicacions Matematiques, Vol. 38 (1994), 433–439.

R. Wolak, Ehresmann connections for lagrangian foliations, J. Geom. and Physics 17 (1995), 310–320.

R. Wolak, The graph of a totally geodesic foliation, Ann. Polon. Math. 60 (1995), 241–247.

J. A. Wolf, Growth of finitely generated solvable groups and curvature of Riemannian manifolds, J. Diff. Geom. 2 (1968), 421–446.

C. M. Wood, A class of harmonic almost-product structures, J. of Geometry and Physics 14 (1994), 25–42.

J. C. Wood, Harmonic morphisms, foliations and Gauss maps, Complex Differential Geometry and Non-linear Differential Equations, Brunswick, 1984, Contemp. Math. 49 (1986), 145–185.

J. C. Wood, Harmonic morphisms, conformal foliations and Seifert fibre spaces, Geometry of low-dimensional manifolds, 1 (Durham, 1989), 247–259, London Math. Soc. Lecture Note Ser., 150, Cambridge Univ. Press, Cambridge, 1990.

J. W. Wood, Foliations on 3-manifolds, Doct. Diss. Berkeley Univ. Calif. (1968).

J. W. Wood, Foliations on 3-manifolds, Ann. of Math. 89 (1969), 336–358.

J. W. Wood, Foliations of codimension one, Bull. Amer. Math. Soc. 76 (1970), 1107–1111.

J. W. Wood, Bundles with totally disconnected structure group, Comment. Math. Helv. 46 (1971), 257–273.

J. W. Wood, Foliated S^1-bundles and diffeomorphisms of S^1, Dynamical Systems, Bahia 1971, Academic Press, 671–681.

J. Wróblewski, Singular foliations of differential spaces, Demonstratio Math. 21 (1988), 805–813.

H. Wu, On the de Rham decomposition theorem, Ill. J. of Math. 8 (1964), 291–311.

H. Wu, A remark on the Bochner technique in differential geometry, Proc. Amer. Math. Soc. 78 (1980), 403–408.

Y. Wu, Essential laminations in surgered 3-manifolds, Proc. AMS 115 (1992), 245–249.

S. Yamagami, Cech cohomology of foliations and transverse measures, Proc. Japan Acad. Ser. A Math. Sci. 58 (1982), 258–261.

S. Yamagami, A note on Hilbert C^*-modules associated with a foliation, Publ. Res. Inst. Math. Sci. Tyoto 20 (1984), 97–106.

S. Yamagami, On a continuous decomposition of the foliation C^*-algebra, Bull. College Sci. Univ. Ryukyus 38 (1984), 1–4.

S. Yamagami, Modular cohomology class of foliations and Takesaki's duality, Geometric methods in operator algebras (Kyoto, 1983), Pitman Res. Notes Math. 123 (1986), 415–539.

K. Yamato, Qualitative theory of codimension one foliations, Proc. Japan. Acad. 48 (1972), 356–359.

K. Yamato, Qualitative theory of codimension one foliations, Nagoya Math. J. 49 (1973), 155–229.

K. Yamato, Examples of foliations with nontrivial exotic characteristic classes, Proc. Japan. Acad. 50 (1974), 127–129.

K. Yamato, Examples of foliations with nontrivial exotic characteristic classes, Osaka J. Math. 12 (1975), 401–417.

K. Yamato, On exotic characteristic classes of conformal and projective foliations, Osaka J. Math. 16 (1979), 589–604.

K. Yamato, Sur la classe caractéristique exotique de Lazarov-Pasternack en codimension 2, C. R. Acad. Sci. Paris 289 (1979), 537–540.

K. Yamato, The Lazarov-Pasternak exotic characteristic class in codimension 2, II, Japan J. Math. 7 (1981), 227–256.

K. Yamato, Exotic characteristic classes of compact Riemannian foliations, Foliations (Tokyo, 1983), 211–227, Adv. Stud. Pure Math. 5, North-Holland, Amsterdam, 1985.

K. Yano, Topological entropy of foliation preserving diffeomorphisms, Proc. Amer. Math. Soc. 85 (1982), 293–296.

K. Yano, Non-singular Morse-Smale flows on 3-manifolds which admit transverse foliation, Foliation (Tokyo, 1983), 341–358, Adv. Stud. Pure Math. 5, North-Holland, Amsterdam, 1985.

K. Yano, Asymptotic cycles on two-dimensional manifolds, Foliations (Tokyo, 1983), 359–377, Adv. Stud. Pure Math. 5, North-Holland, Amsterdam, 1985.

K. Yano, C^1-foliations which cannot be approximated by C^2-foliations, Tôhoku Math. J. 40 (1988), 95–99.

R. Ye, Foliations by constant mean curvature spheres, Pacific J. Math. 147 (1991), 381–396.

H. Yoo, Existence of complete metrics of a Riemannian foliation, Math. J. Toyama Univ. 15 (1992), 35–38.

S. Yorozu, Notes on square-integrable cohomology spaces on certain foliated manifolds, Trans. Amer. Math. Soc. 255 (1979), 329–341.

S. Yorozu, Nonexistence of nonzero foliated harmonic forms on a compact foliated manifold, Ann. Sci. Kanazawa Univ. 18 (1981), 11–17.

S. Yorozu, Behavior of geodesics in foliated manifolds with bundle-like metrics, J. Math. Soc. Japan 35 (1983), 251–272.

S. Yorozu, The non-existence of Killing fields, Tôhoku Math. J. 36 (1984), 99–105.

S. Yorozu, The second fundamental form of a foliation with parallel mean curvatures along the leaves, Ann. Sci. Kanazawa Univ. 22 (1985), 1–10.

S. Yorozu, A_ν-operator on complete foliated Riemannian manifolds, Israel J. Math. 56 (1986), 349–354.

S. Yorozu, Notes on space-like foliations of codimension one in Lorentz manifolds, Ann. Sci. Kanazawa Univ. 23 (1986), 15–26.

S. Yorozu and T. Tanemura, Green's theorem on a foliated Riemannian manifold and its applications, Acta Math. Hungar. 56 (1990), 239–245.

G. Yu, Cyclic cohomology and index theory of transversally elliptic operators, Contemp. Math. 120 (1991), 189–192.

C. Yue, Integral formulas for the Laplacian along the unstable foliation and applications to rigidity problems for manifolds of negative curvature, Ergodic Theory Dyn. Syst. 11 (1991), 803–819.

C. Yue, Brownian motion on Anosov foliations and manifolds of negative curvature, J. Diff. Geom. 41 (1995), 159–183.

A. Zeggar, Nombre de Lefschetz basique pour un feuilletage riemannien, Ann. Fac. Sci. Toulouse Math. 1 (1992), 105–131.

A. Zeghib, Laminations et hypersurfaces géodésiques des variétés hyperboliques, Ann. Sci. Ecole Norm. Sup. 24 (1991), 171–188.

A. Zeghib, Feuilletages géodésiques des variétés localement symétriques, Lyon: ENS (preprint 1992).

A. Zeghib, Sur les feuilletages géodésiques continus des variétés hyperboliques, Invent. Math. 114 (1993), 193–206.

A. Zeghib, Feuilletages géodésiques appliqués, Math. Ann. 298 (1994), 729–759.

A. Zeghib, An example of a 2-dimensional no leaf, Geometric Study of Foliations, Tokyo 1993, 475–477, World Scientific, 1994.

A. Zeghib, Ensembles invariants des flots géodésiques des variétés localement symétriques, Erg. Theory Dyn. Syst. 15 (1995), 479–412.

A. Zeghib, Subsystems of Anosov systems, to appear.

A. Zeghib, Sur une notion d'autonomie de systèmes dynamiques, appliqué aux ensembles invariants des flots d'Anosov algébriques, to appear.

N. Zhukova, On the stability of leaves of Riemannian foliations, Ann. of Global Analysis and Geometry 5 (1987), 261–271.

N. Zhukova, Submersions with Ehresmann connections, Soviet Math. (Izv. Vuz) 32 (1988), 27–38.

N. Zhukova, Foliations that are compatible with systems of paths, Izv. Vyss. Uchebn. Zaved. Mat. 1989, no. 7, 5–13.

N. Zhukova, Global stability of foliations with second-order differential equations on leaves, Soviet Math. 34 (1990), 91–94.

N. Zhukova, Local stability of leaves of Riemannian foliations with singularities, Methods in qualitative theory and bifurcation theory, 93–105, Nizhegorod. Gos. Univ., Nizhnii Novgorod, 1990.

N. Zhukova, Stability criterion of leaves of Riemannian foliations, Russ. Math. 36 (1992), 86–89.

N. Zhukova, Foliations that are compatible with systems of differential equations of arbitrary order, Izv. Vyssh. Uchebn. Zaved. Mat. 1992 (1993), 42–48. Translation: Russian Math. (Iz. VUZ) 36 (1992), 38–44.

N. Zhukova, The graph of a foliation with an Ehresmann connection and the stability of leaves, Izv. Vyssh. Uchebn. Zaved. Mat. (1994), 78–81.

R. Zimmer, Ergodic theory, semisimple Lie groups, and foliations by manifolds of negative curvature, Inst. Hautes Etudes Sci. Publ. Math. 55 (1982), 37–62.

R. Zimmer, On the Mostow rigidity theorem and measurable foliations by hyperbolic space, Israel J. Math. 43 (1982), 281–290.

R. Zimmer, Curvature of leaves in amenable foliations, Amer. J. Math. 105 (1983), 1011–1022.

R. Zimmer, Ergodic theory and semisimple groups, Birkhäuser, Boston, 1984.

R. Zimmer, Amenable actions and dense subgroups of Lie groups, J. Funct. Anal. 72 (1987), 58–64.

R. Zimmer, Arithmeticity of holonomy groups of Lie foliations, J. Amer. Math. Soc. 1 (1988), 35–58.

R. Zimmer, Orbit equivalence, Lie groups, and foliations, Proc. Centre Math. Anal. Austral. Nat. Univ. 16 (1988), 341–349.

R. Zimmer, Positive scalar curvature along the leaves, Appendix C to Global Analysis on Foliated Spaces by C. C. Moore and C. Schochet, MSRI Publ. 9 (1988), 316–321.

A. Zorich, The S. P. Novikov problem on semiclassical motion of electron in homogeneous magnetic field, Russian Math. Surveys 39 (1984), 287–288.

A. Zorich, Asympotic flag of an orientable measured foliation on a surface, Geometric Study of Foliations, Tokyo 1993, 479–498, World Scientific, 1994.

E. Zuzoma, A topological classification of foliations described by one-dimensional Pfaffian forms on an n-dimensional torus, Differential Equations 17 (1981), 898–904.

E. Zuzoma, Singular Reeb foliations on an n-dimensional torus, Mat. Zametki 30 (1981), 123–128. Translation: Math. Notes 30 (1981), 549–551.

Appendix D
Numbers of Papers on Foliation published during consecutive five Year periods up to 1995

The numbers of papers on foliations published during consecutive five year periods are as follows:

Up to 1935:	8	1966–1970:	86
1936–1940:	4	1971–1975:	339
1941–1945:	6	1976–1980:	400
1946–1950:	12	1981–1985:	505
1951–1955:	14	1986–1990:	497
1956–1960:	16	1991–1995:	510
1961–1965:	29		

These add up to 2426 dated publications. Another hundred titles or so are undated preprints, about three quarters of which are to appear in 1996 or later.

Index of Subjects

adapted connection 20
basic form 33
basic Laplacian 85
Bott connection 19
bundle-like metric 43
characteristic form 37
De Rham-Hodge
    decomposition 85
flat bundle 16
flat connection 18
flow 69
foliation (definition) 2,3
geodesible flow 71
Godbillon-Vey class 11
Gray and O'Neill tensors 49
harmonic foliation 27
harmonic basic forms 85
holonomy 18
holonomy invariant
    transversal volume 41
infinitesimal automorphism ... 36
isometric flow 72
Killing vector field 72

leaf 1
Lie foliation 107
Mauer-Cartan form 107
mean curvature form 25
mean curvature vector field ... 25
normal bundle of a foliation ... 4
projectable vector field 4
Reeb foliation 8
Riemannian foliation 43
Roussarie foliation 10
Rummler's formula 38
second fundamental form 24
spectral sequence of $\mathcal{F}$ 34
symplectic foliation 4
tangent bundle of a foliation .. 3
tangential geometry 3
taut foliation 13
Tischler's theorem 30
transition function 3
transversally symmetric 53
trivial $(p+r)$-form 34
twisted duality 95
Weingarten map 25

Index of Notations

$\tilde{B} \times_\Gamma F$ 19
d_∇ 23
d_B 33
d_κ 84
$\operatorname{div}_B Y$ 42
$F^r \Omega^m$ 34
g_Q 21
$H_B \equiv H_B(\mathcal{F})$ ... 33
L 3
$L^\perp$ 20
Q 4
$V(\mathcal{F})$ 5

$W(Z)$ 25
Y 5
α 24
$\langle \alpha, \beta \rangle_B$ 81
$\|a\|$ 30
$\Gamma L, \Gamma(U, L)$ 3
ΓQ^L 5
Δ_B 85
Δ_κ 85
δ_∇ 26
δ_B 83
κ 25

ν 41
τ 25
$\chi \mathcal{F}$ 37
$\Omega(M)$ 34
$\Omega_B \equiv \Omega_B(\mathcal{F})$ 33
$\overset{\circ}{\nabla}$ 19
∇^M 21
∇ 21
∇^L 59
$\mathfrak{g}$ 107
$*$ 26
$\bar{*}$ 81

LHS ($\equiv$ left-hand side)   RHS ($\equiv$ right-hand side)

MMA • Monographs in Mathematics

Managing Editors
H. Amann / K. Grove / H. Kraft / P.-L. Lions

Editorial Board
H. Araki / J. Ball / F. Brezzi / K.C. Chang / N. Hitchin / H. Hofer / H. Knörrer / K. Masuda / D. Zagier

The foundations of this outstanding book series were laid in 1944. Until the end of the 1970s, a total of 77 volumes appeared, including works of such distinguished mathematicians as Carathéodory, Nevanlinna and Shafarevich, to name a few. The series came to its name and present appearance in the 1980s. According to its well-established tradition, only monographs of excellent quality will be published in this collection. Comprehensive, in-depth treatments of areas of current interest are presented to a readership ranging from graduate students to professional mathematicians. Concrete examples and applications both within and beyond the immediate domain of mathematics illustrate the import and consequences of the theory under discussion.

Published in the series since 1983

Volume 64	**R.B. Burckel, An Introduction to Classical Complex Analysis, Vol. 1** 1979, 570 pages, hardcover, ISBN 3-7643-0989-X
Volume 78	**H. Triebel, Theory of Function Spaces I** 1983, 284 pages, hardcover, ISBN 3-7643-1381-1.
Volume 79	**G.M. Henkin /J. Leiterer, Theory of Functions on Complex Manifolds** 1984, 228 pages, hardcover, ISBN 3-7643-1477-X.
Volume 80	**E. Giusti, Minimal Surfaces and Functions of Bounded Variation** 1984, 240 pages, hardcover, ISBN 3-7643-3153-4.
Volume 81	**R.J. Zimmer, Ergodic Theory and Semisimple Groups** 1984, 210 pages, hardcover, ISBN 3-7643-3184-4.
Volume 82	**V.I. Arnold / S.M. Gusein-Zade / A.N. Varchenko, Singularities of Differentiable Maps – Vol. I** 1985, 392 pages, hardcover, ISBN 3-7643-3187-9.
Volume 83	**V.I. Arnold / S.M. Gusein-Zade / A.N. Varchenko, Singularities of Differentiable Maps – Vol. II** 1988, 500 pages, hardcover, ISBN 3-7643-3185-2.
Volume 84	**H. Triebel, Theory of Function Spaces II** 1992, 380 pages, hardcover, ISBN 3-7643-2639-5.
Volume 85	**K.R. Parthasarathy, An Introduction to Quantum Stochastic Calculus** 1992, 300 pages, hardcover, ISBN 3-7643-2697-2.
Volume 86	**M. Nagasawa, Schrödinger Equations and Diffusion Theory** 1993, 332 pages, hardcover, ISBN 3-7643-2875-4.
Volume 87	**J. Prüss, Evolutionary Integral Equations and Applications** 1993, 392 pages, hardcover, ISBN 3-7643-2876-2.
Volume 88	**R.W. Bruggeman, Families of Automorphic Forms** 1994, 328 pages, hardcover, ISBN 3-7643-5046-6.
Volume 89	**H. Amann, Linear and Quasilinear Parabolic Problems, Volume I, Abstract Linear Theory** 1995, 372 pages, hardcover, ISBN 3-7643-5114-4.

Mathematics with Birkhäuser

DIFFERENTIAL GEOMETRY • SYMPLECTIC GEOMETRY • LIE GROUPS

PM 145 • Progress in Mathematics

C. Albert / R. Brouzet / J.P. Dufour, Université Montpellier II, France (Eds)

Integrable Systems and Foliations
Feuilletages et Systèmes Intégrables

1997. 218 pages. Hardcover
ISBN 3-7643-3894-6

The articles in this volume are an outgrowth of a colloqium "Systèmes Intégrables et Feuilletages", which was held in honor of the sixtieth birthday of Pierre Molino.

The topics cover the broad range of mathematical areas which were of keen interest to Molino, namely, integral systems and more generally symplectic geometry and Poisson structures, foliations and Lie transverse structures, transitive structures, and classification problems.

CONTRIBUTORS: *Y. Benoist, G. Cairns, V. Cavalier, H. Flaschka, E. Ghys, A. Haefliger, B. Jessup, A. Lichnerowicz, G. Meigniez, I. Moerdijk, J. Pitkethly, T. Ratiu, J. Turiel, I. Vaisman, and P. Vanhaechke*

For orders originating from all over the world except USA and Canada:
Birkhäuser Verlag AG
P.O Box 133
CH-4010 Basel/Switzerland
Fax: +41/61/205 07 92
e-mail: farnik@birkhauser.ch

For orders originating in the USA and Canada:
Birkhäuser
333 Meadowland Parkway
USA-Secaurus, NJ 07094-2491
Fax: +1 201 348 4033
e-mail: orders@birkhauser.com

Birkhäuser Verlag AG
Basel · Boston · Berlin

VISIT OUR HOMEPAGE http://www.birkhauser.ch

LM • Lectures in Mathematics, ETH Zürich

Department of Mathematics
Research Institute of Mathematics

Each year the Eidgenössische Technische Hochschule (ETH) at Zürich invites a selected group of mathematicians to give postgraduate seminars in various areas of pure and applied mathematics. These seminars are directed to an audience of many levels and backgrounds. Now some of the most successful lectures are being published for a wider audience through the **Lectures in Mathematics, ETH Zürich** *series. Lively and informal in style, moderate in size and price, these books will appeal to professionals and students alike, bringing a quick understanding of some important areas of current research.*

R.J. LeVeque
Numerical Methods for Conservation Laws
2nd Edition, 3rd Printing 1994.
1992. ISBN 3-7643-2723-5

R. Narasimhan
Compact Riemann Surfaces
1992. ISBN 3-7643-2742-1

A.J. Tromba
Teichmüller Theory in Riemannian Geometry
1992. ISBN 3-7643-2735-9

M. Yor
Some Aspects of Brownian Motion, Part I
1992. ISBN 3-7643-2807-X

Part II
1997. ISBN 3-7643-5717-7

G. Baumslag
Topics in Combinatorial Group Theory
1993. ISBN 3-7643-2921-1

M. Giaquinta
Introduction to Regularity Theory for Nonlinear Elliptic Systems
1993. ISBN 3-7643-2879-7

O. Nevanlinna
Convergence of Iterations for Linear Equations
1993. ISBN 3-7643-2865-7

R.-P. Holzapfel
The Ball and Some Hilbert Problems
1995. ISBN 3-7643-2835-5

J.F. Carlson
Modules and Group Algebras
Notes by Ruedi Suter
1996. ISBN 3-7643-5389-9

L. Simon
Theorems on Regularity and Singularity of Energy Minimizing Maps
based on lecture notes by Norbert Hungerbühler
1996. ISBN 3-7643-5397-X

M. Freidlin
Markov Processes and Differential Equations: Asymptotic Problems
1996. ISBN 3-7643-5392-9

J. Jost
**Nonpositive Curvature
Geometric and Analytic Aspects**
1997. ISBN 3-7643-5736-3

Mathematics with Birkhäuser

GEOMETRY • DIFFERENTIAL GEOMETRY • MANIFOLDS

V. Rovenskii, Pedagogical Institute, Krasnoyarsk, Russia

Foliations on Riemannian Manifolds and Submanifolds

1997. Approx. 280 pages. Hardcover
ISBN 3-7643-3806-7

The ideas and methods of foliations are very popular in mathematics and its applications. The key problem of this volume is the role of a Riemannian curvature in studies of manifolds and submanifolds with foliations. Rovenskii discusses the results of many geometers, but the book principally focuses on the author's own investigations into the Riemannian geometry of foliations and submanifolds with generators having nonnegative curvature. The main idea is that such manifolds are decomposed into a direct product when the dimension of leaves is sufficiently large.

Part 1 starts with a short introduction (Chapter 1) to the geometry of foliations and continues with local and global results on Riemannian manifolds with foliations, including rigidity, splitting, and integral formulas (Chapters 2 - 4). In Part 2, Rovenskii gives a survey of submanifolds with generators (Chapter 5) and then combines variational methods (developed in Part 1) with synthetic procedures to obtain rigidity and Segre type decomposition of such submanifolds (Chapters 6, 7). Appendix A, written jointly with V. Toponogov, contains facts on geodesic foliations of a round sphere in relation to manifolds of positive sectional curvature bounded from above. The main result generalizes the minimal diameter theorem by M. Berge.

This book is intended for students and researchers with a basic knowledge of differential and Riemannian geometry.

For orders originating from all over the world except USA and Canada:
Birkhäuser Verlag AG
P.O Box 133
CH-4010 Basel/Switzerland
Fax: +41/61/205 07 92
e-mail: farnik@birkhauser.ch

For orders originating in the USA and Canada:
Birkhäuser
333 Meadowland Parkway
USA-Secaurus, NJ 07094-2491
Fax: +1 201 348 4033
e-mail: orders@birkhauser.com

Birkhäuser
Birkhäuser Verlag AG
Basel · Boston · Berlin

VISIT OUR HOMEPAGE http://www.birkhauser.ch